Adaptation to Climate Change

ASEAN and Comparative Experiences

Adaptation to Climate Change

ASEAN and Comparative Experiences

Editor-in-Chief

KOH Kheng-Lian

National University of Singapore, Singapore

Editors

Ilan KELMAN

University College London, UK and
Norwegian Institute of International Affairs, Norway

Robert KIBUGI

University of Nairobi, Kenya

Rose-Liza EISMA OSORIO

University of Cebu, Philippines

World Scientific

NEW JERSEY • LONDON • SINGAPORE • BEIJING • SHANGHAI • HONG KONG • TAIPEI • CHENNAI • TOKYO

Published by

World Scientific Publishing Co. Pte. Ltd.
5 Toh Tuck Link, Singapore 596224
USA office: 27 Warren Street, Suite 401-402, Hackensack, NJ 07601
UK office: 57 Shelton Street, Covent Garden, London WC2H 9HE

Library of Congress Cataloging-in-Publication Data
Adaptation to climate change : ASEAN and comparative experiences / [edited by] Koh Kheng-Lian (National University of Singapore, Singapore), Ilan Kelman (University College London, UK), Robert Kibugi (University of Nairobi, Kenya), Rose-Liza Eisma Osorio (University of Cebu, Philippines).
pages cm
Includes bibliographical references.
ISBN 978-9814689731 (hardcover : alk. paper)
1. Climatic changes--Government policy--Southeast Asia. 2. Climatic changes--Law and legislation--Southeast Asia. 3. Climatic changes--Asia. I. Koh, Kheng Lian, editor. II. Kelman, Ilan, editor. III. Kibugi, Robert, editor. IV. Eisma-Osorio, Rose-Liza, editor. V. Kelman, Ilan. Asia-Pacific Islander responses to climate change Container of (work).
QC903.2.S645A33 2015
363.738'745610959--dc23

2015023672

British Library Cataloguing-in-Publication Data
A catalogue record for this book is available from the British Library.

In house editor: Rajni Gamage

Typeset by Stallion Press
Email: enquiries@stallionpress.com

Contents

Foreword

I thank my good friend and colleague, Emeritus Professor Koh Kheng-Lian, the editor-in-chief of this book, for inviting me to contribute a Foreword.

The essays in this book were originally presented at a closed-door workshop in July 2014. It was organized by the APCEL Specialist Group on Climate Change Adaptation of the Faculty of Law, National University of Singapore.

Professor Koh Kheng-Lian and Associate Professor Lye Lin Heng had founded APCEL and continue to inspire it with their ideas and passion.

This book addresses the important question of adaption to climate change and, in particular, the role of environmental law, lawyers, and experts from other disciplines in this process. Although the regional focus is Southeast Asia (ASEAN), the experiences of other countries and regions have also been included.

In recent months, we have received both good news and bad news. I would classify the 45th Assessment Report of the International Panel on Climate Change (IPCC) as bad news. The report warns us that things are much worse than had previously been predicted. On the other hand, I consider the massive turn out in New York, in September, at the Climate Summit 2014 as good news. Another good news was the recent joint announcement, in Beijing, by President Barack Obama and President Xi Jinping, of their new and specific commitments on emission reduction.

Let us all hope that the ongoing negotiations will be brought to a successful conclusion next year in Paris.

Tommy Koh
Ambassador-at-Large
Ministry of Foreign Affairs
Co-Chairman
Advisory Committee, APCEL

Preface and Acknowledgments

Preface

As I write this *Preface*, many places around the world are experiencing extreme weather conditions causing typhoons, storms, blizzards, and heat waves. Not all such hazards are impacted by climate change, but climate change can add to the numerous factors that are rapidly changing the planet's weather, as pointed out in the Fifth Assessment Report from the Intergovernmental Panel on Climate Change (www.ipcc.ch) which was published in 2013–2014. These hazards, coupled with ever-increasing vulnerabilities — such as population growth, continuing overconsumption and overexploitation of natural resources, and the rich–poor gap, further exacerbate disasters involving widespread human, material, economic, or environmental impacts.

To lessen the detrimental impacts of the climate change factor, continuing efforts at climate change mitigation are needed. That means increasing uptake of greenhouse gases such as through land use practices, while reducing greenhouse gas emissions such as through reduced use of fossil fuels. However, climate change mitigation is not enough. We also need to adapt to the new realm of disasters, including reducing vulnerabilities as well as learning to live with the changing hazard regime. Both mitigation and adaptation are part of the UNFCCC (www.unfccc.int) climate change negotiations, from which a bleak message about the future is emanating — especially given the weakness of the agreement reached at the Lima negotiations in December 2014. We can simply hope for a more positive outcome from the climate change negotiations in Paris in December 2015. And, from the United Nations World Conference on Disaster Risk Reduction to be held from 14 to 18 March 2015 in Sendai City, Japan, it will determine the modalities of a post-2015 framework for disaster risk reduction.

Adaptation to climate change is entering into the mainstream of policy- and decision-making of many countries, particularly in Asia including the ASEAN Member Countries. The process is still slow as the knowledge and capacity for mainstreaming leaves much to build upon. That is surprising, as adaptation provides little which is fundamentally new, yet the word is of recent "vintage" for practitioners in this field given that mitigation was at the forefront in the years following the initial meetings of UNFCCC and the drafting of the Kyoto Protocol. Adaptation forces countries to think globally but act locally — and act locally they must.

This is evident from the numerous discussions in the Asia-Pacific Adaptation Network Forum 2010 (APAN Forum) held in Bangkok, in Incheon, 2012 and in Kuala Lumpur, 2014. I was privileged to have attended all three forums, and was amazed by the overwhelming response of no less than 550 participants in each of the forums, representing policy-makers, government agencies, international organizations, the private sector, academics, NGOs and CSOs from all over Asia including ASEAN Member Countries. Asia and the world are beginning to wake up to the stark reality of the vulnerabilities to climate change and are finding ways to build resilience in order to attain the objectives of the Sustainable Development Goals. Adaptation is now the buzzword of climate change.

After I attended the APAN Forum in Incheon in 2012, I was inspired to establish the APCEL Specialist Group on Climate Change Adaptation. It was surprising that in the APAN Forums (which cover the Asia–Pacific region), there were hardly any environmental lawyers represented. The APCEL Specialist Group came to fruition in February 2013. The main aim is to encourage environmental lawyers to contribute to the "adaptation" discourse. The objectives of the Specialist Group, with a focus on ASEAN, include:

- Building knowledge on adaptation — how and when to adapt from existing experience of climate change impacts; also projections of adaptation in the future.
- Promoting dialogue and action among policy-makers, legal experts, and other stakeholders, such as physical and social scientists, economists, and vulnerable groups.
- Formulating legal and policy frameworks of adaptation to climate change.
- Assessing key concepts of adaptation, vulnerabilities and resilience, and its possible standardization and harmonization in ASEAN and other countries.
- Managing climate-related disasters and emergency response, rehabilitation, recovery, and reconstruction.
- Understanding the role of science and traditional knowledge in adaptation to climate change.
- Green growth and adaptation.
- Role of insurance companies in adaptation to climate change.
- Decision-making in the context of uncertainties including mainstreaming of community participation.
- Climate change justice and human rights.
- Case studies of current impacts of climate change requiring adaptation in Brunei, Cambodia, Indonesia, Laos, Malaysia, Myanmar, Philippines, Singapore, Thailand, Vietnam, and other parts of the world.
- Scenarios of future adaptation to climate change in ASEAN and the world.
- Governance in ASEAN at multilevels on adaptation to climate change.
- Innovative mechanisms: for example, eco-cities/carbon-free cities as adaptation nodes and survival funds.

- Nature-based solutions including the planting of mangrove trees, oyster beds, etc., and early warning systems.
- The Warsaw International Loss and Damage Mechanism: Beyond Adaptation.
- Promoting and encouraging research among ASEAN Member Countries and other countries.

This publication consists of the papers presented at the "Closed-Door Workshop on Climate Change Adaptation: ASEAN and Comparative Experiences", held at the Faculty of Law of the National University of Singapore (Bukit Timah Campus) from 17–18 July 2014. It was the first workshop organized by the APCEL Specialist Group on Climate Change Adaptation. This Workshop focused on ASEAN, both at the sub-regional, as well as at the national level. In addition, comparative experiences were drawn from other jurisdictions. The papers were excellent and all writers were invited to update and revise their papers in light of the workshop discussions and submit them for this publication.

Why the focus on ASEAN one might ask? It was not because of being exclusive. Indeed, the title of this publication includes not only ASEAN but "comparative experiences" from outside the ASEAN region. This publication is intended to be comparative of the experiences in other regions and countries. ASEAN was chosen as the focal point simply because I teach a course on "ASEAN Environmental Law, Policy and Governance" at the Faculty of Law, National University of Singapore. As all environmental law teachers know, our task in the teaching of environmental law is to constantly update our syllabus to keep up with the dynamics of change in the environment. The subject of adaptation to climate change is relatively recent as mentioned and there is a paucity of research on the legal aspects of adaptation in ASEAN. I included the topic of adaptation in my course which meant a survey of the UNFCCC negotiations from Cancun to Warsaw — and now the negotiations beyond those.

There are also various networks in which many of the ASEAN Member Countries as well as others in the region and beyond are participating: e.g., the Nairobi Work Programme on Impacts, Vulnerability and Adaptation to Climate Change; the Regional Climate Adaptation Knowledge Platform for Asia (SEI, SIDA, and UNEP); Global Climate Adaptation Initiatives and Networks; and Asian Cities Climate Change Resilience Network Rockefeller Foundation. Judging by the overwhelming response in the Asian region and the adaptation discourse in the global arena, this will continue to be one of the important issues in the global climate change negotiations and of the century.

Within Asia, the focus is on the ASEAN region, both at the regional level, as well as at the national level. We are doing a sampling of a few ASEAN member countries: Indonesia, Malaysia, the Philippines, and Thailand. Other Asian countries, not within ASEAN, represented are Bangladesh, People's Republic of China, India, Sri Lanka, and the Republic of Taiwan. We have included other comparative experiences outside the ASEAN region, as climate change is *sans frontières*. We can learn from the experiences of other countries and regions.

The topics covered include the IPCC climate change assessment 2014 findings on impacts, adaptation, and vulnerability in Asia; ASEAN experiences on adaptation to climate change, formulating new legal and policy frameworks for climate change adaptation; crucial issues in specific areas such as geoengineering; water, energy, and industry; coastal and islander populations; the question of the Warsaw International Loss and Damage Mechanism; economic interconnections; internal displacements and external migration; disaster management and risk reduction; and climate change justice and human rights.

Adaptation to Climate Change: ASEAN and Comparative Experiences surveys some of the crucial issues, abovementioned, in countries facing adaptation challenges and in building resilient societies — the very meaning of resilience is not apparent. It is a publication that would provide some insights for policy-makers, administrators, the private sector, scientists, academic scholars, climatologists, NGOs and CSOs both in ASEAN and the world to find solutions. A new governance should be in place to take on the challenges. For example, one of the most important mechanisms to deal with adaptation is disaster management and reduction including humanitarian assistance. ASEAN itself has to take stock of its disaster management and disaster risk reduction in light of Typhoon Haiyan. How can the "whole-of-the-world" approach to deal with massive disasters be achieved both by individual countries as well as collectively within the United Nations, the World Bank, the Asian Development Bank, and others?

Some chapters were updated when the manuscript was in press.

I would like to thank the editors and writers for their cooperation to meet the target of bringing out this publication in time to celebrate Singapore's 50th anniversary of independence, 9 August 1965.

Koh Kheng-Lian, Chair
APCEL Specialist Group on Climate Change Adaptation
Honorary Director
Asia-Pacific Centre for Environmental Law (APCEL)
Emeritus Professor
Faculty of Law
National University of Singapore

Acknowledgments

Apart from those who contributed chapters to this publication, the editors would also like to thank the following persons, including those who delivered welcome remarks and messages at the opening ceremony, presenters, panellists, and moderators for their support for the workshop, mentioned in the *Preface*:

Antonio Benjamin, Judge, Brazilian High Court; Chair, IUCN World Commission on Environmental Law

Lovleen Bhullar, Visiting Faculty, National Law University, Delhi; Adjunct Research Fellow, Faculty of Law, National University of Singapore; Research Associate, Institute of Water Policy, Lee Kuan Yew School of Public Policy

Simon Chesterman, Dean and Professor, Faculty of Law, National University of Singapore

Burhan Gafoor, Ambassador and Singapore Chief Negotiator for Climate Change; Senior Director, Ministry of Foreign Affairs, Singapore

Saleemul Huq, Director, International Centre for Climate Change and Development, Bangladesh

Edwin Khew, Managing Director, Anaergia Pte Ltd; Chairman Sustainable Energy Association of Singapore; Member, APCEL Advisory Committee

Kishan Khoday, UNDP Regional Practice Leader for Environment & Energy in the Arab Region, United Nations Development Programme

Evelyn Khoo, Director (Policy & Planning), National Climate Change Secretariat, Singapore

Tommy Koh, Ambassador-at-Large, Ministry of Foreign Affairs; Special Adviser, Institute of Policy Studies; Chairman, Centre for International Law, National University of Singapore; Co-Chair, APCEL Advisory Committee

Laode M. Syarif, Senior Lecturer, Hasanuddin University, Faculty of Law, Makassar, Indonesia

Lee Sheu Jeen, Communications and Administration Officer, Disaster Law Programme, Asia Pacific, International Federation of Red Cross and Red Crescent Societies

Lye Lin Heng, Associate Professor, Director of APCEL, Faculty of Law, National University of Singapore

Mallika Naguran, Consultant and Publisher, Gaia Discovery

Joy Jacqueline Pereira, Professor and Principal Research Fellow, Southeast Asia Disaster Prevention Research Institute, Universiti Kebangsaan Malaysia

Gloria Estenzo Ramos, Professor, University of Cebu College of Law (UC), Philippines; Member, Governing Board of the IUCN Academy of Environmental Law

Nicholas A. Robinson, Emeritus Professor, Gilbert and Sarah Kerlin Distinguished Professor of Environmental law; Co-Director, Center for Environmental Legal Studies, Pace Law School, US; Member, APCEL Advisory Committee

Administrative Staff and Others

Melinda Tan, Manager, Faculty of Law, National University of Singapore
Khor Shu Kim, Management Assistant Officer, Faculty of Law, National University of Singapore
Shirley Mak, Management Assistant Officer, Faculty of Law, National University of Singapore

Singapore Ministry of Education Academic Research Fund Tier 1 — official grant number: R-241-000-119-112

About the Authors

Professor Dr Abdul Haseeb ANSARI is Head, Research Management Unit (RMU) at the Faculty of Law of the International Islamic University Malaysia, Adjunct Professor in Management and Science University, Malaysia; member of World Commission on Environmental Law, IUCN; and member of the Specialist Group Energy Law and Climate Change, IUCN. His areas of specialization are: Environmental Law, Taxation, and Comparative Jurisprudence. His sub-specialization is Islamic Environmental Law. Prof Abdul Haseeb Ansari has extensive teaching experience in Allahabad University (AU), Allhabad, India; Banaras Hindu University (BHU), Varanasi, India; Punjab University (PU), Chandigarh, India; and International Islamic University Malaysia (IIUM), Malaysia. The courses that he has introduced in IIUM include Environmental Law, Law and Society, Revenue Law, Environmental Law of the Sea, and International Environmental Law. He is a course coordinator for the courses Introduction to Law, Law and Society, Environmental Law, Jurisprudence, Comparative Environmental Law, Environmental Law of the Sea, and Comparative Jurisprudence. He has supervised a number of PhD and Masters dissertations and is a member of the Faculty's Postgraduate Committee (2000– date); Harmonization of Shari'ah; Civil Law Committee (2004–2008); Land Tenure and Environmental Management (2004– date); Globalization and WTO Unit (2004– date); International Law and Maritime Affairs Unit (2008– date); Editorial Board of the *IIU Law Journal* (4 January 2009–August 2011); and Kulliyyah Board. He has published over

100 articles and 11 books. He has bagged several national and international awards.

Ms Caroline COMPTON is currently undertaking her PhD at the Australian National University (ANU) College of Law. Her research examines property law, development and disaster with an interest in understanding how exogenous shock influences institutional development. She received an Endeavour Postgraduate Scholarship to undertake doctoral fieldwork in the Philippines, and is a Visiting Research Associate at the Institute of Philippine Culture, Ateneo de Manila University. Prior to commencing her studies at ANU, Caroline spent a number of years working on development projects in Vietnam. Caroline holds a Bachelor of Arts (Hons.) from the University of New South Wales, a Master of Education from the University of Technology, Sydney, and a Juris Doctor (Hons.) from ANU.

Atty. Rose-Liza EISMA OSORIO is one of the few environmental lawyers in the Philippines. She is one of the founders and managing trustee of the Philippine Earth Justice Center, Inc. (PEJC) and currently the CEO/Chair of the Board of Trustees of the Foundation for the Philippine Environment (FPE). Her expertise is in the areas of governance, institutional development, monitoring and evaluation, environmental policy and advocacy, and technical legal support for natural resource conservation utilizing science-based tools and approaches. She is presently the Project Advisor of the Post-Yolanda Ecosystem Rehabilitation of Kinatarkan Island in Bantayan, Cebu conducted by the Santo Nino de Cebu Augustinian Social Development Foundation. After graduating from the University of the Philippines-Diliman with a Bachelor of Science in Chemistry degree, she earned her Bachelor of Laws at the Siliman University, Philippines and a Certificate of Attendance in Coastal Management at the University of Rhode Island, Coastal Resource Center, US. Currently, she is a faculty member of the College of Law of University of Cebu, teaching environmental law, legal research and writing, and private international law, among others.

Professor FAN Chien-Te is a law professor, founder and current director of the Institute of Law for Science & Technology (ILST) in National Tsing Hua University. The Institute is the first graduate academic institute established in Taiwan which focuses on science- and technology-related legal trainings, energy, and environment conservation. He is also the director of the Bioethics & Law Centre in National Tsing Hua University since 2005. Before he joined National Tsing Hua University, Professor Fan was a professor in Soochow University, School of Law. In addition to his academic achievements, Professor Fan is a consultant to various government agencies, including the President Hall, the Ministry of Foreign Affairs, the Environment Protection Agency, and several other industries. Professor Fan specializes in energy and environmental law, and the areas of biotechnology- and ethic-related issues. He is also an expert in the UNFCCC negotiation process and has participated in a number of its COP meetings. Professor Fan received his LLB degree from Soochow University, LLM degree from the University of Washington, and JD Degree from Puget Sound University.

Professor Daniel FITZPATRICK writes on property rights in a development context. In 2007 he won the Hart Article Prize from the UK Socio-Legal Association for an article entitled: Evolution and Chaos in Property Rights Systems: The Third World Tragedy of Contested Access. He has published in the *Yale Law Journal*, *Law and Society Review*, the *Yale Journal of International Law*, and *Law and Social Inquiry*. He has been a Global Visiting Professor at New York University School of Law (2011); a Visiting Professor at the National University of Singapore (2006–2009); a Visiting Professor at the University of Muenster (2002); and a Distinguished Visitor at the University of Toronto (2007). Currently he is an Australian Research Council Future Fellow (2012–2016) researching on the resilience of property: inundation, displacement and local relocation in the Asia-Pacific. In 2014, he returned to NYU Law as a Global Senior Research Fellow (non-teaching role).

Professor Fitzpatrick has extensive experience in the public policy of property rights and development. He was the UN's land rights adviser in post-conflict East Timor (2000) and post-tsunami Aceh (2005–2006). He is the primary author of the *UN's Land and Disasters: Guidance for Practitioners* (2010). He has undertaken professional consultancies on law and development with the World Bank; AusAID; the Asian Development Bank; Oxfam International; the OECD; UNDP; and UN-Habitat. His work with AusAID includes co-authoring the 2008 "Making Land Work" report for its Pacific Land Program. In 2011, he established the Masters in Law, Governance and Development at the Australian National University. Currently, he is writing a report for Oxfam International on land tenure security after Typhoon Haiyan in the Philippines.

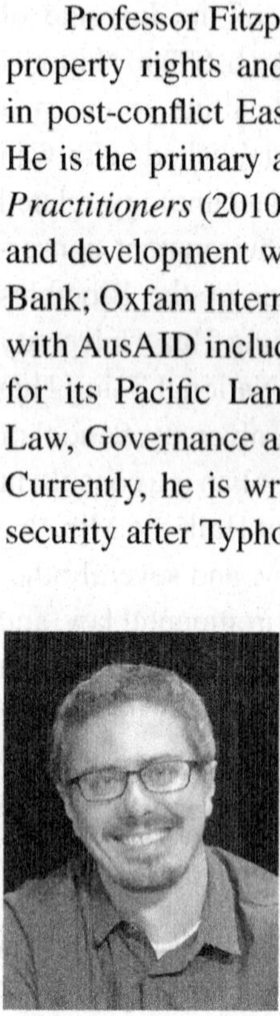

Dr Daniel A. FRIESS is an Assistant Professor and the Principal Investigator of the Mangrove Lab (www.themangrovelab.com), a research group based at the Department of Geography, National University of Singapore (NUS). He is a tropical wetland scientist and his research is focused on how best to incorporate stable ecosystems into broader coastal management strategies over the long term. Dr Daniel and his students have most recently been monitoring the vulnerability of mangroves in Singapore and Thailand to sea level rise, quantifying mangrove ecosystem services for Payments for Ecosystem Services (PES), and conducting mangrove restoration in Sulawesi in conjunction with local NGOs. This research is conducted throughout Southeast Asia, including Singapore, Thailand, Indonesia, the Philippines, and Myanmar. Prior to joining NUS, Daniel was a Research Fellow with the Singapore-Delft Water Alliance and the Applied Plant Ecology Lab in Department of Biological Sciences, NUS. He received his PhD in wetland geography from the

University of Cambridge, UK. He is also a member of the IUCN Mangrove Specialist Group.

Dr GAO Ming-Zhi Anton is an Associate Professor of the Institute of Law for Science and Technology at the National Tsing Hua University, Taiwan. He received one Master's degree (MA) from the National Taiwan University and another Master's degree in energy and environmental law (LLM) from the Institute of Environmental and Energy Law of K.U. Leuven, Belgium, in July 2005. He began the PhD program at K.U. Leuven in October 2005, and received his PhD in Law in December 2009. His main research areas include energy law and policy, European environmental law, renewable energy, feed in tariff, and strategic environmental assessment.

Dr GAO Qi is currently a lecturer of environmental law at KoGuan Law School of Shanghai Jiao Tong University, China. She received her LLB degree from Wuhan University and passed the Chinese bar examination in 2007. She then graduated from the Research Institute of Environmental Law (RIEL) of Wuhan University in 2009 (LLM). Supervised by the late Professor Michael Jeffery, QC. (Principal) and Professor Donna Craig from University of Western Sydney, she completed her PhD study in 2013. Her PhD thesis is scheduled to be published by Brill in April 2014. The book title is *A Procedural Framework for Transboundary Water Management in the Mekong River Basin.*

Her research and publications focus on transboundary water management, climate change, environmental procedural rights, and market-based environmental policy instruments. She has actively worked with the late Professor Michael Jeffery, QC., Professor Ben Boer (University of Sydney), Professor Wang Xi (Shanghai Jiao Tong University), and Professor Ke Jian (Wuhan University) in the above fields. Her research highlights the significance of procedural justice and the key role of civil society in promoting a shift towards environmental good governance in transitional societies, and emphasizes the social, cultural, geopolitical, and economic impact on shaping environmental legal regimes in different countries and regions. She has published several articles in reputable environmental law journals, such as the *Pace Environmental Law Review* and the *Environmental Law Reporter*. She has served as a speaker and panelist at several national and international conferences. She also maintains a close relationship with NGOs and public interest lawyers in China and Southeast Asia.

Dr Asanga GUNAWANSA holds a PhD in law from the National University of Singapore (NUS) and an LLM in International Economic Law from University of Warwick, England. He is an Attorney-at-Law of the Supreme Court of Sri Lanka and has over 20 years. of experience as a legal counsel. Whilst practicing in all areas of general commercial litigation, Dr Gunawansa is an expert in legal aspects of project financing, foreign investment disputes, arbitration, and environmental law. In addition to being engaged in private legal practice, Dr Gunawansa is currently a Legal Consultant to the Asian Development Bank on Public–Private Partnerships. He is also a Legal Advisor to the Ministry of Justice of Sri Lanka and a Member of the Legal Advisory Committee of Sri Lanka Cricket Board. He is also a Member of the Board of Studies of the Council of Legal Education of Sri Lanka. Dr Gunawansa taught construction law, water law, and legal aspects of sustainable development at NUS during the period 2005–2012.

He is currently an Adjunct Assistant Professor attached to the School of Design and Environment of NUS and an Associate Member of the Executive Committee of the Asia Pacific Centre of Environmental Law of NUS. He is also an International Research Associate of the Institute of Water Policy of the Lee Kuan Yew School of Public Policy, NUS. Dr Gunawansa had worked as a State Counsel attached to the Department of the Attorney General of Sri Lanka during the period 1993–2000 and as a Legal Officer attached to the United Nations Compensation Commission and United Nations Relief and Works Agency during the period 2000–2007. Outside his legal practice and academic activities, Dr Gunawansa is a Non-Resident Associate of the Asia Society and a Member of the Editorial Review Board of the *International Journal of Law in the Built Environment* and the *Journal of Built Environment Project and Asset Management*. Dr Gunawansa has also authored several books including the *Asia Pacific Construction Law Case Book Series* published by Lexis-Nexis and *Water Governance: Evaluation of Alternative Architectures* published by Edward Elgar.

Assistant Professor HSU Yun-Hsiang is from the Institute of Law and Government at the National Central University, Taiwan. He received one Master's degree from National Chengchi University and another Master's degree in demography from the University of California, Berkeley, in 2006. After serving at the Council for Economic Planning and Development, Taiwan, for three years, he received a Fulbright Grant from the Institute of International Education and began his PhD program in Public Policy and Management in Ohio State University in 2009. He received his PhD in 2013. His research interests include policy and institutional analysis (IAD framework) and performance management, with a focus on workforce and development policy.

Dr Alejandro IZA is Head of the IUCN Environmental Law Programme and Director of the IUCN Environmental Law Centre since January 2005, five years after joining the Law Programme as a Legal Officer.

From a combined academic and practical background, his area of expertise includes a broad knowledge of international environmental law, with special focus on water resources and marine law, environmental protection, and regional integration. In his academic career, he has taught public international law, the law of European integration, and international environmental law. He holds a Master of Laws degree from the University of London, and a Doctor of Laws degree from the University of Buenos Aires, where he was also a lecturer and research fellow.

Dr Ilan KELMAN (http://www.ilankelman.org and for twitter @IlanKelman) is a Reader in Risk, Resilience, and Global Health at University College London, England and a Senior Research Fellow at the Norwegian Institute of International Affairs, Oslo. His overall research interest is linking disasters and health, including the integration of climate change into disaster research and health research. He has three main areas of focus: (1) disaster diplomacy and health diplomacy (http://www.disasterdiplomacy.org); (2) island sustainability involving safe and healthy communities in isolated locations (http://www.islandvulnerability.org); and (3) risk education for health and disasters (http://www.riskred.org).

Mr M. Hafijul Islam KHAN is an environmental lawyer, working for the last 12 years on legal issues relating to the environment, natural resource management, and climate change. Currently he is leading an organization called Centre for Climate Justice-Bangladesh (CCJ-B). He is also teaching Environmental Law in School of Environmental Science and Management at Independent University-Bangladesh (IUB) as an adjunct faculty member. Mr Khan is a climate negotiator and is working with LDCs climate negotiator group as a core team member, on loss and damage associated with climate change. He worked with BELA, an NGO based in Bangladesh and also with Germanwatch, an NGO based in Germany. He is a member of IUCN-World Commission on Environmental Law. As a consultant, he provided services to UNDP-Bangladesh, IUCN-Bangladesh, CNRS, Oxfam-Bangladesh, USAID-Bangladesh, Asian Development Bank, Environ-Consulting

Services-Malaysia, Germanwatch, Legal Response Initiative, and Department of Environment, Ministry of Environment and Forest (MoEF). Mr Khan completed his LLB and LLM from the University of Dhaka and the second LLM was obtained from the Central European University. He also completed a number of diplomas and short courses on environment, climate change, and natural resource management in various reputed universities including University of California, US; Copenhagen Law School, Denmark; Oxford University, UK; Washington College of Law, US; Jounsuu University, Finland; and Siena University, Italy. One of his recent publications, "Legal and Policy Responses to Loss and Damage Associated with Climate Change", was published in the book *Climate Change: International Law and Global Governance, V-II*, edited by O. C. Ruppel, C. Roschmann, and K. Ruppel-Schlichting, published by Nomos Publishers, Germany, 2013.

Dr Robert KIBUGI is a lecturer in law at the University of Nairobi's Centre for Advanced Studies in Environmental Law (CASELAP) and School of Law. He previously taught at the Faculty of Law, University of Ottawa, Canada. His legal and policy research agenda focuses on, among other subjects, public participation in natural resource governance; land use law for sustainable development; climate change, including the role of law and policy in the adaptation and mitigation of climate change; energy law; water resources management and rights; and water and sanitation. Ongoing research includes examination of legal regulation of low-carbon investments in Tanzania, Zambia, and Mozambique with CIFOR and IDLO; and research on enhancing community benefits in large-scale land acquisitions for investments, supported by IDRC. He holds an LLB and an LLM from the School of Law, University of Nairobi, and an LLD from the Faculty of Law, University of Ottawa. He is an advocate of the High Court of Kenya. Dr Kibugi has published various chapters and articles in peer-reviewed books and journals; more details at: http://profiles.uonbi.ac.ke/rmkibugi/content/dr-robert-kibugi-lecturer-environmental-law.

Dr KOH Kheng-Lian is Emeritus Professor of the Law Faculty, National University of Singapore. She was Director of the Asia-Pacific Centre for Environmental Law (APCEL) from 1996–June 2013 and is currently its Honorary Director, and Chair of the APCEL Specialist Group on Climate Change Adaptation. She was the IUCN-CEL Regional Vice-Chair for South and East Asia, and a member of its Steering Committee (1996–2004). She served as a co-director and resource person of the Ministry of Foreign Affairs, Singapore/World Bank Institute seminars on Urban and Industrial Environmental Management in Singapore from 1999–2001, and as APCEL course director from 2002–2014. She was ADB consultant for the project *Enhancing Effective Regulation of Water Infrastructure in Singapore* (2010).

She was a member of the ADB/APCEL/IUCN/UNEP project on Capacity Building for Environmental Law in the Asia -Pacific Region in 1997 and 1998, which conducted courses for environmental professors, and a co-writer with D. Craig and N. A. Robinson (eds.), *Capacity Building for Environmental Law in the Asian and Pacific Region: Approaches and Resources* (ADB, 2002, in two volumes). She is author of many publications and conference papers in environmental law, including *ASEAN Environmental Law, Policy and Governance: Selected Documents* (2009–2013, in two volumes). She was a member of the IBA (International Bar Association) Task Force which published *Achieving Justice and Human Rights in an Era of Climate Disruption* (2014). She has published books and articles on criminal law, commercial transactions, and straits in international navigation.

She has participated in numerous international environmental law conferences. As a Visiting Professor at University of Mexico and the University of Hawaii, she gave a series of lectures in environmental law. She is in the international resource team of the Konrad Adenauer Stiftung Rule of Law Programme Asia on "Environmental Law Talks". She currently teaches the course on ASEAN Environmental Law, Policy and Governance. She served as a legal officer in the secretariat of the United Nations Commission on International Trade Law (UNCITRAL) in Vienna, 1980–1986.

She is the 2012 Laureate, Elizabeth Haub Prize in Environmental Law conferred jointly by the University of Stockholm, Université Liber de Bruxelles, and the International Council of Environmental Law — http://juridicum.su.se/ehp/news.html. She is an inductee to the Singapore Women's Hall of Fame 2014 for her pioneering work in the development of environmental law in the region — http://www.swhf.sg/the-inductees/17-environment-conservation/138-koh-kheng-lian She holds a PhD LLM (Singapore), Diplôme de Hautes Études Internationales (HEI, University of Geneva), LLB Hons. (University of Malaya in Singapore), and Advocate & Solicitor (Singapore).

Dr Raman LETCHUMANAN joined the S. Rajaratnam School of International Studies (RSIS), Nanyang Technological University (NTU), Singapore, in November 2014 as a Senior Fellow primarily in charge of environment/conservation and related issues. Dr Raman was the Head of the Environment Division at the Secretariat of the Association of Southeast Asian Nations (ASEAN Secretariat) from 2000–2014, where he was overall in charge of promoting regional cooperation on a whole range of environmental and conservation issues for the benefit of ASEAN Member States. From 2002 to 2008 he was also concurrently in charge of disaster management where, among others, he was responsible for drafting, finalizing, and implementing the ASEAN Agreement on Disaster Management and Emergency Response. He was also in charge of the Initiative for ASEAN Integration (IAI) in 2008 and was responsible for the drafting and implementation of the IAI Strategy and Work Plan for 2009–2015.

Prior to this assignment, Dr Raman served the Government of Malaysia for 20 years, since 1981, in various capacities in the Ministry of Science, Technology and the Environment. The last position he held was Director of the Conservation and Environmental Management Division, where he was responsible for formulation and coordination of environmental and conservation polices and strategies for the country. In that capacity, he represented the Government of Malaysia in numerous international negotiations on environmental issues, particularly the climate change and biodiversity conventions. Dr Raman's research expertise includes all areas of environment/ conservation and related issues, disaster management, and sustainable development; particularly from the strategic and policy perspective.

Dr Raman holds a PhD specializing in trade and environment from the University of Tokyo, Japan. He has a first degree majoring in Mathematics and Physics, and a Masters Degree in Science Policy. He is also qualified professionally as a Chartered Management Accountant of the Chartered Institute of Management Accountants, United Kingdom, and as a Chartered Accountant of the Malaysian Institute of Accountants.

Associate Professor Jolene LIN's research interests lie in global environmental governance. Her most recent research focuses on climate change law as a matter of public international and domestic law. She teaches International Environmental Law and Tort Law. Prof Lin is a member of the editorial boards of Climate Law and Transnational Environmental Law. She is an associate member of the Asia-Pacific Centre for Environmental Law, National University of Singapore, and a member of the advisory board for the University of Strathclyde's LLM in Climate Change Law and Policy. She received the 2013 IUCN Academy Junior Scholarship Prize for her environmental law research. Jolene also serves on the Hong Kong Telecommunications (Competition Provisions) Appeal Board and the Hong Kong Appeal Tribunal Panel (Buildings Ordinance).

Associate Professor Benoit MAYER is based at the International Law Institute and Environmental Law Institute of the University of Wuhan, China. He previously studied in the National University of Singapore, McGill University, and Sciences Po (Lyon). He has published a dozen articles and book chapters on climate migration, human rights, and international law, including in the *European*

Journal of International Law, the *Asian Journal of International Law* and the *Asia and Pacific Migration Journal*. He received the 2010 CISDL-IDLO Award of Excellence in Legal Scholarship on Sustainable Development.

Ms Kristen MOTEL is a law clerk with the Connecticut Judicial Branch for the judges of the Connecticut Superior Court. She is a graduate of Pace University School of Law with a certificate in Environmental Law. Kristen has worked as a legal intern for the Pace Environmental Litigation Clinic, the U.S. Environmental Protection Agency, and a private land-use law firm.

Ms Alexandra MUELLER received a Bachelor's degree in Political Science from the University of Bonn, Germany, and a Master of Science degree in Sustainability (Climate Change) from the University of Leeds, UK. She gained working experiences related to climate protection and adaptation as well as nature conservation in several countries in South America before joining the IUCN as project officer, where she primarily works in topics related to transboundary water governance.

Ms Valentina NOZZA is currently enrolled in a master in law program at the University of Trento, Trento, Italy. She is passionate about Environmental Law and is currently writing her thesis on plant genetic resources for food and agriculture as well as working in a law firm in Bolzano, Italy.

Emeritus Professor Richard OTTINGER came to Pace Law School when he retired from Congress in 1984. As a professor he taught in the environmental law program (1984–94). He served as Dean (1994–99). While Dean he had constructed a new modern classroom building which was dedicated as Richard Ottinger Hall in 2013. He was also responsible for bringing to Pace the NY State Judicial Institute. As co-director of the Pace Global Center for Environmental Legal Studies, he founded the Pace Energy & Climate Center, which raised more than $1 million dollars per year, advocating for promotion of clean energy resources. In his 16 years as a member of the United States House of Representatives, he authored a substantial body of energy and environmental laws. He was one of the earliest environmentalists in Congress in 1965. As chairman of the Energy Conservation and Power Subcommittee, Energy & Commerce Committee, he was instrumental in adopting key energy and environmental legislation. Dean Ottinger was a founding staff member of the Peace Corps, serving it during 1961–1964. Currently, Dean Ottinger serves as Dean Emeritus, continuing to advise J.D., LL.M and SJD students and involve them in national and international environmental an energy efficiency and renewable energy research and presentations.

Dr Euston QUAH is Professor and Head of the Department of Economics and Chairman of the Senate Committee on University Policy Matters at Nanyang Technological University (NTU), Singapore. He was formerly Chair, School of Humanities and Social Sciences at NTU; Vice-Dean, Faculty of Arts and Social Sciences; Deputy Director of the Public Policy Program (now called the Lee Kuan Yew School of Public Policy); and headed the economics department at the National University of Singapore (NUS). A prolific writer, Professor Quah had published over 60 papers in journals such as *World Development, Environment and Planning*, *Applied Economics*, *Journal of Economics*, *Journal of*

Public Economic Theory, *Journal of Environmental Management*, *Impact Assessment and Project Appraisal*, *American Journal of Economics and Sociology*, and *International Review of Law and Economics*, among others. His most recent work is a paper in an international publication on cost-benefit analysis by Oxford University Press, 2013. Two books on Cost-Benefit Analysis were published by Routledge, UK in 2007 and 2012 respectively. He was co-author of an Asian Edition of the bestselling *Principles of Economics Text* with Gregory Mankiw published by Cengage (formerly Thomson, US), now a Second Edition in 2013. Professor Quah advises the Singapore Government in various Ministries and was a Member of the recent Prime Minister's Economic Strategies Sub-Committee on Energy and the Environment. He presently sits on the Boards of Energy Market Authority, Fare Review Mechanism Committee of the Ministry of Transport, and the Energy Studies Institute at NUS. He is also a Review Panel Member for the Bill and Melinda Gates Foundation project hosted by the Overseas Development Institute, London. Professor Quah is Editor of the *Singapore Economic Review* (since 2002), and the 16th President of the Economic Society of Singapore. He has been invited by Stanford University, Princeton University, the USA Inter-Pacific Bar Association, WWF for Asia, UNESCAP, Earth Institute of Columbia University (Asian Meetings), ADBI and ADB to speak at their functions and conferences, and he is one of the most highly cited and influential university economists in Singapore.

Dr Kanongnij SRIBUAIAM is an Assistant Professor at Chulalongkorn University Faculty of Law and Research Fellow at the Law and Development Research Center / LDRC. She holds LLB and LLM degrees from Chulalongkorn University, an LLM from Harvard Law School, a BA in International Environmental Studies from International Pacific College (New Zealand), and a Doctorate in environmental public law from the University of Nantes, France. Her interests include the environment, law and society, law reform, and law and development. She is a frequent contributor to various Thai governmental studies and projects.

Dr Kanongnij teaches Advanced Environmental Law; Legal Philosophy; Legal Research Methodology; Seminar on Law and Social Problems; etc. Currently, she serves as a member of the Sub-Committee on Making Framework for Constitutional Drafting responsible for issues on reforming the country for reducing inequality and maintaining justice.

Mr TAN Tsiat Siong graduated from Nanyang Technological University with a Bachelor of Arts in Economics with First Class Honors, and is currently completing his postgraduate degree at the National University of Singapore. He has worked in the fields of Behavioural Economics, Energy Economics, Environmental Economics, and sustainability issues.

Ambassador Amado TOLENTINO is the Executive Governor (for developing countries), International Council of Environmental Law (Bonn, Germany) which, among others, promotes the international exchange of environmental law information. A member of the Philippines Bar and the New York (US) Bar, he received his law education at the University of the Philippines, the Universidad de Madrid (Spain), and the Academy of International Law (Dallas, Texas, US). He was a Visiting Fellow at Hawaii's East West Center and a participant (1976 and 1990), US State Department's International Visitor Program (IVP). An environmental law pioneer, he was elected Delegate to the 1971 Philippines Constitutional Convention. Thereafter, he was appointed first Director of the Philippines' Environmental Management Bureau; Coordinator, ASEAN Experts Group on the Environment; Vice Chair, IUCN Commission on Environmental Law; and environmental law consultant to UNEP and UNESCAP. He participated in the development of UNEPs Montevideo (Environmental Law) Programme in 1982 and 1993 and was also involved in the formulation of ICEL's internationally binding draft Covenant on

Environment and Development. He later joined the Philippines Foreign Service and served as the Philippine's Ambassador to Papua New Guinea and Qatar. He has written extensively for law journals and environmental publications and represented his country in numerous environmental conferences. Presently, he is a professor of environmental law.

Ms WANG Pianpian has a Master of Laws in Environmental Law from Pace University School of Law, New York. Prior to that, she obtained a Juris Master with focus on environmental law from Sun Yat-sen University, China. She also holds a Bachelor of Law and passed the Chinese Bar Exam in 2009. She has consistently focused on carbon markets, carbon accounting associated with US corporate GHG emissions, and sustainable practices in the context of climate change as a research manager of a private company based in New York and a research fellow of Pace Energy and Climate Center.

Dr Professor WANG Xi is the Director of the Environmental and Resources Law Institute of the Law School, Shanghai Jiao Tong University. He is a Deputy Director of the Center for Polar and Deep Sea Development of Shanghai Jiao Tong University. He is a member of the Academic Committee of the Key Institute of Education Ministry of PRC for the Field of Environmental Law. He is a Vice Chairman of the Chinese Society of Environmental and Resources Law. He served as a Deputy Public Prosecutor General of Hubei Province from 2000–2002. He is a member of the National Committee of the Chinese People's Political Consultative Conference (CPPCC) from 1997 to 2012. He received his BA degree from Wuhan Normal College in 1981. He graduated from the Law School of Wuhan University in 1984 (ML) and from Washington University Law School in 1987 (MJS). He got PhD in Law from Wuhan University Law

School in 2000. He teaches courses of International Environmental Law, Comparative Environmental Law, and Chinese Environmental Law. His university textbook of International Environmental Law was recommended by the Ministry of Justice and got a National Award of Excellent University Textbook of Law in 2006 conferred by the Ministry of Education of PRC.

Professor Wang provides legal consultation on environmental law for Chinese national and local legislatures and international organizations. One of his recent research projects is the research on amendment of the Environmental Protection Law of PRC (1989). Based upon the research, he and his team provided a train of thought and a full text of the amendment to the National People's Congress — both were accepted and reflected to a large degree in the new Environmental Protection Law of PRC adopted in April, 2014. He served as the Chinese Coordinator for the Ceremony of Launching the IUCN Academy of Environmental Law and the Chairman of the First Colloquium of the Academy in Shanghai in 2003. He was awarded the Six Annual Senior Scholar Award by the IUCN Academy of Environmental Law in 2014. He has been invited as visiting professor and conference participant by many environmental law institutions in North America, Europe, Australia, Southeast Asia, and Africa. His major publications include a university textbook entitled *International Environmental Law* published in 1998 and 2005, a book entitled *Introduction to Environmental Law of USA* published in 1992, and a book entitled *Environmental Law in China*, published in 2012.

Keynote Address

ASEAN Experiences on Adaptation to Climate Change

*Raman Letchumanan**

KEY MESSAGES

(1) *The ASEAN region — with its geographical, climatic, social, economic and developmental conditions — is very vulnerable to climate change, the devastating effects and, the increased frequency and intensity of which have been witnessed in recent years.*

(2) *While ASEAN, by necessity and as a matter of priority if not survival, has to act to adapt and build resilience to the effects of climate change, mitigating climate change through urgent real reduction of greenhouse gas emissions is essential by all parties consistent with the principle of common but differentiated responsibility.*

(3) *In essence, mitigation is prevention, while adaptation only addresses the symptoms and impacts of climate change. At the implementation level these two aspects are complementary and go hand-in-hand and cannot be distinguished, or dealt with separately, from one another.*

(4) *Unfortunately, mitigation and adaptation have been artificially constructed as distinct issues based on the global climate change negotiation dynamics which is played out as north-south divide, developed versus developing country, and availability of capacity, resources, and technology. In reality, actions on climate change do not recognize*

*Senior Fellow, S. Rajaratnam School of International Studies, Nanyang Technological University, Singapore; formerly Head of the Environment Division, ASEAN Secretariat.

such distinctions but should be based on overall contributions to global environmental sustainability, social development, and economic growth through a multisectoral approach.

(5) *ASEAN has in place a robust legal, institutional, policy, and programmatic framework working simultaneously in all interrelated sectors, which can be further enhanced focusing on climate change, as ASEAN prepares for its post-2015 ASEAN Community building efforts.*

(6) *The proposed establishment of an "adaptation climate change" group in APCEL which brings together policy-makers, the private sector, academic scholars, researchers, NGOs and CSOs both in ASEAN and the world to work together to find solutions is necessary to support the efforts of ASEAN. The immediate task is to develop an effective ASEAN Climate Change Adaptation Work Plan anchored by local experts and based on local solutions and supported by the international community.*

Climate Change — Exacerbating Human-Induced Disasters in ASEAN

Even before the global climate change debate began in 1992, the ASEAN region was, and still is, well-known as the most disaster prone region in the world, both natural- and human-induced. Climate change has further compounded these negative effects, especially human-induced disasters, in the region. These negative impacts are not only devastating to the economies and progressive development of the countries, but have brought unprecedented damage to lives and property. In 2014 alone, Typhoon Haiyan, the prolonged dry weather in the earlier part of the year, and heavy rain and widespread flooding in the later part, have all surpassed records seen over the last few decades or even centuries. Unfortunately, such negative impacts are expected to continue in the coming years and decades, as the global community wrangles painstakingly to come to a new global agreement on addressing climate change in 2015, much less to act on reducing greenhouse emissions to an acceptable level in a timely manner to keep global temperature fluctuations within 2°C.

It is apparent why the ASEAN region is easily prone to escalation of negative impacts of climate change due to its geography, climatic factors, demography, socio-economic activities, and state of development.

ASEAN is particularly vulnerable to the impacts of climate change due to the concentration of people and economic activities in the coastal areas, its rich natural ecosystems and biological diversity, resource-based economies, and the increased vulnerability of the people especially the poor. Due to its geological and geographical factors, the region is also one of the world's most vulnerable regions to suffer from a range of climatic and natural hazards such as earthquakes, typhoons, sea level rise, volcanic eruptions, droughts, heat waves, and tsunamis which are becoming more frequent and severe. In addition, the geophysical and climatic conditions shared by the region have also led to common and trans-boundary environmental concerns such as air and water pollution, urban environmental degradation, and transboundary smoke haze pollution.

The ASEAN region lies within the waters of the Pacific Ocean, Indian Ocean, Andaman Sea, and South China Sea, and stretches more than 3,300 kilometers from north to south (latitudes 30° North to 11° South) and 5,600 kilometers from west to east (longitudes 92° West to 142° East). The countries in the region make up a long coastline, measuring about 173,000 kilometers in total, and are surrounded by major seas and gulfs such as the South China Sea, the Andaman Sea, and the Gulf of Thailand. Any change in global temperatures will certainly have an impact on the region in terms of storms, typhoons, heavy rainfall, and drier seasons.

The ASEAN region is endowed with rich natural resources that sustain essential life support systems both for the region and the world. Even though the region has only about 4% of the world land mass, it contains about 20% of the world's indigenous species of terrestrial biodiversity. The rich marine life and abundant mineral resources support important economic activities such as oil exploration, commercial and small-scale fisheries, and tourism. Apart from providing water, food, and energy, these natural resources play an important role in sustaining a wide range of economic activities and livelihoods.

On the other hand, the region consists of about 600 million people, and several countries and regions have a high population or population density. The level of development varies greatly across countries consisting of least developed and highly developed nations. Poverty levels in the region remain high at about 45%.

While the rich natural endowment provides opportunities for accelerated socio-economic development, climate change impacts can reverse or even destroy the fledging livelihoods of a majority of people who lead subsistence lifestyles supported by the natural ecosystems and resources. Even more developed countries in the region such as Singapore are not spared the impacts of climate change such as, for example, the impact from sea level rise.

In that sense, adaptation to climate change has become an urgent and critical issue to ASEAN and its member states.

Mitigation or Adaptation — The Global Response to Climate Change

Climate change has become a defining and most challenging sustainable development issue of the twenty first century. It is defining in the sense that it is now dictating the pace and nature of economic growth, development, and social progress, while potentially becoming the greatest threat to humankind and survival if left unchecked. It is challenging because of its multifaceted nature, affecting almost all sectors and the basic means and lifestyle of human existence.

The climate change crisis aptly amplifies the consequences of not acting in a holistic and integrated manner. It cannot be fixed by technology or finite human and capital resources alone. It is also the most urgent in the sense that, if remedial measures are not taken, a point of no return would be reached in the next few decades. The climate change crisis is the embodiment of what can go wrong, if action is not taken globally based on the principle of common but differentiated responsibility.

The authoritative Intergovernmental Panel on Climate Change (IPCC) to the United Nations Framework Convention on Climate Change (UNFCCC) has estimated that globally, emission reductions of 25–40% of 2000 levels are needed if the global temperature fluctuations are to be maintained within a range of 2°C, a level which is generally taken to sustain life without any irreversible damage. The Stern Review on the Economics of Climate Change has estimated that the cost of inaction could cost up to 20% or more of global GDP, while in contrast the cost of action undertaken now to address climate change to avoid the worst impacts can be limited to 1% of global GDP each year.

In the ongoing negotiations for renewed and urgent actions on climate change, the debate on mitigating emissions of greenhouse gases, essentially carbon dioxide, is fast turning into a development issue — how much growth each nation can pursue given the constraints on the limiting capacity of atmospheric carbon space. The key argument is that the developed countries, with their historically unfettered industrial growth, have enjoyed these rights and have crowded out the carbon space, which is now limiting the development potential of developing countries, given that they too should have the right to equitable development. This draws attention to the stark reality that the environment cannot be separated from economic growth as well as social development. Each has to support and complement the other.

Much hope is being put on the proposed new global climate change treaty hopefully to be agreed to in 2015 and implemented from 2020. The debate centers around how much each country will commit to reduce greenhouse gases, and how to adapt to the impact of climate change, which invariably is tied to capacity, resources, and technology being provided.

However, the real world does not operate in that way. Even if a global agreement is reached with ambitious commitment and sufficient resources pledged, past experience since 1992 shows that the performance and fulfillment of agreements reached on climate change is dismal. At best, such agreements can only galvanize further support and incentivize collective action — only if countries supported by their regional groups take charge and tackle the climate change issue themselves.

Climate Change — Action at the Multi-Sectoral Level Based on the Concept of Sustainable Development

More importantly, climate change should be addressed in a holistic manner and at the multisectoral level based on the concept of sustainable development. While planning and programming can be done in terms of adaptation and mitigation, at the program/project level it is hard to distinguish between mitigation and adaptation, more so the impact on social development, economic growth, and of course environmental protection.

ASEAN has in place enhanced coordination mechanisms to synergize efforts across the corresponding sectoral bodies, its three community pillars, namely, socio-cultural, economic, and political-security, and across thematic issues such as climate change, disaster management, and health. Climate change impacts on almost all sectors and this is illustrated in the context of ASEAN regional cooperation areas as follows:

- Environment — all areas particularly conservation of biodiversity, water resources management, coastal and marine environment, public awareness and environmental education, forest fires, and transboundary haze pollution.
- Agriculture and Forestry — food security, forest degradation.
- Disaster Management — more and severe weather-related disasters such as floods, droughts, typhoons.
- Science and Technology — weather monitoring and forecasting, research and development on climate-related impacts and responses, vulnerability studies, etc.
- Health — more and severe weather-related diseases.
- Human Development — impact on poverty, sustainable livelihoods, social development, and generally on the achievement of the 2015 MDG goals.
- Economic Growth — affects production and transaction costs through internalizing environmental costs, change in products, services and markets, drag on GDP growth, carbon taxes, allocation and trade of environmental rights, new market opportunities for environmental goods and services.
- Energy/Transportation — renewable energy, energy conservation, mass and efficient transportation, etc.
- Political and Human Security — environmental security issues, political debate influenced more and more by environmental issues, in particular climate change.

ASEAN's Response to Climate Change

(1) *Leadership and Institutional Framework*

The ASEAN Heads of State/Government have proactively led ASEAN's efforts to address climate change issues in the region and beyond. They have consistently issued declarations/statements articulating ASEAN's concerns and position prior to almost every annual Climate Change Summits, the latest being issued in November 2014 addressed to the 21st Climate Change

Summit in Paris in 2015. They view the protection of the environment and the sustainable use and management of natural resources as essential to the long-term economic growth and social development of countries in the region. The ASEAN Vision 2020 calls for "a clean and green ASEAN" with fully established mechanisms to ensure the protection of the environment, sustainability of natural resources, and high quality of life of people in the region.

The Roadmap for an ASEAN Community 2009–2015 adopted by the ASEAN Leaders situates the ASEAN climate change agenda in the context of sustainable development outlining strategies and actions in the ASEAN Socio-Cultural Community Blueprint, ASEAN Economic Community Blueprint, ASEAN Political-Security Community Blueprint, and the Initiative for ASEAN Integration Second Work Plan. ASEAN is therefore addressing climate change, not just through a policy framework on climate change, but through the multisectoral framework of ASEAN community building, with strategies and actions rooted in the various development and sectoral areas.

The ASEAN Charter entered into force on 15 December 2008 bestowed a legal personality upon ASEAN, which for the past 42 years has operated as a coalition of nations born out of the Bangkok Declaration of 1967. Under the Charter, a newly constituted ASEAN Coordinating Council comprising the ASEAN Foreign Ministers, among others, coordinates with the three ASEAN Community Councils to enhance policy coherence, efficiency, and cooperation among them. Each of the three ASEAN Community Councils, among others, ensures the implementation of the relevant decisions of the ASEAN Summit, and coordinates the work of the different sectors under its purview and on issues which cut across the other Community Councils. The ASEAN Socio-Cultural Community Council oversees the work of the ASEAN Environment Ministers.

Environmental issues are helmed by the ASEAN Environment Ministers who are supported by the ASEAN Senior Officials on the Environment (ASOEN), which in turn oversees the work of several technical subsidiary bodies in the areas of climate change, biodiversity and nature conservation, environmentally sustainable cities, multilateral environmental agreements, coastal and marine environment, environmental education, and integrated water resources management.

Transboundary smoke haze pollution issues are helmed by the ASEAN Environment Ministers acting as the Conference of Parties to the ASEAN Agreement on Transboundary Haze Pollution which is supported by a committee of senior officials. Sub-regional ministerial steering committees also operate, respectively, in the Mekong Region and the southern ASEAN region in view of the different seasons and conditions in these two regions.

The work of these environmental technical subsidiary bodies is guided by the strategies and actions enumerated in *Section D — Promoting Environmental Sustainability* of the ASEAN Socio-Cultural Community Blueprint 2009–2015.

(2) *ASEAN Strategies and Actions on Climate Change*

Specifically on climate change, section D10 of the ASCC Blueprint provides and prioritizes collective actions in responding to climate change and addressing its impacts. The strategic objective for addressing climate change is to enhance regional and international cooperation to address the issue of climate change and its impacts on socio-economic development, health, and the environment in ASEAN Member States through implementation of mitigation and adaptation measures, based on the principles of equity, flexibility, effectiveness, common but differentiated responsibilities, respective capabilities, as well as reflecting on different social and economic conditions. The same section identifies 11 priority actions to be implemented by ASEAN Member States, namely:

- Encourage common understanding within ASEAN on climate change issues and where possible, engage in joint efforts and common positions in addressing these issues;
- Encourage the efforts to develop an ASEAN Climate Change Initiative (ACCI);
- Promote and facilitate exchange of information/knowledge on scientific research and development, deployment and transfer of technology and best practices on adaptation and mitigation measures, and enhance human resource development;
- Encourage the international community to participate in and contribute to ASEAN's efforts in afforestation and reforestation, as well as to reduce deforestation and forest degradation;

- Develop regional strategies to enhance capacity for adaptation, low-carbon economy, and promote public awareness to address the effects of climate change;
- Enhance collaboration among ASEAN Member States and relevant partners to address climate-related hazards, and scenarios for climate change;
- Develop a regional systematic observation system to monitor impact of climate change on vulnerable ecosystems in ASEAN;
- Conduct regional policy, scientific, and related studies, to facilitate the implementation of climate change conventions and related conventions;
- Promote public awareness and advocacy to raise community participation on protecting human health from the potential impact of climate change;
- Encourage the participation of the local government, private sector, non-governmental organizations, and communities to address the impacts of climate change; and
- Promote strategies to ensure that climate change initiatives lead to an economically vibrant and environment-friendly ASEAN Community taking into account a win-win synergy between climate change and the economic development.

(3) *Major Climate Change Challenges and ASEAN Initiatives*

The ASEAN Agreement on Transboundary Haze Pollution

Wild fires, especially on carbon rich peatlands, directly release carbon into the atmosphere and destroys biodiversity. ASEAN continues to implement concerted and focused on-the-ground actions to tackle wild fires and the resulting smoke haze problem through the ASEAN Agreement on Transboundary Haze Pollution. The Conference of the Parties (COP) comprising the ASEAN Environment Ministers is responsible for the implementation of the Agreement. In addition, considering the different circumstances and weather patterns in the southern ASEAN and Mekong sub-regions, sub-regional institutional frameworks have been established to address the fire and haze situations in the respective regions.

ASEAN is also pursuing concrete on-the-ground activities, including bilateral initiatives by Malaysia and Singapore with Indonesia. Regional and national plans of actions and targets have been set to reduce fires through monitoring of hotspot activities and preventive actions. The ASEAN Peatland Management Strategy is in place, and a major regional peatland project with activities in several pilot peatland sites across the region is being implemented to address the major source of fires. As forest fires destroy ecosystems and biodiversity, and contribute to climate change by releasing carbon, it is important that the global community becomes more engaged and work with ASEAN to address this recurring problem.

Project on Rehabilitation and Sustainable Use of Peatland Forests in Southeast Asia

This Project which was implemented from 2009–2014 aimed to demonstrate, implement, and scale up sustainable management and rehabilitation of peatland forests in Southeast Asia. This project focused on four main areas: (1) strengthening institutional capacity and frameworks; (2) reducing the rate of degradation of peatlands in Southeast Asia; (3) demonstrating integrated management and rehabilitation of peatlands at target sites; and (4) engaging the private sector and local communities in sustainable peatland management. It fits within the framework of the ASEAN Peatland Management Initiative, and directly supports the ASEAN Peatland Management Strategy, and associated National Action Plans on Peatlands in the participating ASEAN Member States, namely: Indonesia, Malaysia, the Philippines, and Vietnam. ASEAN Member States were able to leverage GEF funding and mainstream the implementation of the project through national mechanisms and budgetary allocations to work with various stake holders at the sites. Based on the success of the project, the ASEAN Environment Ministers agreed to upscale the project to all peatland sites in the region, with the ultimate aim of eliminating any uncontrolled wild fires by the year 2020. The new program has received very encouraging support from the EU, GEF, and other development partners.

ASEAN Heritage Parks (AHP) Programme

The ASEAN Heritage Parks Programme is one of ASEAN's environmental flagship program, whereby member countries designate their selected national protected areas and reserves as ASEAN Heritage Parks (AHP). To date, 33 sites have been designated as AHPs. The AHP Programme aims to concentrate attention on better management of protected areas and serve as a model and learning platform to enhance sustainable management of national protected areas throughout the ASEAN region, in addition to encouraging member states to reserve more biologically rich areas as protected areas. In terms of climate change, natural forests act as carbon sinks, and help to mitigate and build resilience to the damaging effects of climate change.

ASEAN Strategic Plan of Action on Water Resources Management

Recognizing the importance of freshwater resources, ASEAN has embarked on cooperation in integrated water resources management through the ASEAN Working Group on Water Resources Management (AWGWRM) in 2002. Following the endorsement of the ASEAN Long Term Strategic Plan for Water Resources Management in 2003, ASEAN has also adopted the ASEAN Strategic Plan of Action on Water Resources Management (2005) which aims to tackle issues relating to demand and supply allocation, water quality and sanitation, extreme events, and governance and capacity building. A key focus is related to addressing the impact of climate extreme events such as prolonged dry weather and heavy rainfall which is a major concern of the region in recent years.

ASEAN Marine Water Quality Management Guidelines and Monitoring Manual

ASEAN has implemented the ASEAN Marine Water Quality Criteria (AMWQC) initiative in cooperation with Australia, through the Project on "Capacity Building for the Implementation of AMWQC — Phase I (2004–2005)" and "Capacity Building for the Implementation of AMWQC — Phase II (2007–2008)". "The ASEAN Marine Water Quality Criteria: Management Guidelines and Monitoring Manual" was published as the outcome of the Project, which aims at providing a reference document for ASEAN Member States in coordinating marine water quality management policies and monitoring approaches within each of and among the countries. In future the focus will be on the impact of climate change on marine ecosystems and species diversity, which also acts as the sink for carbon emissions.

ASEAN Initiative on Environmentally Sustainable Cities (AIESC)

With the increasing trend of urbanization in the region, ASEAN is implementing the ASEAN Initiative on Environmentally Sustainable Cities

(AIESC) which focuses on addressing urban environmental challenges such as air pollution due to vehicular emissions, solid waste management, and water pollution along with access to clean water. Currently, 25 ASEAN cities are participating in the program. Cities are major consumers of natural resources, energy, and generators of waste, all of which is directly related to climate change. A particular focus will be on environmental sustainability in cities promoting environmental sustainable practices in city infrastructure, services, and urban lifestyles.

The ASEAN Environmentally Sustainable City (ESC) Award aims to recognize and encourage cities to promote Clean Air, Clean Water, and Clean Land initiatives in order to promote more environmentally sustainable cities in the ASEAN region. The Award, given every three years by the ASEAN Environment Ministers, has been bestowed in 2008, 2011, and 2014.

ASEAN Environmental Education Action Plan (AEEAP)

The five-year ASEAN Environmental Education Action Plans (AEEAP), now in its third phase (2014–2018), were adopted by the ASEAN Environment Ministers to serve as regional frameworks to undertake coordinated action at the national and regional levels to promote environmental education (EE), in order to ensure that the people develop from their own social and cultural values a sense of civic responsibility to care for the environment. The action plans focus on four areas, namely: (1) Formal Education; (2) Non-formal Education; (3) Capacity Building; and (4) Networking and Partnerships. The AEEAP 2008–2012 also contributed to the UN Decade on Education for Sustainable Development (2005–2014). One of the key outcomes is the adoption of the ASEAN Guidelines on Eco-Schools and the ASEAN Award for exemplary eco-schools in ASEAN. Awareness and actions on climate change will be a key feature of environmental education at all levels, particularly for the younger generation.

Heart of Borneo Initiative (HoB)

The HoB is a collaborative regional partnership program among the three ASEAN Member States: Indonesia, Malaysia, and Brunei Darussalam in promoting effective conservation of protected areas, sustainable management of productive forests, and sustainable land use. This initiative involves the protection and sustainable management of the key central spine of the natural ecosystem in the Borneo Island.

Coral Triangle Initiative (CTI)

The CTI is a collaborative partnership between Indonesia, the Philippines, and Malaysia working together with Papua New Guinea, Solomon Islands, and Timor Leste in implementing various interventions to protect the diverse marine and coastal resources covering the six participating countries.

ASEAN Agreement on Disaster Management and Emergency Response (AADMER)

Disaster risk reduction measures are immediate response measures to adapt to climate change. ASEAN has included climate change adaptation (CCA) as part of the AADMER Work Programme 2010–2015. As a cross-cutting issue, the intention is to take into account climate change adaptation (CCA) in all the relevant strategic components such as Prevention and Mitigation; Risk Assessment; Early Warning and Monitoring; Preparedness and Response; and Recovery and Rehabilitation. These core components represent the disaster risk reduction (DRR) framework and provide opportunities to mainstream DRR within AADMER. The AADMER Work Programme serves to operationalize the provisions of AADMER in order to realize the ASEAN vision of a disaster-resilient and safer community by the year 2015.

Energy

Climate change has increasingly become an important element in energy policy in ASEAN. More attention has been given to the need to adopt integrated approaches to energy policy-making by taking into account climate change issues to achieve greater energy security in the future. In fact, the ASEAN Plan of Action for Energy Cooperation (APAEC) 2010–2015 carries the theme "Bringing Policy to Actions Towards a Cleaner, More Efficient and Sustainable ASEAN Energy Community".

The three key initiatives of the ASEAN energy security policy/plan within which environment and climate change appear as important elements include:

(1) Energy Efficiency and Conservation (EE&C)

EE&C is considered as one of the most cost-effective ways of enhancing energy security in addressing climate change as well as promoting competitiveness in the ASEAN region.

The strategic goals of ASEAN's EE&C in APAEC 2010–2015 include: (i) pursuing the aspirational goal of reducing regional energy intensity of at least 8% by 2015 based on the 2005 level; (ii) achieving higher end-use energy efficiency for all sectors through regulatory and market approaches, where appropriate; (iii) enhancing institutional and human capacity emphasizing the development of energy efficiency technology and service providers in the ASEAN region; and (iv) encouraging private sector participation, especially financial institutions to support EE&C investment and implementation.

(2) Renewable Energy (RE)

In 2004–2009, ASEAN achieved its 10% target to increase the installed RE-based capacities for power generation. For 2010–2015, ASEAN commits that renewable energies are crucially needed to increase the diversity of energy supply and to reduce the environmental impact of energy use in the region. The strategic goals include: (i) achieving a collective target of 15% for regional renewable energy in the total power installed capacity by 2015; (ii) strengthening of regional cooperation on the development of renewable energy and alternative energy including hydropower and biofuels; (iii) promoting the development of centers of research and development on renewable energy in the region; and (iv) promoting open trade, facilitation, and cooperation in the renewable energy sector and related industries as well as investment in the requisite infrastructure for renewable energy development.

(3) Coal and Clean Coal Technology

In the ASEAN region, coal is well-known as the fastest growing energy source due to the increasing demand of fuel for electricity generation as well as in the industrial sector since 2005. The energy plans of the ASEAN Member States indicate the rapid growth of coal utilization for power generation which presents an opportunity to promote and increase cleaner coal use and trade that could bring in mutual economic benefits towards regional energy security.

Despite growing environmental controls, more coal power projects are moving forward with increasing preference for clean coal technologies. Collective action is therefore needed to strengthen cooperative partnerships in the promotion and utilisation of coal and clean coal technologies among the ASEAN Member States. The Strategic Goals of ASEAN's Coal and CCT in APAEC 2010–2015, include: (i) promoting and increasing cleaner coal use and trade for regional energy security; (ii) encouraging the use of clean coal technologies through regional cooperation; and (iii) building the perception of coal as a cleaner fuel as a result of clean coal technologies.

Agriculture and Forestry

Responding to the urgent need to address the growing threat and impact of climate change, the ASEAN Multi-Sectoral Framework on Climate Change (AFCC) and Food Security (AFCC-FS) has been established as a comprehensive program to address the emerging issues of climate change.

The overall goal of the AFCC is to contribute to food security through sustainable, efficient, and effective use of land, forest, water, and aquatic resources by minimizing the risks and impacts of climate change. To achieve its goal, sectoral activities will be carried out under the two major objectives of: (1) coordination on the development of adaptation and mitigation strategies and (2) cooperation on the implementation of integrated adaptation and mitigation measures.

AFCC pursues a cross-sectoral program approach for effective policy-making and implementation, and also provides an arena for ASEAN to better coordinate the support from its partners. Along with its focus on the agriculture, fisheries, and forestry sectors, it will coordinate with the environment, health, and energy sectors.

Conclusion

Responses to climate change should not be viewed in a narrow perspective, because every sector has a contribution to make to address climate change. The discussion here therefore has taken a broad-based approach, highlighting how actions in various sectoral and thematic areas could contribute to addressing climate change, though the immediate linkage may not be apparent.

ASEAN is committed to pursuing this broad based approach, to play its role in taking voluntary and appropriate mitigation actions, and adaptation measures enabled by technology transfer, provision of concessionary financial assistance, and capacity building to address climate change issues in a proactive and responsible manner.

ASEAN has in place a robust legal, institutional, policy, and programmatic framework working simultaneously in all interrelated sectors, which can be further enhanced focusing on climate change as ASEAN prepares for its post-2015 ASEAN community building efforts.

The proposed establishment of an "adaptation climate change" group in APCEL which brings together policy-makers, the private sector, academic scholars, researchers, NGOs and CSOs in ASEAN as well as the world to work together to find solutions is necessary to support the efforts of ASEAN. The immediate task is to develop an effective ASEAN Climate Change Adaptation Work Plan comprising regional collective actions and national actions, anchored by local experts and local solutions, and supported by the international community.

Theme I

Coastal Warming and Sea Level Rise

Chapter 1

Asia-Pacific Islander Responses to Climate Change

Ilan Kelman

1.1. Introduction

Climate change is affecting the entire planet Earth, but many locations might not experience direct, discernible impacts for a few decades. Other areas are experiencing changes now and need to determine immediately how to deal with those changes alongside those expected in the future. Within this latter group sits the low-lying, smaller islands of Asia, most notably the Pacific Small Island Developing States (SIDS).

SIDS comprise several dozen sovereign countries and non-sovereign territories which are mainly islands, scattered around the tropics and subtropics, but primarily in the Caribbean Sea, Indian Ocean, and Pacific Ocean. Examples of SIDS in non-Pacific locations are Cape Verde and Singapore. Examples of non-island SIDS are Belize and Guyana. The SIDS group formed under United Nations auspices due to facing common development and sustainability problems with solutions that are frequently transferable amongst different SIDS.[1] Currently, the Pacific SIDS which are sovereign are Fiji, Kiribati, Marshall Islands, Federated States of Micronesia, Nauru, Palau, Papua New Guinea, Samoa, Solomon Islands, Timor-Leste, Tonga, Tuvalu, and Vanuatu. The Pacific SIDS also include two New Zealand territories — the Cook Islands and Niue.

[1]UN, *Report of the Global Conference on the Sustainable Development of Small Island Developing States* (Document A/CONF.167/9 United Nations, New York, 2005); UN, *Draft Mauritius Strategy for the Further Implementation of the Programme of Action for the Sustainable Development of Small Island Developing States* (Document A/CONF.207/CRP.7 United Nations, New York, 2005).

There are more Asia-Pacific islanders than just the Pacific SIDS. Indonesia, the Philippines, and Japan are archipelagic countries with many communities facing similar climate change challenges as the SIDS. Malaysia, Cambodia, and Vietnam have numerous island communities along their coastlines. India's islands include Lakshadweep, the Andamans, and the Nicobars while China has Hainan and Hong Kong. Amongst the most famous of Asia's islands affected by climate change are the Sundarbans in Bangladesh and India.[2] The importance of these islands cannot be neglected, but the focus of this chapter is on the Pacific SIDS to scope this study.

The islanders of all these countries recognize that their countries face governance and livelihoods challenges, which are not assisted by the looming threat of climate change — nor by many of the narratives and myths which have emerged with respect to islands and climate change. Yet they all display rich cultures and histories, leading to some level of ability to face the challenges and try to solve them, drawing on their local knowledge, the tight kinship of many of their communities, and a strong desire to retain their identity. How much can be retained, and even the feasibility of inhabiting many of the communities and islands long into the future, is an open debate.

This chapter examines these debates, exploring the vulnerabilities and resiliences to climate change of the Pacific SIDS peoples and communities. Difficulties are inevitably present, but those should be balanced with potential opportunities, even while noting that sacrifices might be necessary to make use of the opportunities. Good practice examples here are presented alongside communities which feel they have few options.

Section 1.2 presents a discussion of climate change's potential impacts on Pacific SIDS, followed by Section 1.3 jointly describing mitigation and adaptation options. The topics, deliberately framed to bring mitigation and adaptation together, are ecosystems, energy, environmental extremes, and mobility (migration). Legal and ethical questions are then

[2]Rajarshi Dasgupta and Rajib Shaw, "Cumulative Impacts of Human Interventions and Climate Change on Mangrove Ecosystems of South and Southeast Asia: An Overview," *Journal of Ecosystems*, 2013, Article ID 379429, 15 pages. Retrieved 12 June 2015, from http://dx.doi.org/10.1155/2013/379429.

explored for climate change mitigation and adaptation in Pacific SIDS communities, covering aspects of responsibility, funding, sovereignty, decision-making, and decision implementation. The conclusions highlight the stark disconnect between who creates climate change impacts and who experiences the worst impacts.

1.2. Climate Change Impacts

Table 1 summarizes the dominant climate change impacts for Pacific SIDS.

Table 1: Dominant Climate Change Impacts for Pacific SIDS

Sphere of Impact	Specific Parameter or Area of Impact	Specific Impacts
Atmosphere	Air temperature	• Increasing globally with widely varying regional and local increases, affecting ecosystems, species, and weather.
	Cycles (e.g., ENSO, NAO, PDO)	• Changing intensities, frequencies, and durations with large uncertainties regarding the directions and magnitudes of the changes.
	Droughts	• Changing intensities, frequencies, and durations. • Changing terrestrial species and ecosystems.
	Humidity	• Changing baseline values at a given location. • Changing soil. • Changing terrestrial species and ecosystems.
	Extreme temperatures	• Changing intensities, frequencies, and durations.
	Precipitation	• Changing baseline volume and type at a given location. • Changing soil. • Changing terrestrial species and ecosystems.
	Storms	• Changing intensities, frequencies, and durations in terms of wind, freshwater flooding, coastal flooding, and ocean waves affecting fishing and coastal zones. • Changing terrestrial species and ecosystems. • Near-shore species and ecosystems could be affected by freshwater and sediment run-off changes.

(Continued)

Table 1: (*Continued*)

Sphere of Impact	Specific Parameter or Area of Impact	Specific Impacts
	Wind patterns	• Changing baseline speeds and direction at a given location. • Changing terrestrial species and ecosystems.
Land	Coastal zone exposure to ocean	• Likely to experience more wave power and changing near-shore species and ecosystems as surrounding coral reefs and coastal vegetation are impacted. • Eroding, retreating, and accruing coastal zones.
	Salt water intrusion	• Salinising of fresh groundwater.
	Terrestrial ecosystems	• Shifting biome. • Changing species composition, including invasive alien species.
	Terrestrial species	• Increased disease susceptibility. • Extinctions. • Migrations. • Prey switching, changing ecosystem balance.
Ocean	Acidification	• Changing ecosystems and species. • Possible geomorphological changes, especially to coral islands and shingle beaches.
	Current changes	• Changing ecosystems and species.
	Marine ecosystems	• Shifting biomes. • Coral bleaching. • Changing species compositions, including invasive alien species.
	Marine species	• Increased disease susceptibility. • Extinctions. • Migrations. • Prey switching, changing ecosystem balance.
	Sea level	• Irrespective of the global mean rise, local rises could be over one meter. • Possible extreme scenarios could lead to an eventual global rise of over five meters. • King tides many centimeters higher. • Wave regimes with higher power and deeper inland penetration.

(*Continued*)

Table 1: (*Continued*)

Sphere of Impact	Specific Parameter or Area of Impact	Specific Impacts
	Sea surface temperature	• Increasing globally, affecting ecosystems, species, and weather.
Settlements	Agriculture	• Affected by atmospheric, water, ecosystem, and species changes. • Changing land, coastal, and water-based agriculture.
	Buildings	• Might be less comfortable or less safe if constructed for a specific climate.
	Extreme events	• Changing intensities, frequencies, and durations of atmospheric-based extreme events (storms and droughts) affect settlement safety and design. • Changing impacts on other extreme events, such as landslides, avalanches, epidemics, earthquakes (e.g., due to soil changes), and volcanic events (e.g., due to tephra dispersion and explosivity intensity).
	Health	• Diseases might migrate into new locations and change their virulence, frequency, and mutation rates. • Heat-, humidity-, and dehydration-related illnesses and productivity losses are likely. • Extreme event casualties will vary according to extreme event frequencies, intensities, locations, and durations.
	Migration	• Some islands evacuated. • Some settlements moved inland. • Some inland peoples moving towards the coasts, especially cities.
Direct policy links	Aid	• Development aid might be affected as donors deal with their own impacts.
	Heritage sites	• Damage and losses such as through inundation, forced migration away from or onto sites, and the changing environment affecting natural heritage.
	Livelihoods	• Changing natural resource-based livelihoods. • Changing tourism-based livelihoods.

(*Continued*)

Table 1: (*Continued*)

Sphere of Impact	Specific Parameter or Area of Impact	Specific Impacts
	Local knowledge	• Could it become less useful for understanding the local environment if changes are rapid or unusual enough? Or can local knowledge adjust at the same speed as the changes?
	Resource use	Conflict and cooperation over: • Terrestrial and marine resources of evacuated islands. • Fresh water, due to precipitation changes. • Marine living resources, especially fisheries, as ecosystems and species change. • Timber, as ecosystems and species change. • Mineral resources, especially fossil fuels, as accessibility of deposits changes. • Exclusive economic zones change with coastlines.

Source: Updated from CICERO and UNEP/GRID-Arendal, *Many Strong Voices: Outline for an Assessment Project Design* (CICERO Report, 2008: 05), CICERO (Center for International Climate and Environmental Research, Oslo); Ilan Kelman and Jennifer West, "Climate Change and Small Island Developing States: A Critical Review," *Ecological and Environmental Anthropology*, 5(1) (2009), 1–16.

Additionally, around the world, many climate change impacts are observed regarding snow, ice, glaciers, and cold weather. Because SIDS are located in the tropics and sub-tropics, and the Pacific SIDS in particular are all in hot climates, these impacts have limited relevance here. Some communities in Papua New Guinea are at a high enough altitude to experience frost, with snow being common on some of the highest locations. These locations would need to consider the impacts of climate change on cold weather phenomena.

Volcanic eruptions, tsunamis, and earthquakes also have prospects for being augmented by climate change,[3] which would affect all Pacific SIDS.

[3]Bill McGuire, *Waking the Giant: How a Changing Climate Triggers Earthquakes, Tsunamis, and Volcanoes* (UK: Oxford University Press, 2013).

The simplistic geophysical explanation is that, as sea levels rise and as glaciers melt, a shifting of the weight on the Earth's crust has the possibility for affecting seismicity and underwater landslides, with subsequent effects on tsunamis. Similar impacts could affect volcanic eruptions. There is much speculation and many uncertainties regarding the interaction between climate change and these geological hazards. In contrast, since climate change by definition affects only planet Earth, there seems little scope for impacts on astronomical hazards, such as solar flares and comets or asteroids impacting the Earth.

1.2.1. *Temperature and precipitation*

Table 2 indicates examples of projected ranges of temperature and precipitation changes for the Pacific region. The values are not as high as for other SIDS locations, but are nonetheless projected to bring many significant changes, as described in Table 1. The regions covered in Table 2 are large, with downscaling to smaller areas — especially specific SIDS and specific islands — being difficult with poor precision and accuracy. That makes it difficult to plan nationally and locally since the mean projected changes over a large area for one scenario offers limited scope for local decision-making.

1.2.2. *Sea level rise*

For the Pacific SIDS, sea level rise for 2081–2100 under one medium emissions scenario is expected to be 0.5–0.6 meters above 1986–2005

Table 2: Projected Change for 2081–2100 Compared to 1986–2005 for a Medium Emissions Scenario

	Mean Annual Temperature Percentiles (°C)			Mean Annual Precipitation Percentiles (%)		
	25th	50th	75th	25th	50th	75th
Northern tropical Pacific	1.2	1.4	1.7	0	1	4
Southern tropical Pacific	1.1	1.2	1.5	0	2	4

Source: IPCC, *Fifth Assessment Report* (IPCC, 2013–2014).

levels.[4] The projected sea level rise occurs partly due to melting ice, such as glaciers and ice sheets, but mainly from the rise of ocean temperature. As water warms above 4°C, it decreases in density. The current mass of water in the oceans will remain the same, meaning that ocean volume will increase, manifesting as a rise in the sea level.

So far, sea level has not been rising uniformly around the world, with some parts of the Pacific experiencing a rise of nearly 10 times the global mean.[5] In addition to the thermal expansion, it is now suggested as being inevitable that a major ice sheet in Antarctica (called the Western Antarctic Ice Sheet, WAIS) will collapse, adding 3–4 meters onto sea level over the next several centuries.[6] If that occurs, then many Pacific SIDS would be entirely flooded, while the capital cities of most Pacific SIDS would experience major damage and disruption.

Even without WAIS collapsing, several of the Pacific SIDS are expected to lose a large proportion of their land, perhaps all of it, especially when adding sea level rise from climate change to sea level variation from storms, waves, and temporary rises due to phenomena such as El Niño. Where Pacific SIDS have large expanses of higher-altitude land — with Fiji, the Solomon Islands, Timor-Leste, Samoa, and Vanuatu, as examples — they still have numerous settlements and infrastructure (such as international airports) along coasts while the inland regions would experience their own ecological alterations due to climate change.

The low-lying Pacific SIDS would have the most problems. Examples are the Marshall Islands with its population of 60,000,[7] Tuvalu with its population of 11,000,[8] Kiribati with its population of 105,000, and Tonga

[4]IPCC (2013–2014), *op. cit.*

[5]A. Cazenave and R.S. Nerem, "Present-day Sea Level Change: Observations and Causes," *Reviews of Geophysics*, 42 (2004), RG3001. DOI: 10.1029/2003RG000139.

[6]Ian Joughin, Benjamin E. Smith, and Brooke Medley, "Marine Ice Sheet Collapse Potentially under Way for the Thwaites Glacier Basin," *West Antarctica Science*, 344 (2014), 735–738; J. Mouginot, E. Rignot, and B. Scheuchl, "Sustained Increase in Ice Discharge from the Amundsen Sea Embayment, West Antarctica, from 1973 to 2013," *Geophysical Research Letters* (2014). DOI: 10.1002/2013GL059069.

[7]Peter Rudiak-Gould, *Climate Change and Tradition in a Small Island State: The Rising Tide* (UK: Routledge, 2013).

[8]John Connell, "An Atoll State in Peril," *Tiempo* (2001). Retrieved 19 May 2015, from http://tiempo.sei-international.org/portal/archive/issue42/t42a1.htm; Bradley C. Parks and

with its population of 115,000.[9] Kiribati and Tonga each has volcanic islands with relatively high topography, but they are small tracts and the vast majority of the population and infrastructure are in low-lying locations, meaning major disruption for the entire country.

Yet complete island or country destruction is not definite, because the expected physical changes to low-lying islands under various sea level rise scenarios have not been well-studied and there is limited long-term empirical evidence to draw upon. Significant geomorphological changes are likely, but that does not necessarily mean inundation, drowning, or eroding away.[10] Additionally, numerous interacting factors might be difficult to anticipate, such as the death of coral reefs due to ocean acidification (in which seawater absorbs carbon dioxide from the atmosphere, leading to carbonic acid formation in the ocean water) and sea warming reducing the protection from wave and current action, thereby accelerating erosion processes. Even with the continuing existence of land, however, it would not necessarily be possible to retain communities in the affected islands due to possibilities for sudden changes.

On 21 October 1972, Cyclone Bebe struck the Ellice Islands which later became Tuvalu. Funafuti Atoll experienced a storm surge which led to the building up of a coral rubble wall with a mean height of 3.5 meters along its 18–19 kilometer length and 30–40 meter width, with these dimensions exceeding the size of some islets within the atoll.[11] If similar events

J. Timmons Roberts, "Globalization, Vulnerability to Climate Change, and Perceived Injustice," *Society and Natural Resources*, 19(4) (2006), 337–355; Holley Ralston, Britta Horstmann, and Carina Holl, *Climate Change Challenges Tuvalu* (Germany: Germanwatch, 2004).

[9]James Lewis, "The Vulnerability of Small Island States to Sea Level Rise: The Need for Holistic Strategies," *Disasters*, 14(3) (1990), 241–248.

[10]Nick Harvey and Bill Mitchell, "Monitoring Sea-level Change in Oceania," *Tiempo*, 50 (2003), 1–6; Paul Kench and Peter Cowell, "Erosion of Low-lying Reef Islands," *Tiempo*, 46 (2002), 6–12; Arthur Webb and Paul Kench, "The Dynamic Response of Reef Islands to Sea-level Rise: Evidence from Multi-Decadal Analysis of Island Change in the Central Pacific," *Global and Planetary Change*, 72 (2010), 234–246; Eugene Rankey, "Nature and Stability of Atoll Island Shorelines: Gilbert Island Chain, Kiribati, Equatorial Pacific," *Sedimentology*, 58 (2011), 1831–1859.

[11]G.B.K. Baines and R.F. McLean, "Re-Surveys of 1972 Hurricane Rampart of Funafuti Atoll, Ellice Islands," *Search*, 7 (1976), 1–2, 36–37; J.E. Maragos, G.B.K. Baines, and P.J. Beveridge, "Tropical Cyclone Bebe Creates a New Land Formation on Funafuti Atoll," *Science*, 181 (1973), 1161–1164.

occurred frequently or if the atolls experienced several decades of major geomorphological reconfiguring, then it would not be possible to establish long-term communities in those locales.

1.2.3. *Additional details on other impacts*

As alluded to above, climate change is likely to induce many changes on the Pacific SIDS with significant resulting impacts on the inhabitability of low-lying communities. Ocean acidification is expected to lead to major mortality amongst coral reefs, but many islands are formed of dead coral rubble and the acidification chemistry of shingle beaches is not fully understood.[12]

Meanwhile, Pacific SIDS including Tonga, Kiribati, and Tuvalu are dependent on freshwater lenses underneath atolls and those are projected to experience major changes. First, rainfall patterns will shift, meaning that some locations will receive more precipitation over the course of a year and some will receive less, but the annual distribution of rainfall might change as well irrespective of the total received. Second, sea level rise and storms will increasingly salinize these water supplies.

For tropical cyclones, it appears as if climate change will lead to reduced frequency of formation but increased intensity once a storm forms.[13] There are feedbacks amongst sea surface temperatures, stratospheric winds, and the lessening temperature differential between the poles and the tropics (the poles warm faster), in addition to other factors, which make analysis challenging, especially for specific SIDS or even specific SIDS regions such as the Pacific.

[12]Andreas J. Andersson, Kiley L. Yeakel, Nicholas R. Bates, and Samantha J. de Putron, "Partial Offsets in Ocean Acidification from Changing Coral Reef Biogeochemistry," *Nature Climate Change*, 4 (2013), 56–61; Katharina E. Fabricius, Chris Langdon, Sven Uthicke, Craig Humphrey, Sam Noonan, Glenn De'ath, Remy Okazaki, Nancy Muehllehner, Martin S. Glas, and Janice M. Lough, "Losers and Winners in Coral Reefs Acclimatized to Elevated Carbon Dioxide Concentrations," *Nature Climate Change*, 1 (2011), 165–169.

[13]Thomas R. Knutson, John L. McBride, Johnny Chan, Kerry Emanuel, Greg Holland, Chris Landsea, Isaac Held, James P. Kossin, A. K. Srivastava, and Masato Sugi, "Tropical Cyclones and Climate Change," *Nature Geoscience*, 3 (2010), 157–163.

1.3. Responding to Climate Change by Joining Mitigation and Adaptation

Pacific SIDS communities should enact climate change mitigation, not because it will slow down climate change, but because the climate change mitigation actions are needed for many other reasons — including overlapping with climate change adaptation. The two processes should not be separate, but should be conducted in tandem.[14]

Regarding mitigation, Pacific SIDS have historically not caused much contemporary climate change, from either greenhouse gas emissions or land use changes.[15] Even greenhouse gas emissions per person and normalized by economic indicators do not place the Pacific SIDS high in country rankings.[16] Nonetheless, Pacific SIDS peoples are not necessarily absolved of all culpability. Many citizens of the Marshall Islands believe that, if the Marshallese had the chance, they would pollute as much as possible, thereby contributing to climate change with limited concern.[17] As such, these Marshallese do not accept that their country is innocent from creating climate change, just that they have not had the opportunity to do so.

This is substantiated by the Pacific SIDS which are causing problems related to climate change such as Papua New Guinea's significant deforestation.[18] Meanwhile, the Pacific SIDS of East Timor has been exploring

---

[14] Hanh H. Dang, Axel Michaelowa, and Dao D. Tuan, "Synergy of Adaptation and Mitigation Strategies in the Context of Sustainable Development: The Case of Vietnam," *Climate Policy*, 3S1 (2003), S81–S96; Sally Kane and Jason F. Shogren, "Linking Adaptation and Mitigation in Climate Change Policy," *Climatic Change*, 45 (2000), 75–102; Evan Mills, "Synergisms between Climate Change Mitigation and Adaptation: An Insurance Perspective," *Mitigation and Adaptation Strategies for Global Change*, 12 (2007), 809–842; A. Nyong, F. Adesina, and B. Osman Elasha, "The Value of Indigenous Knowledge in Climate Change Mitigation and Adaptation Strategies in the African Sahel," *Mitigation and Adaptation Strategies for Global Change*, 12 (2007), 787–797.

[15] IEA, *CO2 Emissions from Fuel Combustion* (International Energy Agency, 2009); Tom Roper, "The Global Sustainable Energy Island Initiative," *New Academy Review*, 3(1) (2004), 147–150.

[16] IEA (2009), *op. cit.*

[17] Rudiak-Gould (2013), *op. cit.*

[18] Phil L. Shearman, Julian Ash, Brendan Mackey, Jane E. Bryan, and Barbara Lokes, "Forest Conversion and Degradation in Papua New Guinea 1972–2002," *Biotropica*, 41(3) (2009), 379–390.

extraction of its fossil fuel reserves for the income to support development and state-building.[19]

The ethics of SIDS using their natural resources can be debated. The Pacific SIDS are experiencing significant problems due to climate change and those are projected to become worse. Does that deny the SIDS the rights to use their forests or fossil fuels to generate income for themselves where those resources are available? A small amount of either could make a big difference to their economies, including helping to fund climate change mitigation and adaptation actions. Since much of the rest of the world has gained substantial wealth from fossil fuels and land use changes — wealth which supports climate change mitigation and adaptation — should the SIDS also have that option? Or is the hypocrisy too much and, as stated on the Marshall Islands, must everyone act as if they can make a difference to curtailing climate change?[20]

Notwithstanding the ethical discussions, the Pacific SIDS peoples are being forced to act on climate change adaptation because climate change is affecting them. Those actions intersect with climate change mitigation, meaning that the processes are linked and will inevitably be pursued simultaneously. Some of these processes are briefly explored in this section.

1.3.1. *Ecosystems*

Both mitigation and adaptation entail using the environment — the land, coastal zone, and water — to improve how environmental aspects are factored in at all scales to deal with climate change, in processes called "ecosystem-based mitigation" and "Ecosystem-based Adaptation" (EBA).[21]

---

[19]Jonathan Steele, "Nation Building in East Timor," *World Policy Journal*, 19(2) (2002), 76–87.

[20]Rudiak-Gould (2013), *op. cit.*

[21]N.A. Marshall, P.A. Marshall, J. Tamelander, D. Obura, D. Malleret-King, and J.E. Cinner, *A Framework for Social Adaptation to Climate Change: Sustaining Tropical Coastal Communities and Industries* (Switzerland: IUCN, 2009); Ángela Andrade Pérez, Bernal Herrera Fernández, and Roberto Cazzolla Gatti (eds.), *Building Resilience to Climate Change: Ecosystem-Based Adaptation and Lessons from the Field* (Switzerland: IUCN, 2010); Karen Sudmeier-Rieux and Neville Ash, *Environmental Guidance Note for Disaster Risk Reduction: Healthy Ecosystems for Human Security*. Ecosystem Management Series No. 8, Commission on Ecosystem Management, Revised Edition (Switzerland: IUCN, 2009); Fabrice G. Renaud, Karen Sudmeier-Rieux, and Marisol Estrella (eds.), *The Role of Ecosystems in Disaster Risk Reduction* (Japan: UNU Press, 2013).

Focusing on EBA, it is defined as "an approach that builds resilience and reduces the vulnerability of local communities to climate change … EBA integrates sustainable use of biodiversity and ecosystem services in a comprehensive adaptation strategy".[22]

One EBA approach, often referred to as a toolkit, which has been applied in Pacific SIDS is the Mangrove Ecosystems for Climate Change Adaptation and Livelihoods (MESCAL).[23] As the project title implies, the focus is on restoring ecosystems, particularly mangroves, as part of wider climate change adaptation strategies. MESCAL has been implemented in the Pacific SIDS of Fiji, Samoa, Solomon Islands, Tonga, and Vanuatu.

Also in Tonga, other initiatives are seen.[24] Work with NGOs at the national and local levels demonstrated efforts to engage the populace in better working with ecosystems for local climate change adaptation, a classic EBA approach. While the theory was shown to apply directly to Tonga, especially given the low-lying nature of many of the communities, the practice was not always so straightforward. The Tongans' successes in EBA activities tended to highlight comparatively simple and day-to-day measures and advice which appeared superficially to be small, but which led to significant results on the ground.

For instance, people would frequently dispose of rubbish by tossing it into the mangroves, partly due to lack of awareness of the consequent damage and partly due to the cost of properly disposing of waste. Teaching respect for the mangroves improved the coastline substantially in some locations while, to a certain degree, instilling previously engrained ethical values regarding responsibility for and interaction with the environment around the communities. Notably though, at the day-to-day level, some NGOs lacked sufficient funds to purchase petrol to drive short distances and they were often hamstrung by the length of time between being promised funds by donors and receiving those funds. When funds did arrive,

[22]Perez *et al.* (2010), *op. cit.*, 15–16.

[23]IUCN website. Retrieved 5 March 2015, from http://www.iucn.org/about/union/secretariat/offices/oceania/priorities/priority_naturebasedsolutions/water_wetlands/about

[24]Stavros Mavrogenis and Ilan Kelman, "Lessons from Local Initiatives on Ecosystem-Based Climate Change Work in Tonga," in Renaud *et al.* (2013), *op. cit.*, 191–218.

implementation was challenging because the community had lost interest and the NGO had lost credibility. Streamlining and speeding up funding processes would make a significant difference to local EBA, even while recognizing that financial accountability might be reduced. Conversely, community accountability might be augmented by having access to the funds promptly. Overall, Tongans were generally focused on securing reliable livelihoods, meaning that it was important for EBA activities to emphasize these advantages to the people and to integrate with local livelihoods.

The experience in Tonga is replicated in other climate change adaptation initiatives in Kiribati,[25] the Marshall Islands,[26] and Timor-Leste[27] amongst other Pacific SIDS. Enacting EBA, to a large degree, means protecting forests, mangroves, and coral reefs. That covers mitigation through maintaining vegetation sinks for greenhouse gases. That also covers adaptation, as described above. Dealing with ecosystems in the context of climate change combines mitigation and adaptation, showing how they support each other.

1.3.2. *Energy*

Much of addressing sustainable energy solutions for Pacific SIDS is contextual. There is rarely a single approach suitable for all locations. Consequently, it is important to recognize the local context and to adapt each energy-related idea, approach, policy, legislation, behavioral change, or technology for that local context.

Small-scale projects are usual for SIDS due to their small land area and small population. The projects should be examined for what might and might not be transferable across communities, islands, and SIDS. Samoa uses passive solar water heaters on roofs. That is transferable to, and is already used in, many other Pacific SIDS. Vanuatu has pioneered the use

---

[25] JC Gaillard, "The Climate Gap," *Climate and Development*, 4 (2012), 261–264.
[26] Rudiak-Gould (2013), *op. cit.*
[27] J. Mercer, I. Kelman, F. do Rosario, A. de Deus de Jesus Lima, A. da Silva, A.M. Beloff, and A. McClean, "Nation-Building Policies in Timor-Leste: Disaster Risk Reduction and Climate Change Adaptation," *Disasters*, (2014), forthcoming.

of coconut oil in vehicles, which has been transferred to other Pacific SIDS such as Fiji, Samoa, the Solomon Islands, Papua New Guinea, Kiribati, and the Marshall Islands including its use for generating electricity.[28] The appropriateness of transferability to SIDS without large areas for coconut tree plantations is questionable since, then, the fuel will need to be produced elsewhere and transported there, exactly as with fossil fuels today. As well, care is needed to ensure that any move towards biofuels does not end up harming ecosystems or reducing food availability due to a focus on bioenergy.

Yet not everything about SIDS is small-scale. Their access to the sea tends to be vast in comparison to their land area and population. Sea-based energy such as wind, wave, current, and thermal has not yet been explored on a large scale around Pacific SIDS — or any SIDS at all. As that develops, major contributions will emerge from SIDS communities in terms of how small places can tap into large ocean-based energy resources. So far, this has happened only for the "large ocean-based energy resources" of fossil fuels, in the Pacific SIDS, most notably for East Timor. For climate change, that entrepreneurship could be switched to renewable supply and demand reduction.

One major challenge is tourism. Most tourists reach the Pacific SIDS via fossil fuel intensive commercial flights or cruise ships. Once on the island, tourists tend to use an order of magnitude or more electricity per capita than the locals. That also includes the excessive amounts of freshwater that tourists usually use along with the large amount of waste disposal required. Many tourists expect the luxuries of air conditioning, unlimited supply of hot water, continually clean linen, and hot food — many of which also have health and safety reasons for implementing them. All require significant amounts of energy.

No one can deny SIDS the right to generate livelihoods through tourism. Implementing renewable energy and demand reduction approaches should apply to the tourism sector, from backpacker hostels to five star hotels. Transportation-related energy demand is trickier. While significant progress

[28]Jan Cloin, "Biofuels in the Pacific: Coconut Oil as a Biofuel in Pacific Islands," *Refocus*, 6(4) (2005), 45–48; Jan Cloin, "Coconut Oil as a Fuel in the Pacific Islands," *Natural Resources Forum*, 31(2) (2007), 119–127.

is being made by the airline and shipping industries to improve fuel efficiency and alternative fuel sources, mainly because they seek larger profits, there are currently few solutions to flying or sailing on cruise ships without excessive energy use. This is a dilemma to continue investigating: How can the desire (even if not the need) for tourism-related income be balanced with the energy demands for travelling to and from the destination along with the energy demands at the destination?

Answering that question melds climate change mitigation through energy with climate change adaptation through energy, alongside many other approaches mentioned above. The Pacific SIDS will require funds to support and implement adaptation projects. Those funds could be raised by the SIDS themselves through tourist revenues, patenting and selling innovative energy and transportation approaches, and especially by combining these two through pioneering low-energy tourism approaches. What these are in reality remains an open question for creativity, experimentation, and innovation to answer.

1.3.3. *Environmental extremes*

As discussed in Section 1.2, contemporary climate change influences many hazards, exacerbating some and diminishing others. Climate change is, in effect, a potential hazard driver or potential hazard diminisher, rather than being a hazard itself. The complexities of the interactions between climate change and specific hazards across the variety of environments in Pacific SIDS makes attribution and projections challenging.

Adapting to these changes is a different story and is straightforward compared to trying to project potential changes. The reason is that adapting to environmental extremes influenced by climate change is no different to dealing with environmental extremes without climate change, as discussed by the field of "disaster risk reduction". The United Nations Secretariat responsible is the UN International Strategy for Disaster Reduction (UNISDR) which defines "disaster risk reduction" to be "[t]he concept and practice of reducing disaster risks through systematic efforts to analyse and manage the causal factors of disasters, including through reduced exposure to hazards, lessened vulnerability of people and property, wise management of land and the environment, and improved preparedness for adverse events".[29] In effect, it is about recognizing that disasters arise from a combination of hazard and vulnerability, but the key to reducing disaster risk is to understand and tackle vulnerability, irrespective of the hazard which is often created or exacerbated by human activities.[30] Consequently, climate change has limited relevance for adapting to environmental extremes, since people need to deal with those extremes anyway.

Considering climate-related extremes such as floods and droughts, society frequently implements structural measures such as dams, reservoirs, levees, and river engineering, altering water flow and dampening out flow extremes, such as from spring snow melt or the dry season. That significantly changes the hazard. Inhabitants near rivers tend to become inured to the absence of regular flood and drought cycles. Because few extremes occur, disaster risk reduction activities tend to lapse. Awareness of and coping mechanisms for potential floods and droughts diminish alongside understanding how to predict and react to floods and droughts. The visibility of the structural measures contributes to the psychological feeling of security. Eventually, a large flood or drought must occur, yielding damage which is far greater than would have occurred if the affected community were used to regular, smaller-scale floods and droughts.[31]

Instead, disaster risk reduction measures would envelop both hazard and vulnerability, rather than focusing on only hazard. Examples would be permitting rivers to use floodplains to flood, constructing dwellings with

---

[29]UNISDR, "Terminology." Retrieved 5 March 2015, from http://www.unisdr.org/we/inform/terminology.

[30]Kenneth Hewitt (ed.), *Interpretations of Calamity from the Viewpoint of Human Ecology* (Australia: Allen & Unwin, 1983); James Lewis, *Development in Disaster-Prone Places: Studies of Vulnerability* (UK: Intermediate Technology Publications, 1999); Ben Wisner, Piers Blaikie, Terry Cannon, and Ian Davis, *At Risk: Natural Hazards, People's Vulnerability and Disasters*, Second Edition (UK: Routledge, 2004); Ben Wisner, JC Gaillard, and Ilan Kelman (eds.), *Handbook of Hazards and Disaster Risk Reduction* (UK: Routledge, 2012).

[31]Robert E. Criss and Everett L. Shock, "Flood Enhancement through Flood Control," *Geology*, 29(10) (2001), 875–878; David Etkin, "Risk Transference and Related Trends: Driving Forces towards More Mega-Disasters," *Environmental Hazards*, 1 (1999), 69–75; Maureen Fordham, "Participatory Planning for Flood Mitigation: Models and Approaches," *The Australian Journal of Emergency Management*, 13(4) (1999), 27–34; Graham A. Tobin, "The Levee Love Affair: A Stormy Relationship," *Water Resources Bulletin* 31(3) (1995), 359–367.

flood resistance measures, and ensuring that people have the knowledge to recognize possible flood and drought signs in order to act before the flood or drought becomes a potential hazard. All these measures have long been implemented around the world, are needed as part of climate change adaptation, and should be used irrespective of climate change.

Nevertheless, despite this discussion, Pacific SIDS yield some exceptions because it is possible that environmental extremes might occur under climate change which have never before been experienced in certain locations. In particular, low-lying atolls are likely to experience sea flooding at heights and rates which have not been seen since these atolls were inhabited.[32] Additionally, to assure inhabitation on an island, it is necessary to have land, food, and water. If islands are inundated or undergo severe geomorphological changes, if coral reefs and coastal vegetation die, if freshwater supplies are salinated, if available land-based or sea-based food diminishes, and/or if other changes detrimentally impact basic life needs to a large extent, then no amount of disaster risk reduction can save that community on that atoll — as might have occurred in many Pacific SIDS communities during the 14th century.[33] Mobility, discussed in the next section, might be necessary.

1.3.4. *Mobility (Migration)*

An often-touted suggestion regarding Pacific SIDS people's responses to climate change is that the people would move from their homes.[34] As described earlier, full flooding of communities or islands completely disappearing is not necessarily a definite result from climate change. Yet there is little doubt that many communities on Pacific SIDS will be highly impacted by climate change and might need to consider, or be forced, to move as a consequence. Examples of the SIDS cited as being amongst

---

[32]For background information, see William R. Dickinson, "Pacific Atoll Living: How Long Already and until When?," *GSA Today*, 19(3) (2009), 4–10.

[33]Patrick D. Nunn, Rosalind Hunter-Anderson, Mike T. Carson, Frank Thomas, Sean Ulm, and Michael J. Rowland, "Times of Plenty, Times of Less: Last Millennium Societal Disruption in the Pacific Basin," *Human Ecology*, 35(4) (2007), 385–401.

[34]Karen McNamara and Chris Gibson, "'We Do Not Want to Leave Our Land': Pacific Ambassadors at the United Nations Resist the Category of 'Climate Refugees'", *Geoforum*, 40(3) (2009), 475–483.

the most vulnerable are Tuvalu, Tonga, Kiribati, the Marshall Islands, and the Federated States of Micronesia. It is not just sea level which might precipitate or force mobility, but also food, water, livelihoods, and ecosystem impacts, as discussed in Section 1.2.

If mobility becomes necessary or chosen, then several decisions need to be made. First, the timing of the movement would need to be determined. Should communities move immediately, start preparing immediately to move as soon as they are ready, carefully plan for moving several years in the future which could perhaps even be decades, or wait until it is no longer possible to live in their community? Such decisions are not straightforward, with each choice having advantages and disadvantages. Careful planning or the choice to move long into the future could be stymied if a storm or tsunami causes substantial damage in the interim.

Once the timing of the movement is decided, then decisions are needed regarding where to move to. The people could aim to integrate into communities which exist already, on islands or not, whether those be megacities, small towns, or rural locations. This entails accepting the loss of identity and culture. The people could instead aim to re-establish their original community, perhaps even a SIDS country, elsewhere. In such cases, land would need to be ceded, purchased, or built for the SIDS peoples to settle on. Issues of autonomy or sovereignty would need to be resolved, under international law but in consultation with the host location. While some have suggested the possibility of mobile SIDS states, settling on large-scale platforms or floating islands which roam the world's seas, most discussion is about fixed land and locations.

For instance, if a Filipino island were given to people moving from Tonga, or if Filipino land were extended to accommodate newcomers, then decisions must be undertaken regarding whether or not the independent state of Tonga would still exist. Other possibilities include being enfolded within the Philippines or being an autonomous province, special administrative area, or territory of the Philippines with the Tongans being Filipino citizens. Decisions would be needed regarding services to provide alongside the level of harmonization between Tonga and the Philippines, including official language(s), currencies, healthcare, education, and justice systems. There could be clashes over topics such as religious freedom, capital punishment, and family planning.

Within these discussions, many contested aspects manifest regarding the labeling and the power to decide. Many authors jump on the bandwagon of "climate refugees" or "climate change refugees", yet numerous legal, moral, and practical issues have been raised, indicating a preference to avoid those terms.[35] In international law,[36] no scope exists to assign refugee status to people crossing borders for environmental reasons. Many fear that attempting to redefine the term "refugee" to account for climate change or environmental reasons will lead to further, ostensibly detrimental changes for those currently claiming refugee status. It would certainly be a political minefield to assign international status, such as "refugee", to people displaced within the borders of their own sovereign state.

As such, the mobility debate surrounding Pacific SIDS peoples' responses to climate change is not clear-cut. No definitive evidence exists that large-scale movement would be required, no clear pathway appears to effect this mobility if it is required, and many legal and moral issues remain to be resolved.

1.3.5. *Accepting uncertainty*

These topics of climate change mitigation and adaptation, and joining them, are about practical, on-the-ground action. Yet many uncertainties are articulated, often compounded by the uncertainties detailed in Section 1.2 regarding impacts. In addition to uncertainties, unknowns exist regarding both impacts and specific mitigation and adaptation actions to deal with those impacts. Frameworks are needed for examining how those uncertainties and unknowns could come into play with people making decisions, keeping in mind that such frameworks also include uncertainties and

---

[35]Giovanni Bettini, "Climate Barbarians at the Gate? A Critique of Apocalyptic Narratives on 'Climate Refugees'," *Geoforum*, 45 (2013), 63–72; Betsy Hartmann, "Rethinking Climate Refugees and Climate Conflict: Rhetoric, Reality and the Politics of Policy Discourse," *Journal of International Development*, 22(2) (2010), 233–246; Calum T.M. Nicholson, "Climate Change and the Politics of Causal Reasoning: The Case of Climate Change and Migration," *The Geographical Journal*, 180(2) (2014), 151–160; McNamara and Gibson (2009), *op. cit.*

[36]UNHCR, *Convention and Protocol Relating to the Status of Refugees* (UNHCR, 1951/1967).

unknowns. To explore some such frameworks, the next section examines some legal and ethical questions.

1.4. Legal and Ethical Questions

This section raises some of the legal and ethical questions emerging regarding Pacific SIDS responding to climate change. As discussed in Section 1.3, many construct climate change as being a justice issue, particularly for the Pacific SIDS who have contributed minimally to climate change yet will suffer extensively because of it. Not all Pacific SIDS peoples agree with that framing, such as the Marshall Islanders who suggest that they themselves should do as much as possible to counter climate change.[37] Different conceptualizations of ethical responsibility emerge in terms of some expecting those who primarily caused climate change to solve the challenges and others not wishing to assign blame but, instead, accepting common responsibility for tackling climate change.

Irrespective of the stance adopted, the full answer to "Who is responsible?" is not given. There are still ethical questions in terms of the "who" being institutions, individuals, or most likely some combination. Institutions would cover governments and governing groups (supranational, national, and subnational), non-governmental organizations, the private sector, schools and universities, scientific groups, the United Nations, and international organizations, amongst others. Individuals could be leaders acting in a personal capacity or in a joint institutional-personal capacity, along with all citizens — explicitly including all ethnic groups, religions, linguistic groups, ages, genders, and abilities. Homeless people, transients, and prisoners should also be considered within the category of "individuals".

These comments cover ethical responsibility. Legal responsibility is a subject of much more entangled debate.[38] Other chapters throughout this volume cover many aspects relevant to Pacific SIDS, but few of the

[37] Rudiak-Gould (2013), *op. cit.*

[38] Michael B. Gerrard and Gregory E. Wannier (eds.), *Threatened Island Nations: Legal Implications of Rising Seas and a Changing Climate* (UK: Cambridge University Press, 2013); Lilian Yamamoto and Miguel Esteban, *Atoll Island States and International Law: Climate Change Displacement and Sovereignty* (New York: Springer, 2014).

questions are resolved — not even in terms of which court would have jurisdiction, how disputes would be resolved, or how decisions would be monitored and enforced. Many initiatives regarding law have been explored through national laws, international human rights law, and attempts to make ecocide — the destruction of ecosystems — enshrined in international law as the fifth crime against peace.[39]

Two minor notes are pertinent regarding legal responsibility for Asia-Pacific responses to climate change. First, in 2002, the Prime Minister of Tuvalu indicated the intention to sue the US and Australia over greenhouse gas emissions and failure to sign the Kyoto Protocol, but the lawsuit did not proceed after the Tuvaluan Prime Minister suffered an election defeat.[40] Second, in 2013, a case was brought to court in New Zealand in 2013 from a citizen of Kiribati claiming climate refugee status, but the court ruled against him because there are no current statutes which would permit the court to use climate change as a reason for making him a refugee or protected person.[41]

The "who?" question continues beyond responsibility. Even if ethical and legal responsibility were assigned and accepted, decisions regarding the Pacific SIDS' responses to climate change would require funding to be enacted. Who provides the funds and who monitors the spending? What further justice and equity issues emerge with respect to funding?

For instance, if land is purchased or if compensation is given for resettling people from Pacific SIDS, should local market values be provided or should all sellers (forced or otherwise) be treated equally? Should resource and access rights around abandoned islands be used as part of the compensation or should ownership remain fully with the people who have moved? Some authors have started to explore these questions legally

[39]Polly Higgins, *Eradicating Ecocide: Laws and Governance to Prevent the Destruction of Our Planet* (UK: Shepheard-Walwyn, 2010).

[40]Rebecca Elizabeth Jacobs, "Treading Deep Waters: Substantive Law Issues in Tuvalu's Threat to Sue the United States in the International Court of Justice," *Pacific Rim Law and Policy Journal*, 14(1) (2005), 103–128; Ralston *et al.* (2004), *op. cit.*

[41]J. Priestley, *Reserved Judgment Of Priestley J — In the High Court Of New Zealand Auckland Registry, between Ioane Teitiota, Applicant, and the Chief Executive of the Ministry of Business Innovation and Employment, Respondent* (CIV-2013-404-3528 [2013] NZHC 3125, High Court of New Zealand, 2013).

and technically,[42] and Kiribati even purchased some land in Fiji in 2014, but the social and ethical implications remain an understudied topic, especially with regards to compensation for the loss of the intangible values of the land, such as identity, tradition, ancestors, and home.

One of the principal justice issues, from both ethical and legal perspectives, emerges with respect to sovereignty. Is retaining sovereignty of all Pacific SIDS an attainable goal, irrespective of where the islanders move to? Realistically, how long would notions and desires of sovereignty remain through various generations? What would occur if factions within a current Pacific SIDS differ regarding desired sovereignty, citizenship, and maintaining their SIDS identity?

Most fundamental regarding all these questions is the decision-making process, reprising the "who?" question. Who has the power, authority, legal right, and moral right to control the decision-making process and to enact the decisions? Limited literature exists on such operational realities or ways of moving forward, although many of the citations throughout this chapter describe some aspects and questions in detail. As such, this section represents a research agenda which is also highly relevant for policy and practice.

One example bringing together many of these threads and seeking to find answers to the questions is the Many Strong Voices (MSV) program.[43] MSV was founded in 2005 at the request of peoples from the Arctic and SIDS, including substantial contributions and support from the Pacific SIDS. Program participants are dominated by representatives from Arctic and SIDS communities, but also include indigenous peoples' organizations, international organizations, government agencies, non-governmental organizations, and research institutes all collaborating to support local voices and local capacity for responding to climate change.

The Arctic and SIDS peoples have many differences due to their regions' respective climates, plus the Pacific SIDS have major differences in culture and history from other SIDS and amongst themselves. But all

[42]Gerrard and Wannier (2013), *op. cit.*; Yamamoto and Esteban (2014), *op. cit.*

[43]Many Strong Voices website — http://www.manystrongvoices.org, with scientific publications on "Small Island States (SIDS) and Climate Change." Retrieved 5 March 2015, from http://www.islandvulnerability.org/docs/islandsclimatechange.pdf.

these groups also display numerous similarities. Examples are natural resource dependency and, in many cases, being relatively isolated coastal communities. Their voices, knowledge, and wisdom tend to be sidelined, in terms of not only describing their observations of their changing climate, but also possible solutions from their own knowledge balanced with requests for outside assistance.[44] MSV works to overcome that marginalization to formulate and implement climate change-related decisions in tandem with many other programs and support mechanisms.

MSV, and the answers to the legal and ethical questions, extend beyond climate change. Many communities around the Pacific SIDS have had habitability challenges due to reasons other than climate-related phenomena. Volcanoes forced entire islands to be evacuated in Tonga in 1946[45] and Baliau, Papua New Guinea in 1957 and 2004.[46] Mismanagement of its phosphate resources has led to Nauru having few livelihood opportunities remaining.[47] Should Pacific SIDS threatened by reasons not linked to climate change be accorded the same treatment as those threatened by climate change? Legal and ethical questions pervade the management of the habitability and livelihoods of many Pacific SIDS.

1.5. Conclusions

The stark disconnect between who creates climate change impacts and who experiences the worst impacts has been shown for the case of Pacific SIDS. The reality is that, irrespective of legal or other regimes introduced to try to assist those most vulnerable to climate change, from a Pacific SIDS perspective, most actions will likely need to be taken by the islanders themselves. This approach would be embraced by many of the islanders,

[44] Ilan Kelman, "Hearing Local Voices from Small Island Developing States for Climate Change," *Local Environment*, 15(7) (2010), 605–619; Karen McNamara, "Voices from the Margins: Pacific Ambassadors and the Geopolitics of Marginality at the United Nations," *Asia Pacific Viewpoint*, 50 (1) (2009), 1–12.

[45] James Lewis, "Volcano in Tonga" *Journal of Administration Overseas*, XVIII(2) (1979), 116–121.

[46] Jessica Mercer and Ilan Kelman, "Living alongside a Volcano in Baliau, Papua New Guinea," *Disaster Prevention and Management*, 9(4) (2010), 412–422.

[47] Christopher Weeramantry, *Nauru: Environmental Damage under International Trusteeship* (UK: Oxford University Press, 1992).

especially when not all SIDS peoples worry about justice or blame in the same way as academics do, and especially when many SIDS peoples wish to be in control of their own fates and to make their own decisions. That, arguably, is an ethical stance to ensure that people have the resources and abilities to make decisions about their own lives with respect to climate change, as long as that does not unduly interfere with others — such as through causing climate change.

That is where external assistance plays an important role, with the Many Strong Voices program being an example. Pacific SIDS peoples thereby gain the resources and apply their abilities to make their own climate change-related decisions and to enact those decisions. That assists in redressing some of the injustices raised by many without removing responsibility or dignity from the SIDS. That also becomes practical, such as SIDS peoples joining mitigation and adaptation rather than perpetuating the artificial and unnecessary separation which pervades external approaches.

Wherever one lives, at whatever time period in history, there is always some degree of control regarding influences on one's society and some degree of lack of control. Contemporary climate change impacting Pacific SIDS has a strong imbalance in terms of the islanders not having much control over the situation and the impacts, yet being forced to deal with major impacts without being provided with the necessary resources. Simultaneously, many decisions influence others outside the islands and require cooperation from them, yet the outsiders do not often provide what is needed or become involved in the process.

For instance, consider if islanders are being forced to migrate due to climate change, agree within their community on a location, and are able to effect the move themselves without external resources. In such a case, the host communities and country need to agree to the resettlement and the governance processes for the settlers. No other country has yet volunteered to be involved in such a process in advance of a crisis. A similar situation emerges with environmental extremes and Ecosystem-based Adaptation. An island community might determine, and be supported by their country, that in a specific instance, a structural approach is needed or an ecosystem needs to be altered. External interests might oppose that because it harms or changes an ecosystem. Ethical and legal battles could ensue, especially regarding who has the right to control the decision-making process.

Throughout, it is important to emphasize that the islanders are making decisions due to climate change to which they have contributed little. Their decisions affect others, so others ought to get involved and have a say, but ultimately, the islanders deserve control over their own fates as much as feasible within the context of climate change. Climate change is affecting the entire planet and all of humanity. Approaches to deal with climate change should operate at similar scales rather than neglecting one group, such as Pacific SIDS.

Chapter 2

Managing Southeast Asian Ecosystems to Reduce Coastal Population Vulnerability under Sea Level Rise

*Daniel A. Friess**

2.1. Introduction

The world may see an accelerated sea level rise (SLR) of up to one meter by the end of the 21st century due to warming ocean expansion and glacier/ice sheet melting.[1] The recent IPCC Fifth Assessment Working Group II Report[2] states with *very high confidence* that coastal systems and low-lying areas will experience a range of adverse impacts related to SLR, such as increased submergence, coastal flooding, and coastal erosion.

Coastal ecosystems threatened by SLR coexist with human coastal populations in many parts of the world. At least 1.2 billion people (more than 20% of the world's population) are concentrated along the coast.[3] Urban populations continue to increase in size and density more rapidly in the coastal zone, compared to other locations,[4] especially in the Asian region. For example, China's economic policy of creating Export Processing Zones (EPZs) in coastal areas (to access better shipping and import/export

---

*The research presented herein has been funded by several partners, including the Department of Geography, National University of Singapore (R-109-000-141-133) and the Ministry of Education, Government of Singapore (R-109-000-166-112).

[1]IPCC, *Chapter 13. Sea Level Change*, Intergovernmental Panel on Climate Change, Working Group I, Fifth Assessment Report, 2014a.

[2]IPCC, *Impacts, Adaptation, and Vulnerability*, Intergovernmental Panel on Climate Change, Working Group II, 5th Assessment Report, 2014b.

[3]C. Small and R.J. Nicholls, "A Global Analysis of Human Settlement in Coastal Zones," *Journal of Coastal Research*, 19(3) (2003), 584–599.

[4]K.C. Seto *et al.*, "A Meta-Analysis of Global Urban Land Expansion," *PLoS ONE* 6, e23777 (2011).

links) such as Guangzhao and Shenzhen has encouraged the movement of tens of millions of (predominantly female) inland rural migrants seeking manufacturing job opportunities in the rapidly urbanizing coastal zone. A number of key megacities are also located in Asia's coastal zone.

Urban and rural coastal populations are thus rapidly increasing in areas that are physically and ecologically dynamic, and vulnerable to an array of natural hazards linked to extreme weather events (such as storm surges, cyclones, tsunamis), and the longer-term impacts of global climate change (increased inundation and coastal erosion). This chapter discusses the role of tropical coastal ecosystems (particularly mangroves and seagrasses) in reducing the vulnerability of coastal populations in the ASEAN region to SLR. However, coastal habitats themselves are also uniquely susceptible to SLR; if we are to maximize their effectiveness for coastal adaptation then we need to manage the exposure of coastal ecosystems to SLR and other anthropogenic stressors, and increase their adaptive capacity to respond to stressors that cannot be adequately mitigated. Reducing exposure and increasing adaptive capacity requires (1) coastal ecosystems to be viewed through a lens of coastal vulnerability, similar to how we currently assess human population vulnerability; and (2) a significant change in how we construct broad integrated coastal management frameworks and the scales at which we construct them.

2.2. Concepts of Vulnerability

2.2.1. *The vulnerability framework*

Vulnerability was defined by the IPCC's Third Assessment Report as the degree to which a system is susceptible to, or unable to cope with, the adverse effects of a threat such as climate change.[5] The definition of vulnerability has been further modified in the IPCC's Fifth Assessment Report[2] as a multidimensional concept that involves "a propensity or predisposition to be adversely affected [by a hazard]". This definition places an increased emphasis on the role of structural conditions of poverty and inequality in shaping overall vulnerability.[6]

[5]IPCC, *Climate Change 2001: Impacts, Adaptation, and Vulnerability*, Intergovernmental Panel on Climate Change, Working Group II, Third Assessment, 2001.

[6]IPCC (2014b), *op. cit.*

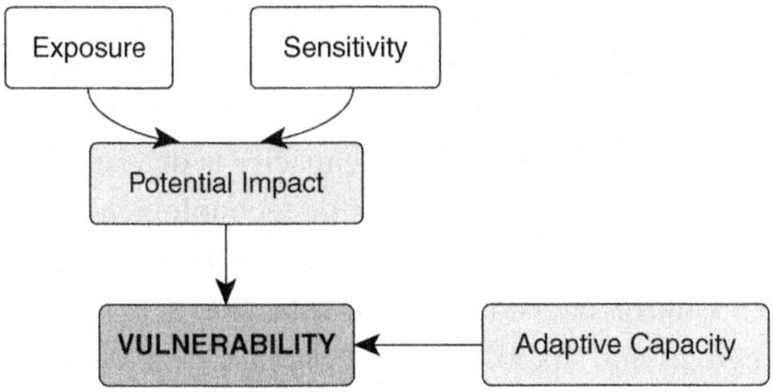

Figure 1. The Components of Vulnerability.

Source: IPCC (2014b), *op. cit.*

Vulnerability is a broad concept that must be clearly defined in the context of which it is being used. Vulnerability is most often used in conjunction with the aspect that is deemed to be vulnerable, such as "socioeconomic vulnerability", "coastal population vulnerability", "infrastructural vulnerability", "political vulnerability" or "ecosystem vulnerability". Regardless of the precise definition, multiple components are key contributors to vulnerability (Figure 1):

- *Exposure* is defined by the IPCC's Fifth Assessment Report as the presence of people, livelihoods, species or ecosystems, environmental services and resources, infrastructure, or economic, social, or cultural assets in places that could be adversely affected.[2] Broader definitions incorporate the degree to which a system is exposed to threats or an event that may harm the system. Threats may refer to extreme weather events such as storm surges, cyclones, tsunamis, or the longer-term impacts of global climate change, such as increased inundation and coastal erosion. Exposure also incorporates elements of threat severity and magnitude. Exposure is probably the most researched and understood aspect of coastal population vulnerability;
- *Sensitivity* as per the IPPC's Fifth Assessment Report refers to the degree to which a system may be affected (positively or negatively) by an event or other threat such as climate change or variability.[5] Both exposure and sensitivity contribute to the potential impact that a system may experience;
- *Adaptive capacity* as per the IPPC's Fifth Assessment Report refers to the ability of systems, institutions, humans, and other organisms to adjust

to potential damage, to take advantage of opportunities, or to respond to consequences.[5] Increasing adaptive capacity allows a system to better prepare for and manage stressors and impacts of varying frequency and magnitude. The degree of adaptive capacity is determined by a number of interlinked factors, such as available technological options, resource availability, social capital, and legislative and institutional mechanisms available to make decisions. Thus, adaptive capacity differs greatly among regions and socio-economic statuses. In coastal populations, social memory can also be a key component of adaptive capacity, providing key points of renewal and reorganization after a disaster or event.[7]

The temporal and spatial scale of observation is key when considering the three components of vulnerability. Firstly, exposure, sensitivity, and adaptive capacity change are not static concepts, but vary through time, for example, as stressors (e.g., SLR) increase, or as a community becomes more economically developed and invests in measures that increase adaptive capacity. Secondly, spatial scale may affect the degree of exposure, sensitivity, or adaptive capacity that can be readily observed. For example, local-scale networks (e.g., a coastal village) may be readily adaptive in some aspects, though not adaptive in other aspects that require large socio-economic investment or larger-scale policy instruments.

2.2.2. *The vulnerability of coastal populations to sea level rise*

We often consider vulnerability in terms of short-term, discrete, low-frequency but high-magnitude events, such as storm surges, cyclones, and tsunamis. Gradual, long-term threats such as those caused by sea level rise are more insidious, and potentially harder to determine in terms of exposure and sensitivity. Such impacts may be cumulatively greater than discrete events over the long term.

Considering coastal climate change impacts in the context of the IPCC vulnerability framework presented in Figure 1, it is clear that we know the most about potential exposure to climate change hazards such

[7] N. Adger *et al.*, "Socio-Ecological Resilience to Coastal Disasters," *Science*, 309(5737) (2005), 1036–1039.

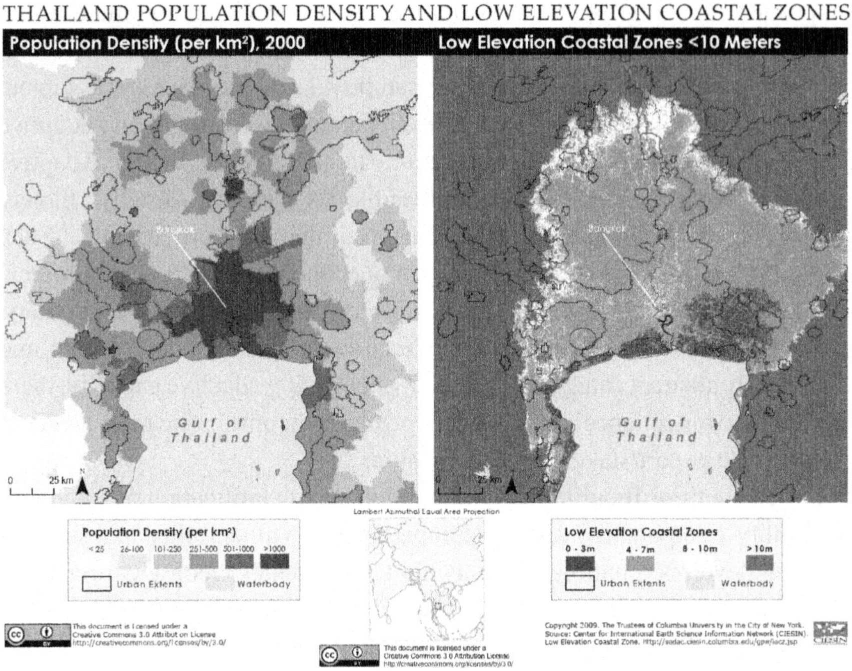

Figure 2. A Confluence of High Population Density and Low Elevation Increase Coastal Population Vulnerability in Bangkok, Thailand.

Source: CIESIN (2009), *op. cit.*

as SLR, and associated/knock-on impacts such as increased inundation and coastal erosion. Exposure can be mapped in its simplest form by combining spatial data on population density and coastal elevation. For example, Figure 2 indicates the potential direct exposure of Bangkok, Thailand to SLR impacts.[8] By extrapolating this data into the future, we may be able to estimate the proportion of Bangkok's population that could be at risk from flooding due to a rate of SLR of varying scenarios. However, indirect exposure is much more difficult to identify and quantify. Bangkok contributes a substantial proportion of Thailand's GDP, so much of Thailand's broader economy will also be indirectly exposed to the impacts of SLR. How should this be incorporated into a measure of

[8]CIESIN (2009), *op. cit.*

exposure? This highlights the role of scale in identifying and quantifying exposure, as introduced previously.

While some aspects of exposure can be readily estimated or assumed, the other two components of vulnerability, namely sensitivity and adaptive capacity, are perhaps more abstract terms to quantify and model. Adaptive capacity in particular may be unfeasible to measure directly, though measuring the characteristics that influence it may be more feasible.[9] Such characteristics include broad categories of economic wealth, technological capacity, knowledge and skills, infrastructure and institutions, and others. The question still remains "how to accurately quantify such broad and potentially abstract categories?". When quantifying adaptive capacity, there is still an over-reliance on simplified proxies (income, education level) or small-scale expert/stakeholder discussion.

Despite the difficulties in identification and quantification of some vulnerability components, several comprehensive vulnerability assessments exist in developed nations, such as the United States Geological Survey's *National Assessment of Coastal Vulnerability to Sea-Level Rise* Report,[10] and the later National Oceanographic and Atmospheric Administration's *State of the Coast* Report,[11] the UK Environment Agency's *Vulnerability of Estuaries to Sea Level Rise* series,[12] and Australia's *Adapting to Climate Change* initiative.[13] Importantly, these assessments are freely available to all stakeholders, both governmental, private and public. Several of these assessments have also produced simple-to-use, and freely available web

---

[9] L. Jones, E. Ludi and S. Levine, *Towards a Characterization of Adaptive Capacity: A Framework for Analysing Adaptive Capacity at the Local Level.* Overseas Development Institute, UK, 2010.

[10] USGS, *National Assessment of Coastal Vulnerability to Sea Level Rise*, United States Geological Survey, 2013. Retrieved 9 March 2015, from http://woodshole.er.usgs.gov/project-pages/cvi/.

[11] NOAA, *State of the Coast*, National Oceanic and Atmospheric Administration, 2011. Retrieved 19 May 2015, from http://www.environment.gov.au/climate-change/adaptation/australias-coast/mapping-sea-level-rise.

[12] EA, *Vulnerability of Estuaries to Sea Level Rise — Stage 1: A Review*, Environment Agency, United Kingdom, 2010.

[13] DoE, *Mapping Sea Level Rise*, Department of the Environment, Government of Australia, n.d. Retrieved 5 March 2015, from www.climatechange.gov.au/climate-change/adapting-climate-change/australias-coasts-and-climate-change/mapping-sea-level-rise.

tools, in order to better communicate vulnerability to potentially impacted populations. However, similar comprehensive and large-scale vulnerability assessments either do not exist, or are not made publically available for the ASEAN region. For example, several coastal vulnerability assessments have been conducted for the tropical island city-state of Singapore by the national government. The results are shared by multiple government agencies and departments, and feed into the national planning process, but are not available to the public.

Interestingly, most of the developed nation vulnerability assessments listed above are not truly comprehensive in nature, but are instead still focused predominantly on the physical, and to a lesser extent, socio-economic characteristics that determine exposure. This is most probably due to issues related to characterizing and quantifying the more abstract and fluid concepts of sensitivity and adaptive capacity. For example, the NOAA *State of the Coast* Report[10] includes a publically accessible web visualization tool, that allows users to investigate exposure parameters such as wave height, tidal range, geomorphology, erosion rate, and SLR rate, though it does not include parameters related to societal sensitivity or adaptive capacity. However, the increased technological and resource means of developed nations to invest in such exposure mapping and modelling could in itself be considered a contribution that increases adaptive capacity.

2.3. Coastal Ecosystems Decrease Sensitivity and Increase Adaptive Capacity

Adaptive capacity is often considered in terms of technological, resource, and institutional terms. However, we could also add the role of socio-ecological structures as an important contributor to adaptive capacity. Important socio-ecological structures to consider in particular are tropical vegetated coastal ecosystems. Mangroves are a group of halophytic (salt-tolerant) trees, shrubs, and associated fauna found on low-energy, muddy coastlines. Mangroves in 2001 covered 137,760 km^2 throughout the tropics, with 42% of the world's mangroves located in Asia.[14] Seagrasses are a group

[14] C. Giri *et al.*, "Status and Distribution of Mangrove Forests of the World," *Global Ecology and Biogeography*, 20(1) (2011), 154–159.

of halophytic grass species, which are found further offshore in the inter- to sub-tidal zone (permanently flooded). The area and distribution of tropical seagrasses is much harder to determine compared to mangroves, though they may cover as much as 177,000 km^2 globally,[15] with a substantial proportion located in the tropics.

Tropical vegetated coastal ecosystems can play a role in reducing coastal population vulnerability to SLR by increasing the adaptive capacity of a local coastal population. The role of coastal ecosystems in adaptive capacity is only now being realized, and is now recognized under the 2011 UNFCCC Cancun Adaptation Framework[16] and recent COP10 Aichi targets of the Convention on Biological Diversity,[17] among others. Thus, the role of coastal ecosystems in the vulnerability framework is now being discussed and debated at the highest levels of international policy.

Coastal ecosystems reduce sensitivity to coastal hazards, and contribute to adaptive capacity by (1) providing ecosystem services to afford coastal protection and increase livelihoods (thus increasing resource access and social capital), and (2) by providing flexibility to threats caused by a changing climate.

2.3.1. *Coastal ecosystems provide crucial ecosystem services*

For millennia, communities have established and thrived in the coastal zone in part due to the broad range of ecosystem services that coastal ecosystems provide. In recent years, a greater understanding of the broad suite of coastal ecosystem services has emerged. The Millennium Ecosystem Assessment[18]

---

[15]E.P. Green and F. Short, *World Atlas of Seagrasses*. (California: University of California Press, 2003).

[16]UNFCCC, Report of the Conference of the Parties on Its Sixteenth Session, held in Cancun from 29 November to 10 December 2010. Part Two: Action Taken by the Conference of the Parties at Its Sixteenth Session (FCCC/CP/2010/7/Add.1). United Nations Framework Convention on Climate Change, 2011.

[17]CBD, COP 10 — 10th Meeting of the Conference of the Parties to the Convention on Biological Diversity Decision X/2. Strategic Plan for Biodiversity 2011–2020 United Nations Environment Programme, Nagoya, Japan. Convention on Biological Diversity, 2010.

[18]MEA, *Ecosystems and Human Well-being*, Millennium Ecosystem Assessment, 2005.

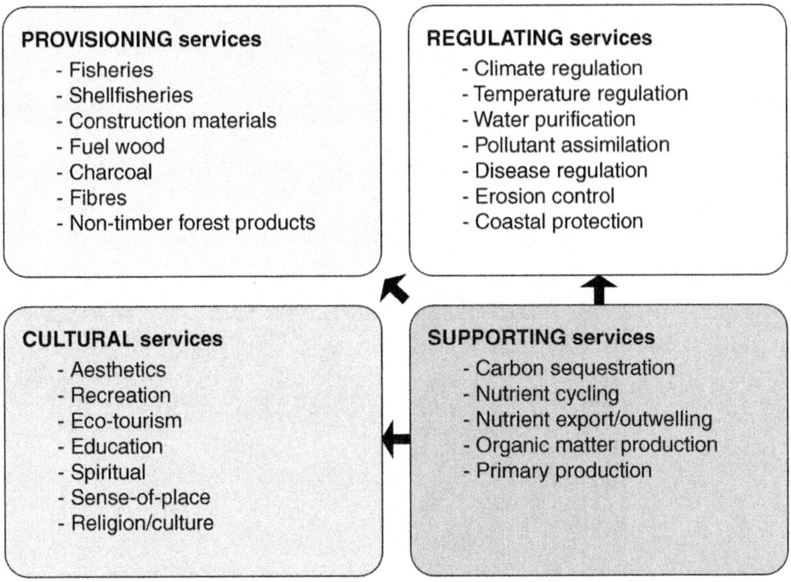

Figure 3. A Hierarchy of Ecosystem Services Provided by Coastal Ecosystems.

characterizes ecosystem services into several functional groups, depending on their role in providing, regulating or supporting ecosystem services, or whether they play a cultural role (Figure 3).[17–19] Ecosystem services from mangroves and seagrasses may be worth as much as US\$5449 $\times$ 10^9 per year, equivalent to a substantial proportion of global GDP[18] (though it is notoriously difficult to quantify the non-monetary value of coastal ecosystem services).

Mangroves and seagrasses play a strong provisioning role for local to regional populations (Figure 4). Almost 50% of the animal protein consumed in Southeast Asia is derived from fish,[21] many species of which

[19] R. Costanza *et al.*, "The Value of the World's Ecosystem Services and Natural Capital," *Nature*, 387 (1997), 253–260.

[20] C. Lacambra *et al.*, "Bioshields: Mangrove Ecosystems as Resilient Natural Coastal Defences," in Renaud and Estrella (eds.), *The Role of Ecosystems in Disaster Risk Reduction* (Japan: UNU Press, 2013), pp. 82–108.

[21] N. Harvey, *Global Change and Integrated Coastal Zone Management: The Asia-Pacific Region* (Berlin: Springer, 2006).

Figure 4. Provisioning Services Provided by Mangroves in Bintuni Bay, West Papua Include: (a) Fisheries; (b) Tobacco Papers and Roof Tiles from Nypa Palm Leaves; (c) Transporting of Recently Harvested Mangrove Fuel Wood; (d) Mangrove Wood Chips.

are closely associated with mangroves and/or seagrasses for parts of their life cycle. These coastal ecosystems play an important nursery function for juvenile fish. Mangroves in West Papua, Indonesia are harvested for wood chips and paper (Figure 4), while mangroves in much of Southeast Asia are converted to charcoal. The myriad of provisioning services afforded by coastal ecosystems increases adaptive capacity by increasing socio-economic status and available resources, allowing a community to remain resilient in the face of a stressor.

Provisioning services are readily identifiable and quantifiable, though the other ecosystem services classes are indirect and intrinsic. Supporting services such as carbon sequestration and nutrient cycling/export play an indirect role in adaptive capacity by supporting the provisioning services described previously. Regulating services can be somewhat directly linked to adaptive capacity. For example, strong focus has been applied to mangroves (and to a much lesser extent, seagrasses) and their role in wave attenuation and coastal protection. Focus was particularly strong after the 2004 Southeast Asian tsunami, where early reports[22,23] described how mangroves in front of coastal villages buffered the incoming waves and increased community resiliency. While we now know that the role of mangroves in attenuating large waves such as tsunamis is much more nuanced than this,[24] a rapidly increasing knowledge base is empirically proving the important role mangroves can play in attenuating day-to-day tidal events, storms and surges through increasing surface roughness and frictional processes. Seagrasses play a similar protection role for human coastal populations, by reducing current velocity, dissipating wave energy, and stabilizing dynamic inter- and sub-tidal surfaces.[25] While it must be made clear that coastal ecosystems on their own will probably not provide a level of coastal protection adequate to fully defend a coastal population (especially a densely populated, urban population), there may be scope to utilize this ecosystem service by hybrid engineering, or growing coastal ecosystems in conjunction with (smaller) hard engineering structures such as sea walls.

The final functional group of services — cultural ecosystem services (Figure 3) — may not, at first glance, seem to play a role in increasing adaptive capacity. However, cultural ecosystem services may contribute to social capital and community cohesion, which are two important contributors that allow a community to adapt and transform to threats and stressors.

2.3.2. *Coastal ecosystems may incorporate some element of "self-repair"*

As natural systems in a dynamic environment, coastal habitats experience a continuous and never-ending life cycle of growth, disturbance, death and

[22]F. Danielsen *et al.*, The Asian Tsunami: A Protective Role for Coastal Vegetation, *Science*, 310(5748) (2005), 643.

[23]K. Kathiresan and N. Rajendran, "Coastal Mangrove Forests Mitigated Tsunami," *Estuarine, Coastal and Shelf Science*, 65(3) (2005), 601–606.

[24]A.M. Kerr, A.H. Baird and S.J. Campbell, Comments on Kathiresan and Rajendran (2005), *op. cit.*, *Estuarine, Coastal and Shelf Science*, 67(3) (2005), 549–541.

[25]B. Ondiviela *et al.*, "The Role of Seagrasses in Coastal Protection in a Changing Climate," *Ecological Engineering*, 87 (2014), 158–168.

renewal. Many coastal plant species exhibit pioneer life cycle strategies, are able to rapidly colonize a suitable area with a large number of seeds and propagules, grow rapidly and die at a relatively young age.[26] These strategies allow many coastal plant species to take advantage of rare "windows of opportunity", where physical conditions are suitable for plant establishment for only a short period of time,[27] and to survive in a dynamic and rapidly changing physical environment with a multitude of disturbances and stressors. Thus, coastal habitats may be considered adaptive to external disturbances in some circumstances.

The ability of a coastal ecosystem to potentially "bounce back" from some types of disturbance depends on the ecosystem in question and its biological traits, the frequency, magnitude, and duration of the disturbance. However, ecosystem resilience of mangroves to disturbances is a key knowledge gap.[19] Encouraging reports have recently been published concerning mangrove regeneration in response to Cyclone Nargis, an event which devastated large parts of the Ayeyarwady Delta in Myanmar in 2008. Here, mangroves in 60 plots rapidly resprouted after the disturbance.[28] However, this response is species-specific (most likely seen in species such as *Avicennia* spp. and *Laguncularia* spp. that can coppice[19,25]), and the areal extent of regeneration is swamped by the huge area of mangrove lost due to ongoing conversion to rice agriculture.[29]

Conflicting reports are emerging from the field in the Philippines in the wake of Typhoon Haiyan (Yolanda) in November 2013. Some reports suggest that as of May 2014, some coastal mangroves and seagrass showed either very little damage, or rapid regeneration and regrowth of disturbed

[26]D.A. Friess *et al.*, "Are All Intertidal Wetlands Naturally Created Equal? Bottlenecks, Thresholds and Knowledge Gaps to Mangrove and Saltmarsh Ecosystems," *Biological Reviews*, 87(2) (2012), 346–366.

[27]T. Balke *et al.*, "Windows of Opportunity: Thresholds to Mangrove Seedling Establishment on Tidal Flats," *Marine Ecology Progress Series*, 440 (2011), 1–9.

[28]T.T. Aung, Y. Mochida and M.M. Than, "Prediction of Recovery Pathways of Cyclone-Disturbed Mangroves in the Mega Delta of Myanmar," *Forest Ecology and Management*, 293 (2013), 103–113.

[29]E.L. Webb *et al.*, "Deforestation in the Ayeyarwady Delta and the Conservation Implications of an Internationally-Engaged Myanmar," *Global Environmental Change*, 24 (2014), 321–333.

mangroves. However, other reports are suggesting that some areas are yet to show signs of re-establishment.

Seagrasses may show a mixed and delayed response to cyclone disturbance, depending on water depth. For example, a 1000 km^2 off the coast of Queensland, Australia, only began to recover after two years, and only in deeper water, as shallow water seagrasses were all destroyed by wave action.[30] One issue for both mangroves and seagrasses is a lack of systematic monitoring after a major disturbance due to a lack of pre-disturbance baseline data, as understandably, monitoring ecosystem response is not a priority after a major disturbance.

Thus, the ability of a coastal ecosystem to self-repair may vary spatially due to habitat type, quality, and prior health and disturbance history. Self-repair is also somewhat dependent on the type of disturbance being experienced. The examples given here concern low-frequency, high-magnitude, and short-duration disturbances such as cyclones. Will similar rules of self-repair still hold true in the case of low-magnitude, long-duration stressors such as SLR?

2.3.3. *Coastal ecosystems provide flexibility*

Nature-, or Ecosystem-based Adaptation measures (EbA) use natural ecosystems to buffer populations against negative climate change impacts through the sustainable delivery of broad ecosystem services relating to coastal protection and disaster risk reduction, food security, and livelihood diversification.[31] A key tenet of EbA is its perceived flexibility. For example, mangroves and seagrasses are able to migrate horizontally and accrete vertically to keep pace with SLR, if rates are not excessive, and if sufficient accommodation space and sediment supply is available (see Section 2.4). Thus, coastal ecosystems may be able to migrate with SLR in some conditions, compared to a rigid, hard engineered coastal defence. Flexibility is key to adaptation, because coastal populations will have to adapt to rates

[30] A.R. Preen, W.J. Long and R.G. Coles, "Flood and Cyclone Related Loss, and Partial Recovery, of More than 1000 km^2 of Seagrass in Hervey Bay, Queensland, Australia," *Aquatic Botany*, 52 (1995), 3–17.

[31] H.P. Jones, D.G. Hole and E.S. Zavaleta, "Harnessing Nature to Help People Adapt to Climate Change," *Nature Climate Change*, 2 (2012), 504–509.

of SLR that are as yet unknown, with predictions varying greatly. What rate of SLR should be used define the building criteria for a seawall over the next 50 years, if the rate of SLR over that time period is unknown?

2.4. Coastal Ecosystems Are Threatened by Climate Change

The sustainable and long-term delivery of ecosystem services (as a contribution to human adaptive capacity) requires a stable and resilient ecosystem.[32] Even with the ability of coastal ecosystems to self-repair to varying degrees, disturbances can still affect ecosystem service provision over the medium to long term. For example, almost 10 years after the Southeast Asian tsunami in 2004, some locations on the west coast of Thailand still show much reduced mangrove biomass,[33] with implications for ecosystem service provision related to carbon sequestration and storage. Threats linked to climate change, such as SLR, also have the potential to dramatically disturb tropical vegetated coastal ecosystems; a rate of SLR of only by 38 cm the year 2080 (a very conservative estimate) could result in the loss of 22% of the world's coastal wetlands.[34]

2.4.1. *Sea level rise is a non-linear threat*

Coastal and marine ecosystems are expected to bear the brunt of climate change impacts, such as sea level rise, increased windiness and storm surges, increased sea surface temperatures, and ocean acidification.[1] The implications of such impacts are severe, and include habitat loss, reduced ecosystem productivity and ecosystem service provision, and irreversible ecological transformation.[35] Such ecological transformations will be inherently non-linear and depend on a series of thresholds that are based on species-specific tolerances to a range of environmental stressors.[25]

[32]T. Elmqvist *et al.*, "Response Diversity Ecosystem Change and Resilience," *Frontiers in Ecology and the Environment*, 1(9) (2003), 488–494.

[33]N.R.A. Jachowski *et al.*, "Mangrove Biomass Estimation in Southwest Thailand Using Machine Learning," *Applied Geography*, 45 (2013), 311–321.

[34]R.J. Nicholls, F.M. Hoozemans and M. Marchand, "Increasing Flood Risk and Wetland Losses Due to Global Sea-level Rise," *Global Environmental Change*, 9 (1999), 69–87.

[35]O. Hoegh-Guldberg and J.F. Bruno, "The Impact of Climate Change on the World's Marine Ecosystems," *Science*, 328(5985) (2010), 1523–1528.

The non-linearity of ecosystem service degradation and ecological transformation poses particular management and policy challenges. If a tipping point has been surpassed where a state change is ensured, though a time lag means that no impact has been observed, then how do you know when to plan and enact a management intervention?

2.4.2. *Response of vegetated coastal ecosystems to sea level rise*

The concept of thresholds introduced above is key to the relationship between vegetated coastal ecosystems and SLR. The establishment and long-term survival of coastal ecosystems is strongly dependent on their tolerance to tidal inundation. For example, the mangrove ecosystem is found at or above mean sea level (MSL), where the intertidal surface is dry more often than it is flooded. Seagrasses are found below MSL due to a stronger tolerance to flooding.

SLR is a threat to vegetated coastal ecosystems as it shifts the MSL baseline, requiring the ecosystem to also move in order to maintain its position relative to the tidal frame. Vegetated coastal ecosystems can achieve this via two mechanisms. Firstly, coastal ecosystems can migrate landwards over time to higher elevations, a mechanism that has allowed coastal ecosystems to survive periods of sea level change over millennia. However, the potential for migration is curtailed in much of Southeast Asia, as the landward extent of coastal ecosystems is constrained by artificial coastal defence structures, resulting in "coastal squeeze". For example, Figure 5 shows a mangrove in Johor, Peninsular Malaysia. Successive land drainage has encroached into the mangrove, with the subsequent agricultural land protected by clay dykes. These dykes block the natural landward migration of mangroves under SLR. This example is characteristic of much of the west coast of Peninsular Malaysia.

Seagrasses may show a similar ability to migrate landwards under sea level rise, if artificial barriers do not impede their movement. For example, a recent study[36] in Moreton Bay, Queensland, Australia modelled seagrass migration patterns under an SLR scenario of 1.1 m by the year 2100. Over

[36]M.I. Saunders *et al.*, "Coastal Retreat and Improved Water Quality Mitigate Losses of Seagrass from Sea Level Rise," *Global Change Biology*, 19(8) (2013), 2569–2583.

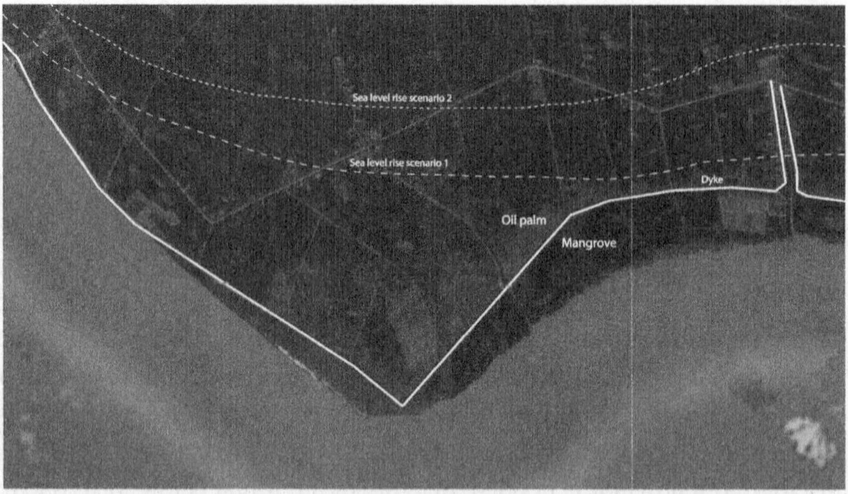

Figure 5. A Mangrove Near Semerah, Johor State, Peninsula Malaysia. Historical Mangrove Area Has Been Drained for Oil Palm and Other Agricultural Land Uses. The Clearly Defined Sea Dyke Prevents the Mangrove from Naturally Migrating Landwards under Different Sea Level Rise Scenarios (Dashed and Dotted Lines).

Source: Google Earth image, 2009.

this time period, seagrass loss was modelled in the order of 17% from coastal squeeze only (e.g., no anthropogenic stressors such as increased development and encroachment). However, the removal of artificial barriers such as dykes, to allow seagrasses to migrate, resulted in seagrass losses of only 5% by 2100. Interestingly, seagrass losses due to sea level rise were lower in areas with high water quality. This suggests that low water quality has a synergistic effect with SLR and could cause impacts that are worse than SLR alone (see Section 2.4.4).

Coastal vegetated ecosystems can also keep pace with SLR vertically, as well as laterally. Several physical and biological processes contribute to positive surface elevation.[25] Sediment is brought in on the tide and deposited onto the mangrove surface, which builds up vertically over time. Below-ground root production can also contribute to increasing surface elevation, especially in systems (such as the neotropics) where incoming sediment supply is limited. Other processes, such as subsidence, organic matter oxidation, and compaction can offset positive elevation gains.

2.4.3. *Monitoring ecosystem vulnerability to sea level rise in Southeast Asia*

A stable ecosystem is required if we are to maximize their contribution to adaptive capacity, though the non-linear response of coastal ecosystems to SLR means it is often difficult to assess whether an ecosystem is actually stable or not. This difficulty is compounded by the fact that changes to SLR and surface elevation operate over millimeter spatial scales, and decadal to millennial temporal scales. The Rod Surface Elevation Table (RSET) is one potential tool to monitor the multitude of sub-surface processes that affect surface elevation change at the millimeter scale. The RSET consists of a solid steel rod that is drilled into the bedrock (e.g., 5–8 m in southwest Thailand, 10–14 m in West Papua, Indonesia, 11–18 m in northwest Singapore). This provides a stable datum for a removable arm, which is used to measure how the whole soil column expands or contracts. The movement of pins attached to the arm are recorded — an accreting surface will push the pins up, while an eroding surface will lower the pins. The RSET can act as a sentinel to identify sites where the soil column may not be expanding as fast as SLR.

The RSET method has been deployed in mangrove and saltmarsh areas around the world. However, there is a distinct geographical bias to SLR vulnerability monitoring, with sites concentrated in North and Central America, Europe, and Australasia. We are data-deficient in many other regions, especially South America, Africa, South Asia, East Asia, and Southeast Asia.[37] This is somewhat surprising, because RSETs are particularly suitable for deployment in the ASEAN region. Compared to other surface elevation measurement methods (which cannot achieve the vertical resolution required for wetland studies), RSETs are very cheap to deploy, simple to measure, and utilize a standardized method to allow for comparisons between sites and countries.[38] Thus, while RSET research has primarily been the domain of academic researchers, it is potentially suitable

[37]E.L. Webb *et al.*, A Global Standard for Monitoring Coastal Wetland Vulnerability to Sea-Level Rise, *Nature Climate Change*, 3 (2013), 458–465.

[38]Donald R. Cahoon, "Surface Elevation Table," U.S. Geological Survey. Retrieved 9 March 2015, from http://www.pwrc.usgs.gov/set/.

for long-term monitoring by government forestry departments and coastal management agencies.

2.4.4. *Interactions between climate change and other anthropogenic stressors*

The threat of SLR is compounded by the myriad of other anthropogenic stressors that act within the densely populated coastal zone. The example of Mandai mangrove, situated on the northwest coast of Singapore, exemplifies the range and scale of different stressors than can impact coastal vegetated ecosystems.[39] This mangrove is suffering from erosion and internal vegetation collapse due to decades of impacts from stressors as diverse as heavy metal pollution, microplastic pollution, insect pests, overharvesting, increased shipping traffic causing erosion, sediment starvation due to nearby dam construction (reducing its resilience to SLR), and the impacts of transboundary reclamation. Thus, SLR in this, and similar examples throughout Southeast Asia, would probably not be the most immediate and pressing concern to coastal vegetated ecosystems in many locations. What may be of more importance in these locations is the potential interaction between these multiple stressors. The nature of the interaction between SLR and other stressors will determine their overall impact. The interaction may be[40]:

(1) *Additive* — purely the combined impacts of both stressors in isolation;
(2) *Antagonistic* — one stressor reduces the impact of another, so the combined impact is less than would be expected of additive stressors;
(3) *Synergistic* — one stressor amplifies the impact of another, so the combined impact is more than would be expected of additive stressors.

Interactions between SLR and other anthropogenic stressors pose real management challenges,[41] not least because the nature of the interaction

[39] D.A. Friess *et al.*, "Mandai Mangrove, Singapore: Lessons for the Conservation of Southeast Asia's Mangroves," *Raffles Bulletin of Zoology*, S25 (2012), 55–65.

[40] C.J. Brown *et al.*, "Managing for Interactions between Local and Global Stressors of Ecosystems," *PLoS ONE*, 8(6) (2013), e65765.

[41] D.A. Friess *et al.*, "Payments for Ecosystem Services (PES) in the Face of External Biophysical Stressors," *Global Environmental Change*, 30 (2015), 31–42.

is largely unknown until the non-linear impact is realized. However, the interactions between SLR (a global, largely uncontrollable stressor) and other anthropogenic stressors (potentially smaller scale and easier to control) also provides real management opportunities. The case study of synergistic interactions between SLR and water quality in Moreton Bay, Australia (Section 2.4.3) provides an example of how stressors could be managed. While it is unfeasible to reduce SLR within a sufficient timescale, management action can be taken to improve aspects of water quality in order to mitigate SLR-driven losses.[35]

2.5. Moving towards a Landscape Approach to Manage Coastal Ecosystems

With key coastal ecosystem services under threat from SLR and other (potentially synergistic) anthropogenic stressors, we urgently need a management framework that can minimize negative impacts of stressors, while increasing coastal ecosystem adaptive capacity. These objectives necessitate a landscape approach to coastal management and decision-making. Coastal ecosystems in the ASEAN region are often (or perceived to be) managed individually and in isolation, and only at the site-scale. For example, habitat restoration is often conducted in a single site (e.g., a set of abandoned shrimp ponds) and for one ecosystem type (mangroves), due to funding, land tenure and decision-making limitations. Larger-scale coastal management plans that encompass multiple ecosystem at the core are rarer in the region, though are required in order to maximize ecosystem service provision, and manage the ecosystem stressors that originate from, and act on scales much greater than the single site. A coastal management plan could be defined using the coastal vulnerability assessment framework outlined previously, particularly focusing on the exposure and adaptive capacity of coastal vegetated ecosystems.

2.5.1. *Reducing the exposure of coastal vegetated ecosystems*

Previously, we have seen how coastal ecosystems are exposed to a multitude of threats, including SLR and multiple anthropogenic stressors of diverse origin and mechanism. Reducing exposure to these (potentially synergistic) threats is key to reducing the vulnerability of coastal ecosystems to SLR, but is such a goal achievable?

Management of stressors is challenging because many of them originate outside the target conservation site, and thus outside the purview of site-specific management plans.[40] Taking the Mandai example previously (Section 2.4.4), mangroves require the input of sediment in order to increase their surface elevations and keep pace with rising sea levels. However, sediment supply has been severely curtailed in northwest Singapore since the early 1970s, when the Kranji estuary was dammed to create a freshwater reservoir, blocking terrestrial and fluvial sediment from entering the coastal zone. Thus, the main mechanism enabling Mandai mangrove to increase its resilience to SLR (sediment) is impacted by a management decision made decades ago and multiple kilometers away from the site. Other stressors, such as water pollution and ship traffic, also occur beyond the boundaries of Mandai mangrove.

Mitigating and managing a diverse array of external anthropogenic stressors requires the identification of the actors causing harm, and modifying their behavior, e.g., through top-down legislative mechanisms, financial incentives, or compensation to proactively reduce a stressor. A landscape approach is required to attempt to encompass actors that are operating at scales substantially beyond the targeted conservation site. Increasing the scale of management, and increasing the number of actors substantially increases the institutional and transaction costs of any management intervention, which may not make it cost-effective to conserve the targeted ecosystem.[40] It also may not be feasible to mitigate some stressors, especially diffuse stressors that act on long temporal and spatial scales, such as SLR. Focus in these instances may have to be less on reducing exposure and more on increasing the adaptive capacity of the targeted ecosystem.

2.5.2. *Increasing the adaptive capacity of coastal vegetated ecosystems*

Previous examples show that coastal vegetated ecosystems can show some resilience to SLR (depending on the rate), if landward accommodation space is available to migrate into. Creating space to migrate increases the adaptive capacity of coastal vegetated ecosystems, potentially giving them scope to adapt to external stressors that cannot meaningfully be mitigated (such as SLR). However, the coastal zone is heavily utilized for urban development, industry and agriculture, so creating space requires a trade-off with economic development to some degree, and will be a contentious planning issue, probably requiring a long-term, top-down approach to plan for future land uses. Thus, the act of creating accommodation space requires a robust, large-scale and long-term legislative framework of coastal management. Such frameworks are currently lacking in much of the ASEAN region, though other regions offer lessons on how such a framework could be constructed and implemented.

In the United Kingdom (UK), accommodation space (e.g., areas where sea defences can be removed and retreated to allow coastal ecosystem migration) has been identified through a series of iterative, national-scale Shoreline Management Plans (SMPs).[42] The entire coastline of England and Wales is divided into management cells and sub-cells, where specific management interventions (increase defences, maintain defences, retreat defences) are planned for the next 50 years. It is also used to identify sub-cells that may not be economically viable to protect over the long term due to climate change, etc. Importantly, SMP cells are delineated by coastal processes, not administrative boundaries. While this adds complications and transaction costs by requiring transboundary cooperation, it ensures that management actions are self-contained, and minimizes knock-on geomorphological impacts further along the coast.

Frameworks for the creation of accommodation space have also been proposed for the coastline of Australia.[43] These frameworks rely less on spatial planning and site identification (as with the UK SMPs), but focus more on stakeholder analysis, including the distribution of public and private benefits, property rights and values, and institutional and decision-making structures.

Ideally, an accommodation space framework would achieve both the spatial planning components (UK SMPs) and the socio-economic conflicts (Australia), because a framework will have multiple objectives to

---

[42]EA, *Shoreline Management Plans*, UK Environment Agency, 2009. Retrieved 9 March 2015, from www.gov.uk/government/publications/shoreline-management-plans-smps.

[43]N. Abel *et al.*, "Sea Level Rise, Coastal Development and Planned Retreat: Analytical Framework, Governance Principles and an Australian Case Study," *Environmental Science and Policy*, 14(3) (2011), 279–288.

achieve. Firstly, a framework has to identify areas that are suitable for the creation of accommodation space — an objective that requires the balanced assessment of multiple physical, economic, and social trade-offs. What is the economic/social/ecological value of the land being sacrificed? What is more important — agricultural production or a seagrass meadow? Is it cost-effective to protect this land under scenarios of climate change or should sea defences be removed to create new intertidal areas? Who owns the land?

Some of these questions may be difficult to answer in the ASEAN context. We may know the economic value of agricultural land, but not its full socio-ecological value. Land tenure is particularly unclear in the coastal zones of many ASEAN countries. The creation of accommodation space and "managed retreat" has been pursued in some European countries for more than 20 years now, though it is a management intervention that is yet to be applied meaningfully in the ASEAN context. Thus, the examples from the UK and Australia provide important experience in how to identify locations of accommodation space, but also how to manage complex land use trade-offs and land tenure conflicts.

2.6. Recommendations for a Holistic Approach to Coastal Vulnerability Reduction

Coastal ecosystems hold substantial promise to increase adaptive capacity and reduce the vulnerability of human populations along the coast. So, how do we better position coastal ecosystems within large-scale coastal management frameworks in Southeast Asia? A recent call for action[44] outlined four key recommendations to increase the use of coastal ecosystems in EbA:

(1) Build the case for natural coastal protection methods, by advancing coastal protection science and quantifying ecosystem service co-benefits in addition to coastal protection;
(2) Bring ecosystems into mainstream decision-making processes by considering ecosystems in vulnerability assessments; developing tools that model complex combined risks; building decision-support systems

[44] M.D. Spalding *et al.*, "The Role of Ecosystems in Coastal Protection: Adapting to Climate Change and Coastal Hazards," *Ocean & Coastal Management*, 90 (2014), 50–57.

to help communities visualize the impacts of, and solutions to, coastal adaptation; engaging stakeholders; and enacting environmental policies;

(3) Incorporate proven management interventions such as hybrid engineering solutions, habitat restoration, and Marine Protected Areas (MPA);

(4) Build capacity for implementation at the local and national level.

These recommendations are important starting points, though the discussion in this chapter suggests several additional recommendations that consider the vulnerability of coastal ecosystems themselves to SLR:

(5) *Expand coastal vulnerability assessments to incorporate the exposure, sensitivity and adaptive capacity of coastal ecosystems, as well as coastal populations.*
The main tenets of coastal vulnerability assessment — exposure, sensitivity and adaptive capacity — are well-established and provide an important framework with which to estimate the vulnerability of coastal populations. Though substantial data limitations exist, this framework could also be applied to assess coastal ecosystem vulnerability.

(6) *Increase the collection and availability of region-specific information on ecosystem service provision, and the potential vulnerability of ecosystems to climate change-related stressors.*
Expanding the vulnerability framework to coastal ecosystems requires data on ecosystem exposure, sensitivity, and adaptive capacity. Similar to human coastal vulnerability assessments, exposure is probably the easiest component to estimate for coastal ecosystems, though substantial data limitations still exist (in particular related to elevation data at sufficient vertical resolution). Estimating sensitivity will require a concerted effort focused on mesocosm, field, and modelling studies. Research concerning the adaptive capacity of coastal ecosystems may still be at the conceptual stage. For all three components, we require data that is specific to the ASEAN region, as mangroves and seagrasses in Southeast Asia differ greatly from systems in the neotropics and Australasia, in terms of species composition, biomass, geomorphology, and threats.

(7) *Expand research, assessment and implementation of spatial scales in EbA.*
Coastal ecosystems — and the external stressors that impact them — operate over multiple spatial scales, many of which are beyond the confines of the single site that is targeted for adaptation action. Landscape-scale coastal management is required in order to attempt to mitigate stressors (both SLR and other synergistic anthropogenic stressors) that are affecting the target ecosystem.[40]

(8) *Expand research, assessment and implementation of temporal scales in EbA.*
We have an understanding of coastal ecosystem processes over the short term, and how coastal ecosystems respond over millennial timescales, though the non-linear relationship between coastal ecosystems and SLR means we lack an understanding of coastal ecosystem dynamics at decadal-century scales — exactly the timescale that managers and decision-makers require.[45] It is important to incorporate the time lag between a physical forcing event and a geomorphological or ecological response into modelling (and planning) processes,[46] though it is currently difficult to do so due to insufficient knowledge of the various feedback loops within the coastal environment.

These recommendations, aimed at incorporating ecosystem resilience and dynamics into EbA, will help evidence-based management and policy interventions by ensuring that coastal ecosystems are most resilient and can continue to sustain an adequate level of ecosystem service provision.

2.7. Conclusions

Coastal ecosystems contribute essential ecosystem services to human populations in the coastal zone and beyond, and can be considered an important contributor to increased adaptive capacity and reduced vulnerability to climate change-related stressors such as SLR. However,

---

[45]C.D. Woodroffe and C.V. Murray-Wallace, "Sea-Level Rise and Coastal Change: The Past as a Guide to the Future," *Quaternary Science Reviews*, 54 (2012), 4–11.

[46]L. McFadden, T. Spencer and R.J. Nicholls, "Broad-Scale Modelling of Coastal Wetlands: What Is Required?," *Hydrobiologia*, 577(1) (2007), 5–15.

coastal ecosystems are by their very nature dynamic, and exist according to fine thresholds that may tip them into alternative stable states, that may not contribute to adaptive capacity to the same degree. This dynamism means that coastal ecosystems are susceptible to the very climate change-related stressors that they also help coastal populations adapt to in the first place. In addition, ecosystems in the congested and over-utilized coastal zone are more immediately threatened by anthropogenic stressors such as encroachment and pollution. Thus, in their degraded state they may be less able to provide ecosystem services and contribute to adaptive capacity.

This chapter advocates a long-term, large-scale planning framework that can best encapsulate the dynamic nature of coastal vegetation ecosystems. The adaptive capacity of coastal ecosystems to SLR can increase the adaptive capacity of human coastal populations, if we manage our coast in such a way that we allow coastal ecosystems the space and time to adapt. Such a framework in the ASEAN region will allow us to more effectively manage future human population vulnerability and habitat degradation in a dynamic and changing coastal environment.

Theme II

Legal Frameworks/Policies/Governance for Climate Change

Chapter 3

Adaptive Water Governance: Lessons Learned from Implementing an Ecosystem Approach in Mesoamerica

Alejandro Iza, Alexandra Müller†, and Valentina Nozza‡*

3.1. Introduction

The IPCC Report on Climate Change and Water concludes that freshwater resources are highly vulnerable to climate change.[1] Climate change impacts on the hydrological cycle include changes in the amount and quality of available freshwater resources and extremes like floods and droughts, and are projected to intensify in the future.[2] Hence, water is one of the resources that is going to be most seriously affected by climate change, and where climate change manifests its impacts more fully: climate change is water change. Considering the importance of freshwater for both human and natural systems, this poses serious threats to life on earth and brings about the need for urgent solutions to sustainable water management on all scales. Accordingly, effective governance is needed to manage water resources sustainably. In this context, it is important to note that 276 river and lake basins are shared among two or more countries. Competing uses over these resources can lead to over-allocation, pollution and degradation of the ecosystems. Eventually, this can lead to tensions and conflicts

*Alejandro Iza, Director, IUCN Environmental Law Centre and Head, IUCN Environmental Law Programme.

†Alexandra Müller, Project Officer, IUCN Environmental Law Centre Bonn, Germany.

‡Valentina Nozza, intern, IUCN Environmental Law Centre Bonn, Germany.

[1] B.C. Bates, Z.W. Kundzewicz, S. Wu, and J.P. Palutikof (eds.), *Climate Change and Water. Technical Paper of the Intergovernmental Panel on Climate Change* (Switzerland: IPCC Secretariat, 2008).

[2] *Ibid.*

between riparian states. Given the vital importance of freshwater, innovative approaches are needed that take into account these complexities.

Adaptive governance of water resources has evolved as a response to these challenges. The approach is still new and experiences are needed to supplement and give practical meaning to this concept. It is crucial to see whether and how water management and "water governance capacity"[3] are transformed when building a robust governance framework and applying the principles of adaptive water governance on the ground.

IUCN is putting the principles of adaptive governance into practice by implementing an ecosystem approach in several pilot river and lake basins around the world. This chapter presents some lessons learned from the implementation of adaptive water governance at the local level in Mesoamerica.

The chapter is divided into two segments. The first segment introduces the concept of adaptive water governance and explains the principles it relies upon. Further, it analyzes why International Water Law plays a crucial role in supplementing adaptive water governance. The second segment reports experiences from the application of the concept on the ground. It states challenges encountered and lessons learned during IUCN's practical work in Mesoamerica. The chapter concludes that adaptive governance is a good way to enhance both water governance and adaptive capacity on a local level as well as on a transnational one and promotes cooperation and trust building.

3.2. Water, Climate Change and Ecosystems

Freshwater systems are complex, characterized by their own dynamics and organic processes, which are determined by ongoing feedbacks and self-organization.[4] Generally, these systems are considered to possess the

[3]In their book, *RULE — Reforming Water Governance*, Iza and Stein define water governance capacity as "a society's level of competence to implement effective water arrangements through policies, laws, institutions, regulations and compliance mechanisms" — A. Iza and R. Stein (eds.), *RULE — Reforming Water Governance* (Switzerland: IUCN, 2009).

[4]C. Folke, "Resilience: The Emergence of a Perspective for Social-Ecological Systems Analyses," *Global Environmental Change*, 16(3) (2006), 253–267.

capacity to respond and adapt to different stressors,[5] but constant human interactions present factors that are not easily borne.[6] Freshwater therefore has the complex characteristics of a combined social-ecological system,[7] through which greater variation within its organic processes results in increased unpredictability and declined capacity to adapt.[8] Furthermore, climate change alters the hydrological cycle in a non-comparable fashion, posing an additional, significant amount of uncertainty.[9] Population growth will add to this through a rising demand for food, freshwater, and energy, and increases in agricultural and industrial pollution.[10] All of these unsupportive external forces result in a direct impact on the quantity and quality of the freshwater resources available.

Good water governance is a means to address these challenges by implementing effective water management through policies, laws, and institutions.[11] It is intrinsic to health, food security, and economic development of hydrological basins. Poor water governance results in over-allocation and pollution of water resources and in the degradation of essential ecosystem services people rely upon for water storage and management during periods of flood risks and droughts. Recently states have become increasingly aware of the scarcity of this vital resource: up to now, around 29 states suffer from more or less severe water shortages and according to the World Meteorological Organization, if conditions remain unchanged, the number will rise to 34 by 2025.[12]

---

[5]K.R. Young and J. Lipton, "Adaptive Governance and Climate Change in the Tropical Highlands of Western South America," *Climatic Change*, 78 (2006), 63–102.

[6]*Ibid.*

[7]Folke (2006), *op. cit.*

[8]W. Medema, B.S. McIntosh and P.J. Jeffrey, "From Premise to Practice: A Critical Assessment of Integrated Water Resources Management and Adaptive Management Approaches in the Water Sector," *Ecology and Society* (2008). Retrieved 10 March 2015, from http://www.ecologyandsociety.org.

[9]B.C. Bates *et al.* (eds.), *Climate Change and Water. Technical Paper of the Intergovernmental Panel on Climate Change* (Switzerland: IPCC Secretariat, 2008).

[10]*Ibid.*

[11]Iza and Stein (2009), *op. cit.*

[12]G. Aguilar Rojas and A. Iza, *Governance of Shared Waters: Legal and Institutional Issues* (Switzerland: IUCN, 2011).

So far, states have made major efforts in mitigating and adapting to climate variability, but the paradigm of stationarity,[13] where natural system fluctuations occur within a range of certainty, no longer applies. To achieve good water governance that responds to the demands and challenges posed by climate change, it is central to accept and incorporate uncertainty into project plans, policies, and laws related to water. Decision-makers often lack the resources to conduct sound studies and regional modelling on the specific impacts of climate change on their national territory, so there is a need to develop governance frameworks that are functional and designed flexibly enough to allow for periodic changes to the objectives, information, and rights in order to adjust appropriately.[14] Still, most current legislations fail to address climate change adaptation in a holistic manner. This can be related to the fact that climate impacts are frequently seen as hypothetical events happening in the future[15] and that it is seen as more convenient to *react* to them by applying technological measures as coping mechanisms, instead of anticipating and preventing them.

However, *preventive* and *proactive* action is needed to reduce vulnerability and engage successfully in the sustainable development of hydrological basins. In the context of climate change, vulnerability is described as the "state of susceptibility to harm from exposure to stresses associated with environmental and social change and from the absence of capacity to adapt".[16] This means that vulnerability is the extent to which a community is able or unable to cope with adverse climate impacts. Generally, developing countries are considered more vulnerable since they do not possess sufficient social, financial, and technological means to adapt.[17] On the other hand, adaptive capacity is the ability to respond to these challenges through learning how to manage them, developing new knowledge, and being malleable enough to adopt new

---

[13] *Ibid.*

[14] *Ibid.*

[15] International Union for Conservation of Nature, *Climate Change and Governance: Ecosystem-Based Approach in Mesoamerica* (Switzerland: IUCN, 2010).

[16] W.N. Adger, "Vulnerability," *Global Environmental Change*, 16 (2006), 268–281.

[17] IUCN (2010), *op. cit.*

working methods and strategies.[18] The more a community is resilient to changing circumstances — is able to adapt — the less it is vulnerable.

3.3. Adaptive Governance as an Opportunity

In the light of this demand, adaptive governance has evolved as an approach to deal with the multiple challenges that have been associated with the governance of climate change. It is seen as an ongoing process for integrating uncertainty and insufficient knowledge into planning processes by continually reviewing management practices and learning from the outcomes. Decisions need to be informed in order to be prepared to act and to decide how to best modify current intervention strategies.[19] The concept is particularly suitable for water resources, due to their intrinsic biophysical processes, dynamics, and feedbacks, as explained above.

Adaptive governance, in contrast to single adaptation measures or conventional resources management, is an integrated way of managing natural resources across relevant sectors, and places emphasis on flexibility of and learning from applied measures.[20] The ecosystem approach becomes central to adaptive governance, as it favors biodiversity and ecosystem services over inflexible built structures. It promotes a holistic management of water, land, and living resources, striving for sustainable management, conservation, and restoration of water resources in order for them to provide services that facilitate adaptation to climate change of the local population.[21] The resilience of a resource to the adverse impacts of climate change is augmented if its ecological integrity is maintained.[22] The ecosystem approach includes the ideas of sustainability, interconnection of

---

[18] N.A. Marshall *et al.*, *A Framework for Social Adaptation to Climate Change: Sustaining Tropical Coastal Communities and Industries* (Switzerland: IUCN, 2009).

[19] International Union for Conservation of Nature Information Paper, *Recovery from the Indian Ocean Tsunami: Guidance for Ecosystem Rehabilitation Incorporating Livelihoods Concerns* (Switzerland: IUCN, 2005).

[20] Medema, McIntosh, and Jeffrey (2008), *op. cit.*

[21] A. Colls, N. Ash, and N. Ikkala, *Ecosystem-Based Adaptation: A Natural Response to Climate* (Switzerland: IUCN, 2009).

[22] J. Troell and G. Swason, "Adaptive Water Governance and the Principles of International Law," in J. Roberts and J.C. Sanchez (eds.), *Transboundary Water Governance: Adaptation to Climate Change* (Switzerland: IUCN, 2014).

ecological and social systems, adaptive management, and the precautionary approach. It proposes to enable people to adapt to climate change by preserving the structure, functions, and services of ecosystems.[23]

Ecosystems provide vital services to people: they are sources for food and freshwater and fulfill an essential role during extreme climate events such as droughts and floods in storing water and preventing soil erosion. In addition, they provide habitats for plants and animals and are of cultural, scientific, and educational value.[24] The flexibility of the ecosystem approach allows it to account for the great uncertainties regarding the impacts of climate change on water resources. Enhancing the health of the ecosystem instead of installing purely technological measures is therefore a so-called "no regret" or "win-win" strategy.

The cross-sectoral nature of vulnerabilities and adaptive responses further requires a multilevel approach.[25] Other sectors such as agriculture, energy, and health are also projected to be hit hard by climate change and since these are related to water, mutual feedbacks are in a position to enhance negative impacts. A horizontal integration and coordination across sectors and within single sectors will therefore be necessary.[26] Furthermore, vertical integration is necessary, since water management takes place at the local level, while decisions are taken at the national, regional, or even global levels.[27]

Governance is understood as a set of policies, laws, and institutions. As common pool resources, freshwater resources such as river basins are particularly complex to govern and require effective institutional arrangements to ensure their sustainable use.[28] The existence and effectiveness of institutions is therefore crucial, as well as the institutional design. The latter plays a critical role to ensure integration of all levels. For adaptive governance, the focus should be on the river basin level as the scale of decision-making

[23]Aguilar Rojas and Iza (2011), *op. cit.*
[24]*Ibid.*
[25]Troell and Swason (2014), *op. cit.*
[26]*Ibid.*
[27]*Ibid.*
[28]E. Ostrom, *Governing the Commons: The Evolution of Institutions for Collective Action* (UK: Cambridge University Press, 1990).

and management.[29] Adequate water legislation that regulates water management and promotes the establishment of water institutions, for example basin organizations at different levels, is crucial for effective governance. However, there is the need for sufficient capacity to translate national laws and regulations or even global agreements to the basin level.

Adaptive governance accounts for the uncertainties associated with the variations in water quantity and quality due to climate change, which will significantly differ in the short-, medium-, and long-term.[30] This relates to the timing, intensity, and the character of climate impacts, such as changes in precipitation patterns, droughts, and floods.[31] In contrast to conventional governance approaches, adaptive governance can account for these uncertainties in the sense that collaborative, flexible, and learning-based approaches are at the core of the concept, including decision-makers and stakeholders from all societal levels.[32] The objective is to equip decision-makers and stakeholders with an understanding of the social-ecological system and the required skills to deal with variability and uncertainty.[33] Multi-level, broad-based stakeholder participation is therefore another critical aspect of this approach.

3.4. Water Governance and the Principles of International Water Law

To have a functional, flexible, and innovative governance framework in place that facilitates effective and sustainable adaptation of water resources requires policies, laws, and institutions to be in place and operational. Results from the ground indicate that transboundary water cooperation in particular still shows significant governance gaps, despite the fact that 60%

---

[29] D.E. Huitema *et al.*, "Adaptive Water Governance: Assessing the Institutional Prescriptions of Adaptive (Co-)Management from a Governance Perspective and Defining a Research Agenda," *Ecology and Society*. Retrieved 10 March 2015, from http://www.ecologyandsociety.org.

[30] R.S. Bradley *et al.*, "Threats to Water Supplies in the Tropical Andes," *Science*, 312(5781) (2006), 1755–1756.

[31] Troell and Swason (2014), *op. cit.*

[32] C. Pahl-Wostl *et al.*, "Social Learning and Water Resources Management," *Ecology and Society*. Retrieved 10 March 2015, from http://www.ecologyandsociety.org.

[33] *Ibid.*

of the global freshwater resources are shared between two or more states.[34] For these states, good water governance cannot be seen as a purely national matter: Even if states claim sovereign rights over the portions of rivers and lakes located within their territories, certain activities, projects, and programs can potentially affect all riparian states. Consequently, the idea of the hydrological unity of a river or lake basin and the community of interest of all riparian states are crucial elements for a sustainable management of shared resources. Only through cooperation in the management of the basin, overcoming the lack of confidence and the fear of losing sovereignty, is it possible to achieve effective risks prevention and reduction, water security, economic development, and environmental protection on a basin-wide level.[35]

International Water Law provides guidance on how to govern shared rivers and lakes; thus, it supports sustainable and equitable use of the resources. Two major international instruments, the *Convention on the Protection and Use of Transboundary Watercourses and International Lakes* adopted in 1992 (Helsinki Convention) and the *UN Convention on Non-Navigable Uses of International Watercourses* adopted in 1997 (UN Watercourse Convention)[36] set out a series of substantial and procedural obligations states should consider when discussing transboundary water governance.[37] The international community has recognized that these treaties contain a set of principles, most of which are already part of International Customary Law and can therefore be considered binding upon states. Despite being international instruments of multiregional and global character, they are just as important for water governance at the local level. The following section describes briefly two principles of International Water Law that play a central role for invigorating adaptive water governance in a transboundary context.

[34]M. Smith and M. Cartin, *Water Vision to Action: Catalysing Change through the IUCN Water and Nature Initiative* (Switzerland: IUCN, 2011).

[35]A. Drieschove and G. Eckstein, "Cooperative Transboundary Mechanisms," in J. Roberts and J.C. Sanchez (eds.), *Transboundary Water Governance: Adaptation to Climate Change* (Switzerland: IUCN, 2014).

[36]Aguilar Rojas and Iza (2011), *op. cit.*; United Nations Convention on the Law of the Non-Navigational Uses of International Watercourses, 1997, Art. 5.

[37]*Ibid.*

3.4.1. *Equitable and reasonable utilization*

The principle of equitable and reasonable utilization is based upon the idea of the community of interest of all basin states and includes "both the right to utilize the watercourse and the duty to cooperate in the protection and development thereof...".[38] It requires states to use, develop, and protect watercourses in an equitable and reasonable manner. The doctrine in question is flexible: what is considered an equitable apportionment may vary over time with changes in the circumstances of the states concerned. "Equality of rights" does not imply that every state is entitled to the same amount of water, rather it refers to the right to receive an equitable share of the benefits deriving therefrom. Article 6 of the UN Watercourse Convention provides a series of factors that need to be taken into account to determine when an apportionment is equitable and reasonable. However, none of the factors prevails over the others and each one needs to be weighted in comparison with the others: geographical and climatic factors, social and economic needs, effect of the use on other states, existing and potential uses, availability of alternatives of comparable value, etc.[39] Even if a state engages in an activity which causes harm to other riparians, this harm will be just one more factor to be taken into account in establishing equitable utilization. In addition, prior or existing uses do not entitle an absolute protection of that activity,[40] but old activities may give way to new uses to achieve an equitable allocation of benefits. Ultimately, the implementation of this principle depends on good faith and cooperation between states.[41]

3.4.2. *Prevention of significant transboundary harm*

This principle does not absolutely prohibit the causation of any harm to other riparian states, but rather it requires states to avoid significant harm in a way and to the extent that this is reasonable under given circumstances,

---

[38] United Nations Convention on the Law of the Non-Navigational Uses of International Watercourses, 1997, Art. 5, para. 2.

[39] Troell and Swason (2014), *op. cit.*

[40] United Nations Convention on the Law of the Non-Navigational Uses of International Watercourses, 1997, Art. 10.

[41] S.C. McCaffrey, *The Law of International Watercourses* (UK: Oxford University Press, 2007).

recalling the principle of due diligence. The "no harm" rule refers to harm resulting from both activities directly affecting a watercourse, like building a dam, and activities not necessarily connected with the use of a watercourse, such as deforestation. As outlined by the UN Watercourse Convention, environmental pollution or harm is prohibited only if the harm is significant and unreasonable. The adjective "significant" is meant to be a threshold low enough to exclude trivial harm from the discussion and high enough to trigger before substantial or serious damage is done.[42] According to the ICJ in the *Pulp Mill* Case, a transboundary environmental impact assessment (TEIA) would put into practice numerous procedural requirements included in the principle of equitable and reasonable utilization and in the "no harm" rule.[43] Additionally, it should be noted that a strategic environmental assessment (SEA) might be a useful tool to integrate climate change concerns into basin-wide planning processes.[44] Strategic environmental assessments are environmental impact assessments that have a broader scope and can therefore be used to evaluate vulnerabilities of projects to climate impacts, instead of focusing on the environmental impacts of the project only.[45]

3.4.3. *Procedural obligations*

The duty to cooperate is a cornerstone of international environmental law. Under Article 8, the UN Watercourse Convention requires states to cooperate in good faith to adequately protect international watercourses.[46] Moreover, it encourages states to eventually establish joint management mechanisms and institutions. However, even if cooperation holds many beneficial outcomes for states, putting it into practice is a challenging undertaking. The primary obstacles to good cooperation are the lack of political will, shortages of institutional capacities, and the asymmetries between national legislations of riparian states. Procedural obligations

[42] *Ibid.*

[43] Troell and Swason (2014), *op. cit.*

[44] *Ibid.*

[45] *Ibid.*

[46] United Nations Convention on the Law of the Non-Navigational Uses of International Watercourses, 1997, Art. 6.

such as the obligation to regularly exchange data and information,[47] the obligation of prior notification,[48] and the duty to consult with other riparian states[49] are fundamental tools to overcome these challenges. Cooperation over technical issues like water quality or meteorological data enhances trust among riparian states and thus might make it possible for technical expertise to overcome political difficulties[50] and foster basin-wide and regional stability.

At the transboundary level, these principles feed into the adaptive governance framework, supporting it via laying out guidance on transnational cooperation, which is crucial for the sustainable management of shared resources or those resources that are of strategic significance in a given region. Therefore, when talking about adaptive governance in transboundary basins, the referred principles need to be taken into account and adequately factored in national activities and projects as well as in transnational undertakings.

3.5. Testing Adaptive Water Governance on the Ground: Experiences from Implementing the Ecosystem Approach

3.5.1. *Water governance in Mesoamerica*

IUCN has been working on adaptive water governance by applying the ecosystem approach in Mesoamerica in the framework of two different projects. One of these projects is entitled *Building River Dialogue and Governance* (BRIDGE). The other project is known as *Climate Change Governance Capacity: Building Regionally- and Nationally-tailored Ecosystem-based Adaptation (EbA) in Mesoamerica.*

BRIDGE works in four regions — Mesoamerica, the Andes, the lower Mekong, and Africa — and aims to promote good water governance through consensus building, demonstration, leadership, and learning in the context of transboundary river and lake basins. It endeavors to enhance water security and facilitate progress in ecosystem management using

[47] *Ibid.*, Art. 9.
[48] *Ibid.*, Art. 11.
[49] *Ibid.*, Art. 17.
[50] Rojas and Iza (2011), *op. cit.*

the ecosystem approach, water supply, sanitation, and sustainable water management. The objective of BRIDGE is to achieve these goals by catalyzing cooperation among riparian states, establishing or reinforcing transboundary institutional mechanisms, and supporting basin-wide planning and technical coordination.

Within Mesoamerica, BRIDGE has been working in three basins: Coatán, Goascorán (Sumpul sub-basin), and Sixaola, which also served as intervention sites for the above-mentioned EbA project. In these sites, the ecosystem approach theory was put into practice through a three-component integrated methodology that included: water governance capacity building, conservation and restoration of ecosystems, and improving local livelihoods. The EbA project goal was primarily to enhance ecosystem resilience to climate change by catalyzing enhanced governance arrangements and strengthening capacities for developing and implementing legislation and policy development on climate change and water management.

Both projects are implemented in transboundary basins. A fundamental element to the project intervention therefore, is to apply hydro-diplomacy but integrating the principles of multilevel dialogues and broad-based stakeholder participation. The understanding is that diplomacy takes place under the authority of sovereign states, but — because water is a resource used by everyone and managed at multiple scales — for agreements to work effectively, they require the support of water users. Thus, different agreements are needed for effective water governance: These may include both international treaties and a wide range of formal and informal agreements between local authorities, technical agencies, and the private sector.[51] To achieve efficient long-term results, water management issues need to be addressed at the appropriate scales (local, national, regional). National authorities need to support local actions with comprehensive regulations to give them both clear guidance and the necessary policy backstopping.

However, participation of a broad range of stakeholders in planning, implementing, and monitoring plans and strategies can be problematic because water users often lack both water governance capacity and formal

---

[51] *Ibid.*

and legitimate representation. Thus, it is essential to ensure that local stakeholder groups are given the opportunity to be equally represented and that they have the means to participate actively and meaningfully in planning processes. Public involvement calls for a strong commitment from local communities, to put into action water policies and plans that respect local capacities, needs, and vulnerabilities as well as national water management and development policies. Broad stakeholder participation helps people to understand climate science and to overcome popular misconceptions about the nature, degree, and the consequences of climate change causes and impacts.[52]

The following section explains the challenges faced and the lessons learned through applying the ecosystem approach on the ground in Mesoamerica. Using examples from the field, it shows evidence that adaptive governance is an auspicious way to enhance both good water governance and adaptive capacity at the local as well as at the transnational level. Slowly, it is being recognized that political goals such as economic growth, developing sustainable livelihoods, and environmental protection can be realized through more efficient management of water.[53]

For the purpose of this chapter, the three above-mentioned basins in Mesoamerica where both BRIDGE and EbA interventions have taken place have been selected to illustrate the lessons that can be learned from testing adaptive governance on the ground.

3.5.2. *Presentation of the three river basins*

3.5.2.1. *Sixaola basin*

The Sixaola River is a contiguous river that establishes part of the boundary between Costa Rica and Panama, in the southern part of the Central American Isthmus. It originates in the Talamanca Range and drains into the Caribbean. The basin is especially rich in biodiversity and a number of indigenous communities rely heavily on the basin's ecosystems for their livelihoods.

---

[52] *Ibid.*

[53] IUCN Water and Nature Initiative, "What We've Learned: Implementation Principle 7." Retrieved 2 April 2014, from http://www.waterandnature.org/en/results/what-weve-learned.

The basin is characterized by social and economic diversity, which have influenced its governance. While in the lower basin, transnational companies manage large fruit plantations for consumption overseas, indigenous territories located in the middle and upper part rely heavily on their own production. The basin is adjacent to the International Park "La Amistad", a transboundary protected area of significant importance,[54] which has been a Biosphere Reserve since 1982 and was declared a World Heritage Site in 1983.[55] The river is used for navigation and communication between the different communities, as well as for irrigation and provision of drinking water. One major issue is poor water quality due to the use of pesticides, which have resulted in public health problems.[56]

The primary impacts from climate change are projected to be heavy rainfalls, which will cause severe floods, damaging livelihood and resources.[57] Droughts will have a lesser impact.[58] Additional to its high vulnerability to climate change, the Sixaola Basin faces significant power asymmetries among stakeholders, in particular, the agro-industry sector and the local communities, and this has a correlation with the improvement of water governance arrangements in the basin.[59]

As far as the basin governance framework, in 1992, Costa Rica and Panama signed the *Treaty on Cooperation for Border Development* which covers several areas, such as social, economic, commercial, and environmental. The treaty sets out an institutional cooperative structure, the Costa Rica-Panama Permanent Binational Commission (PBC), which was, however, rather dormant during the first years of its existence. The following institutions were created under the treaty: the PBC, the primary decision-making body for the whole basin; two executive secretariats representing each country and monitoring the decisions made by the PBC;

---

[54]The particular importance of the International Park "La Amistad" is further expressed by the fact that a Binational Sectoral Technical Commission was particularly established for "La Amistad" within the framework of the Border Treaty to ensure coherence in the governance of this jointly executed project.

[55]M. Peña Chacón and M. Perez de Madrid, "Case Study: Up-Scaling Adaptation in the Sixaola River Basin," in J. Roberts and J.C. Sanchez (eds.), *Transboundary Water Governance: Adaptation to Climate Change* (Switzerland: IUCN, 2014).

[56–59]*Ibid.*

ordinary and special Binational Sectoral Technical Commissions (BSTCs) that provide technical assistance to particular sectors; and the Binational Executive Technical Units (BETUs) that serve as implementation units.[60] Within the framework of the BCTCs, the Binational Commission for the Sixaola River Basin (BCSRB) was established with a view of responding to environmental threats and climate change impacts. Its primary function was to oversee a binational project entitled Integrated Ecosystem Management of the Sixaola Binational River Basin, funded by the Inter-American Development Bank (IDB). As a binational body, it served as a platform for broad-based stakeholder participation, but had no formal decision-making authority. It is worth noting that while Costa Rica recently passed a new Water Bill, at the start of the BRIDGE and EbA project interventions, there was no law that regulated water management or the establishment of water institutions. Panama, on the contrary, has a water law in place which promotes the establishment of basin organizations.

The aim of IUCN's intervention in the Sixaola Basin was to create an enabling environment for deepening the cooperation between Costa Rica and Panama within the basin, promoting sensitization and awareness raising on climate change impacts as well as specific interventions to demonstrate EbA in practice, and embracing the governance arrangements within the basin through strategic and technical support to the BCSRB. Statutes were drafted which led to a formalization process of the Commission within the PBC, clarified the objectives and activities for the Sixaola Commission, and elucidated the roles and responsibilities of its members (national agencies, local authorities, communities, and local stakeholders). Before this process took place, the lack of clarity coupled with national sensitivities relating to specific border issues compromised the cooperation process and led to the ineffectiveness of the Commission. The trainings (technical, legal, institutional) and the strategic support provided by IUCN to the wide array of basin stakeholders raised awareness about the environmental challenges and also resulted in strengthened water governance capacities

[60] *Ibid.*

to progressively establish the mechanism to manage the Sixaola Basin effectively and sustainably.

Alongside the institutional development, emphasis was given to increasing local capacities in natural resources governance and climate change adaptation strategies, as well as to public participation. The micro-basin Quebrada Rosa (within the Panamean territory) was selected to implement a micro-basin community management plan, as a result of which a water committee was established. Special trainings on Ecosystem-based Adaptation measures and strategies were conducted which triggered agricultural diversification and reforestation within the micro-basin. A diversification of seeds was introduced with a view to motivating local communities to prioritize organic and local seeds instead of introduced ones. The example of the micro-basin Quebrada Rosa could be replicated along the basin and upscaled within the region to demonstrate the benefits of cooperative action to address climate change challenges.

3.5.2.2. *Coatán basin*

The Coatán basin is a shared basin located in the Southwest of Mexico and Northwest of Guatemala. The river originates on the slopes of the Tacaná Volcano and drains into the Pacific Ocean. The basin is of strategic importance for both states due to the number of cities and intensive agricultural activities that depend on its waters and associated resources. Within Guatemala, the waters are mostly used for small-scale irrigation and domestic consumption, whereas in Mexico the main use is irrigation.[61] The local economy primarily depends on agriculture, in particular coffee production and export.[62] An already existing challenge with major impacts for agricultural production is water scarcity in the middle and lower basin during the dry season.[63] A threatening issue across the region is the increase in floods and mudslides due to soil erosion from wide-ranging deforestation and unregulated land use changes.[64] Another challenge consists of the weak institutional framework in the basin: there is limited interrelation between the local and national levels, and consequently limited

[61] M. Cartin *et al.*, *Tacaná Watersheds Guatemala and Mexico* (Switzerland: IUCN, n.d.).
[62–64] *Ibid.*

stakeholder participation.[65] With regard to water legislation, there is an asymmetry between the two basin states. Mexico passed a water law in 1992 and amended it in 2004. While Mexico's water law recognizes basin institutions at different levels and promotes water management at basin level, Guatemala has not yet enacted a law that regulates water resources in particular or recognizes institutional structures for water management.

Based on the local realities within the Coatán basin, IUCN's efforts have been geared towards establishing an effective framework of local community governance. This model was implemented in the Mexican section of the basin, through the establishment of the micro-basin Committee of Buena Vista. This undertaking triggered the organization of a first joint transboundary workshop on water management to provide a platform for having a dialogue and sharing experiences between the two countries. As a result of this, other communities responded by establishing micro-basin councils.

As a result of an initial lack of cooperation at the national level between Guatemala and Mexico, transboundary water governance was predominantly driven from the local level. With a view of promoting further dialogues on this matter, BRIDGE organized a network of local "water champions", promoting cooperation on matters related to integrated water management. This network took the lead in approaching local authorities and engaged one Mexican and two Guatemalan municipalities in dialogues. In addition, trainings were delivered along the basin to raise further awareness and build capacities on Ecosystem-based Adaptation as well as to strengthen water governance capacities of stakeholders and local governments. This process helped the newly established institutional arrangements implement new farming systems, introducing agroforestry and reforesting the basin.

3.5.2.3. *Goascorán basin*

The Goascorán basin is a transboundary basin shared by Honduras and El Salvador. The Goascorán river originates in Honduras and drains into the

[65] *Ibid.*

Pacific Ocean, more precisely in the Gulf of Fonseca, shared by El Salvador, Honduras, and Nicaragua. The basin presents a diversified economy: the middle and lower parts depend mainly on agriculture, livestock, tourism, and fishing, whereas the upper part has the potential for eco-tourism, silvopastoralism, and agroforestry.[66] Honduras enacted a water law in 2009 that establishes different forms of basin organizations at different levels. El Salvador does not yet have a water law in place that regulates water management. In 2006, the EU funded a project for the development of the border between Honduras and El Salvador. Part of this project had the objective to promote the adoption of a basin management plan, as well as a plan for integrated water management. The project further laid out the establishment of a Binational Management Group (BMG) for the Goascorán basin with the intention to promote broad-based stakeholder participation in water management in the border areas. However, with the end of the previously mentioned EU funded project, the BMG stopped its operation, which happened before IUCN started its work in the area.

Having an institutional framework already in place (the BMG), IUCN's aim was to revitalize it and stimulate its operation. At the sub-regional level, work was undertaken with local institutions to ascertain stakeholder's challenges and expectations as well as widen the stakeholder base in order to include all interest groups. Emphasis was also given to those stakeholders not previously included in the BMG, such as local development agencies and national ministries from both basin countries dealing with planning, interior affairs, and agriculture. The existing management plan was revised with a view of matching the interests of all stakeholders and was used as a basis for developing various management actions at the local and transboundary levels. The BMG then served as a mechanism for cooperation in the basin and put into place a new transboundary committee, focusing particularly on economic, financial, and environmental issues. In the absence of a binational treaty to regulate the basin, the BMG concentrated on formalizing its establishment and strengthening its institutional base by setting up national basin institutions according to the national legislation of the two basin states.

---

[66] P. MacQuarrie *et al.*, *Goascorán Basin: Honduras and El Salvador* (Switzerland: IUCN, 2013).

In Honduras a number of meetings took place with the objective of establishing sub-basin councils. However, in El Salvador, the process is more cumbersome in the absence of national water legislation providing for the establishment of basin institutions. Thus, an equivalent legal entity to the sub-basin councils in Honduras was to be identified for the Salvadoran part of the Goascorán Basin. For one of the sub-basins of the Goascorán, IUCN facilitated the establishment of the Sumpul Binational Committee, with representation from local organizations in Honduras. Trainings to strengthen the water management and Ecosystem-based Adaptation capacities have been a central point of departure for IUCN's collaboration with this institution. Capacity-building efforts focused on matters relating to reforestation, monitoring of water quality, and pollution control. The communities living in the Sumpul have started the process of shifting from monocrops to agrodiversification models as a means for reducing vulnerability to climate impacts.

3.5.3. *Lessons learned*

From the IUCN project interventions in Mesoamerica it is possible to outline some experiences and lessons learned deriving from the implementation of adaptive governance and the ecosystem approach on the ground.

An overarching lesson has been that what appear to be intransigent situations may be dynamic underneath, or else the simple act of starting new dialogues can lead to unexpected shifts in transboundary arrangements. These can have unforeseen consequences that must be managed adaptively. Thus, it is clear that in the context of water-diplomacy initiatives, significant impacts cannot be expected in a short period of time. However, incremental steps within a broader influencing strategy can make a difference, especially if that strategy is agile and adaptive. The same is true with regard to uncertainty. Investing in natural, instead of built infrastructure, can be considered as a no-regret measure, since healthy natural infrastructure can more easily adapt to unforeseen changes. Applying the ecosystem approach to water management, livelihoods could be improved and adaptive capacity fostered at the local level, for example, by enhancing the resilience of local ecosystems through agroforestry, such as in the case of the Coatán Basin.

3.5.3.1. *Institutions*

Institutions are absolutely fundamental for effective and sustainable management of water. While this idea has been already proposed within the concept of adaptive governance, IUCN's on-the-ground intervention unfolds other aspects. Legislation becomes particularly crucial for an understanding of mandates of different agencies, as well as roles and responsibilities of various stakeholders. For example, in the absence of a transboundary agreement between Honduras and El Salvador in the Goascorán Basin, national-level governance structures were conceived with a view to progressively developing a binational cooperation agreement (through adhering to the national level constructs). Honduras has a water law in place that provides for the establishment of basin organizations. In El Salvador, such a law does not exist. As a result, it was only possible for an institutional arrangement to be established straightforwardly for the Honduran part of the basin,[67] whereas for El Salvador there was a need to identify other legal settings, in particular local development councils comprising public and private actors. However, these settings do not provide explicit legal competences as far as water management is concerned, something which may compromise coherence of the transboundary water governance arrangements towards the future.

The very existence of institutions, however, is not enough. As shown in the case of the BCSRB between Costa Rica and Panama, there was a need to articulate different institutional levels (that of the commission for the entire border area — PBC — with the transboundary basin commission — BCSRB), to clarify their mandates and responsibilities, and provide for an effective dialogue between different institutions, agencies, and stakeholders represented in both commissions in order to unlock cooperation. For example, enlarging the institutional basis for transboundary water management within the Costa Rica-Panama PBC led to uncertainty among some public and private actors on whether strengthening the Sixaola Basin Commission (BCSRB) will dilute the importance of the PBC. Capacity building and

[67]The Honduran set-up explicitly calls for the integration of national authorities in the case of a transboundary watershed, which certainly fits the purpose of vertical integration and facilitates dialogue at the national level.

training as part of water diplomacy helped to overcome misconceptions and foster trust.

3.5.3.2. *The issue of scale*

Any project seeking to enhance water governance capacities needs the support of water users to be operational at the local level. Without stakeholder participation, plans and institutions will lack legitimacy and basin communities will most likely not support them genuinely. In order to move forward towards achieving concrete solutions and tangible livelihood improvements, it is indispensable to pay particular attention to stakeholders' priority issues and matters of concern. In other words, when states decide to engage in transboundary cooperation, community actors need to be involved in the process, making sure that they "own" the agreements and that their priority issues are truly addressed.[68]

Against the background of restricted cooperation between Guatemala and Mexico at the national level, a bottom-up approach to hydro-diplomacy enabled the establishment of a space for local communities in the Coatán Basin to collaborate with their neighbors living across the international border. At the local level, micro-basin councils were established and started to implement a number of projects such as the rehabilitation of water systems in the Department of San Marcos, Guatemala, and fostering exchange of experiences with the State of Chiapas, Mexico.

The establishment of networks of "water champions" has been one of the fundamental ingredients for supporting the establishment of governance arrangements in the context of transboundary basins and catalyzing local actions to address climate change adaptation. A multiplicity of stakeholders are part of these networks, including municipal leaders, who are particularly engaged in the promotion of actions relating to climate change and water governance. The network members have received a number of trainings on water cooperation, governance, EbA, as well as on leadership skills development. A concrete result of the "water champions" network engagement was the establishment of the Buena Vista Micro-basin Committee in Mexico in 2011. In dialogues and workshops the networks have begun to

---

[68]International Union for Conservation of Nature, *Water Diplomacy* (Switzerland: IUCN, n.d.).

establish local committees and design workplans and advocacy strategies to implement them. While support of micro-basin councils is gaining traction at the national level, interactions with adjacent groups and communities within and beyond national boundaries is starting to increase transboundary cooperation. These types of local level interventions have the ability to move across scales once demonstrations are given and a positive enabling environment is built.

3.5.3.3. *Broad-based stakeholder participation*

As the example of the Goascorán basin illustrates, for water governance to be effective at the transboundary level, it requires fostering dialogues between institutions and a broad base of basin stakeholders. The Binational Management Group established by the previously mentioned EU-funded project was too narrowly focused and participation and decision-making were limited to municipal stakeholders. In a negotiated process which included stakeholders from beyond the water sector, the BMG was revitalized, adopting a new Management Plan in coordination with national and regional plans. The outcome was that several new stakeholders such as NGOs, local development associations, and national ministries joined the BMG, increasing its legitimacy and broadening its base and objectives. The key to success for reforming the BMG was to expand its stakeholder platform.[69]

For effective and adaptive water governance, it is important that all stakeholders, public and private actors, are able to participate. All need to be aware of expected climate change repercussions, vulnerabilities, and other impacts their basin will face in the future to allow for the identification of common needs and search for solutions like the preparation of joint management plans. Finally, the facilitation of spaces where different actors can come together could range from formal round table dialogue sessions to informal training workshops on technical issues or events like seminars, conferences, or symposia.[70]

[69–70] *Ibid.*

3.6. Conclusion

Adaptive water governance is a promising approach that aims for promoting sustainable water governance, in particular under the impacts of climate variability and change. The approach is propitious to serve specific local conditions and account for flexibility. However, certain elements are crucial to its practical implementation, these being primarily flexibility and the integration of uncertainties. They are accounted for through the application of the ecosystem approach; effective institutions including legislation; vertical and horizontal integration and the incorporation of different scales, encompassing broad-based stakeholder participation.

This chapter has investigated if and how the principle of adaptive governance implemented on the ground as part of two IUCN-led projects can facilitate the effective and sustainable management of water resources at multiple scales, taking into consideration the various challenges ahead related to the uncertainty imposed by climate change. To substantiate the argument, on-the-ground experiences were presented and the related lessons learned carved out.

Evidence from the practical experiences in Mesoamerica makes clear that building and/or strengthening governance arrangements requires capacity building and establishment of local level structures such as micro-basin councils to ensure informed participation and decision-making, transparency, and legitimacy. Furthermore, transboundary structures and settings for integrated water management must be flexible enough to identify the appropriate legal formats that would be recognized in each individual basin state both in case of existence and non-existence of a basin-wide agreement. However, this is particularly needed when there are no formal transboundary agreements in place, like treaties. National-level governance structures can then evolve into transboundary platforms.

The identification and operationalization of all possible synergies (local–national–international, public–private, etc.) should also be a requirement for an effective joint management of a transboundary basin. It maximizes the effects of the interventions, allows for more efficient investments, and ensures that proposed strategies are sustainable.

In summary, the experience so far shows that the principles of adaptive water governance can be a useful way to enhance both good water management and adaptive capacity to climate change on a local level as well as on a transnational one, and lead to a more robust and effective governance framework which accounts for different stakeholder priorities. Finally, adaptive water governance is in line with and facilitates collaboration among states, and could also be a good starting point for broader regional cooperation and trust building.

References

Adger, W.N. "Vulnerability." *Global Environmental Change*, 16 (2006), 268–281.

Aguilar Rojas, G. and A. Iza, *Governance of Shared Waters: Legal and Institutional Issues* (Switzerland: IUCN, 2011).

Bates, B.C., Z.W. Kundzewicz, S. Wu, and J.P. Palutikof (eds.), *Climate Change and Water. Technical Paper of the Intergovernmental Panel on Climate Change* (Switzerland: IPCC Secretariat, 2008).

Bradley, R.S., M. Vuille, H.F. Diaz, and W. Vergara. "Threats to Water Supplies in the Tropical Andes." *Science*, 312(5781) (2006), 1755–1756.

Cartin, M., R. Welling, R. Córdoba, O. Rivera, C. Rosal, and F. Arrevillaga, *Tacaná Watersheds: Guatemala and Mexico* (Switzerland: IUCN, n.d.).

Colls, A., N. Ash, and N. Ikkala. *Ecosystem-Based Adaptation: A Natural Response to Climate* (Switzerland: IUCN, 2009).

Drieschove, A. and G. Eckstein. "Cooperative Transboundary Mechanisms." In J. Roberts and J.C. Sanchez (eds.), *Transboundary Water Governance: Adaptation to Climate Change* (Switzerland: IUCN, 2014).

Folke, C. "Resilience: The Emergence of a Perspective for Social-Ecological Systems Analyses." *Global Environmental Change*, 16(3) (2006), 253–267.

Huitema, D., E. Mostert, W. Egas, S. Moellenkamp, C. Pahl-Wostl, and R. Yalcin. "Adaptive Water Governance: Assessing the Institutional Prescriptions of Adaptive (Co-)Management from a Governance Perspective and Defining a Research Agenda." *Ecology and Society*. Retrieved 10 March 2015, from http://www.ecologyandsociety.org.

International Union for Conservation of Nature. *Climate Change and Governance: Ecosystem-based Approach in Mesoamerica* (Switzerland: IUCN, 2010).

International Union for Conservation of Nature Information Paper. *Recovery from the Indian Ocean Tsunami: Guidance for Ecosystem Rehabilitation Incorporating Livelihoods Concerns* (Switzerland: IUCN, 2005).

IUCN Water and Nature Initiative. "What We've Learned: Implementation Principle 7." Retrieved 2 April 2014, from http://www.waterandnature.org/en/results/what-weve-learned.

Iza, A. and R. Stein (eds.), *RULE — Reforming Water Governance* (Switzerland: IUCN, 2009).

MacQuarrie, P., R. Welling, N. Porras, and R. Córdoba. *Goascorán Basin: Honduras and El Salvador* (Switzerland: IUCN, 2013).

McCaffrey, S.C. *The Law of International Watercourses* (UK: Oxford University Press, 2007).

Marshall, N.A., J. Tamelander, D. Obura, D. Malleret-King, and J.E. Cinner. *A Framework for Social Adaptation to Climate Change: Sustaining Tropical Coastal Communities and Industries* (Switzerland: IUCN, 2009).

Medema, W., B.S. McIntosh, and P.J. Jeffrey. "From Premise to Practice: A Critical Assessment of Integrated Water Resources Management and Adaptive Management Approaches in the Water Sector." *Ecology and Society*. Retrieved 10 March 2015, from http://www.ecologyandsociety.org.

Ostrom, E. *Governing the Commons: The Evolution of Institutions for Collective Action* (UK: Cambridge University Press, 1990).

Pahl-Wostl, C., M. Craps, A. Dewulf, E. Mostert, D. Tabara, and T. Taillieu. "Social Learning and Water Resources Management." *Ecology and Society*. Retrieved 10 March 2015, from http://www.ecologyandsociety.org.

Peña Chacón, M. and M. Perez de Madrid. "Case Study: Up-Scaling Adaptation in the Sixaola River Basin." In J. Roberts and J.C. Sanchez (eds.), *Transboundary Water Governance: Adaptation to Climate Change* (Switzerland: IUCN, 2014).

Smith, M. and M. Cartin. *Water Vision to Action: Catalysing Change through the IUCN Water and Nature Initiative* (Switzerland: IUCN, 2011).

Troell, J. and G. Swason. "Adaptive Water Governance and the Principles of International Law." In J. Roberts and J.C. Sanchez (eds.), *Transboundary Water Governance: Adaptation to Climate Change* (Switzerland: IUCN, 2014).

United Nations Convention on the Law of the Non-Navigational Uses of International Watercourses, 1997.

Young, K.R. and J. Lipton. "Adaptive Governance and Climate Change in the Tropical Highlands of Western South America." *Climatic Change*, 78 (2006), 63–102.

Chapter 4

Legal and Policy Framework for Ecosystem-Based Adaptation to Climate Change in Malaysia: A Reform-Oriented Study

Abdul Haseeb Ansari

4.1. Prelude

Average global temperature and its global disastrous impacts, due to unusual weather conditions, are escalating despite greenhouse gases (GHGs) reduction commitments under the Kyoto Protocol, as the mitigation mechanisms alone cannot do away with the suffering of people on a long-term basis. They have to be combined with adaptation.[1] Adaptation to tenacious adverse impacts of climate change is crucial for alleviating them, lest these impacts become a menace for people. Adaptation within the context of UNFCCC is very restrictive and lacks detailed strategy. It is for this reason that mitigation measures combined with adaptation can best be practised at the regional, national, and local levels.[2] It has also been said so in the Fourth Assessment Report of the IPCC, 2007 (IPCC, 2007). This is because although there are a number of strategies developed under the auspices of the UNFCCC, their

[1]R. Klein *et al.*, "Inter-Relationships between Adaptation and Mitigation," (2007), 745–777. Retrieved 12 November 2013, from http://su.diva-portal.org/smash/record.jsf?pid=diva2:180448.

[2]K.L. Koh and B. Lovleen, "Governance on Adaptation to Climate Change in the ASEAN Region," *Carbon and Climate Law Review*, (2011), 82–90; M. Sahraie, "Sustainable Development Law on Climate Change Recognizing or Pro-Actively," (December 2011). Retrieved 10 March 2015, from cisd1.org/public/docs/news/5._Melodie_Sahraie_2.pdf; S.B. Othman, "Adaptation to Climate Change and Reducing Natural Disaster Risk: A Study on Country Practices and Lesson between Malaysia and Japan," (October 2011). Retrieved 10 March 2015, from www.adrc.asia/aboutus/Vrdata/finalreport/2011A_MYS_Surina_FRR.pdf.

implementation depends on viable national adaptation plans and effective programs of action within the Cancun Adaptation Framework and Nairobi Work Program supported with encouraging transfer of environmentally sound technologies (ESTs), research support, financial support, and support in capacity building. A successful program at these levels depends on political will on the part of the state supported by all stakeholders, national and international, including people at the grassroots level. All these have been well spelled out and discussed in the IPCC WGII AR5 entitled *Climate Change 2014: Impacts, Adaptation and Vulnerability*, released on 1 April 2014.[3]

The Fifth IPCC Report *Assessing Impacts of Climate Change on Human Health, Settlements and Natural Resources*, released on 30 March 2014, states that Asia is facing the brunt of climate change and will see severe stress on water resources and foodgrains production. *It warned that the worst is yet to come if no measures are taken to curb the ill effects of global warming* (emphasis added). Chapter 24 of the IPCC Working Group II Assessment Report of 2014 (Art. 5) concludes that the current increase in warm days and higher temperature will lead to lower rice yields, which may result in water scarcity and changes in arid and semiarid areas, loss of mangroves and coral reefs, and increase in mortality rate. The Report clearly emphasizes taking all possible measures for mitigation and Ecosystem-based Adaptation (EbA). The legal aspect of EbA outlines the responses to the impact of climate change, which can enhance adaptive capacity and reduce climate risk vulnerability. In fact, mitigation measures, adaptation practices, and change in consumption patterns are three important tools for sustained living on the planet. Mitigation and consumption practice have to be adopted to do away with the suffering of the people posed by climate change. For example, the April 2014 water crisis in the Klang Valley of Malaysia has prompted the government to resort to water rationing, which will instill among the people the importance of water and not to misuse or waste it, and to harvest rain water, which is broadly practised in India. This may be resorted to in Sao Paulo, Brazil, and other parts of the world

[3]For a detailed account, see IPCC Intergovernmental Panel on Climate Change, "Climate Change 2014: Impacts, Adaptation and Vulnerability," (2014). Retrieved 10 October 2014, from www.ipcc.ch/report/ar5/wg2.

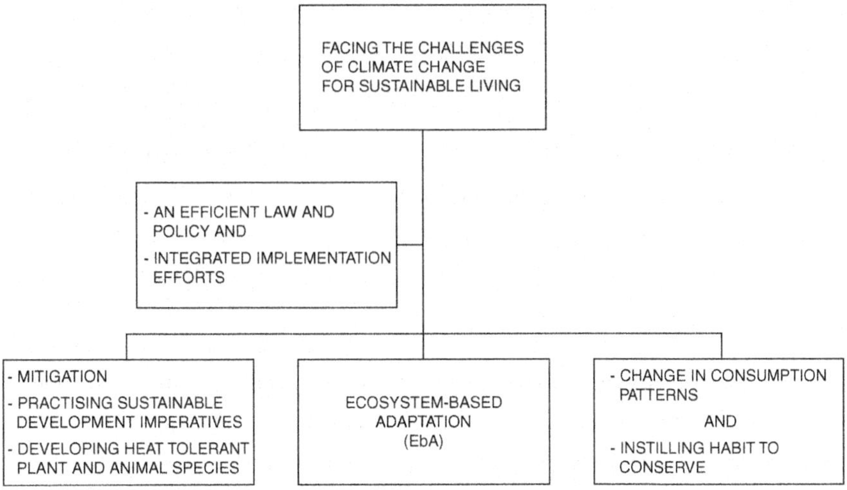

Figure 1. Challenges of Climate Change and Strategies to Mitigate Them.

facing water crisis. Actually, these measures are part of adaptation. The government had to adopt immediate mitigation measures by bringing water from Negari Sembilan to Selangor, to upgrade dams, and to improve the water purification and supply system. For improving consumption patterns, governments do not have to spend money; whereas for mitigation measures it will have to spend money and employ necessary technology. This can be represented graphically, as in Figure 1.

The Convention on Biological Diversity (CBD) defines EbA as: "the use of biodiversity and ecosystem services as part of an overall adaptation strategy to help people adapt to the adverse effects of climate change." The UNDP Draft Working Paper, 3 November 2010, on "Ecosystems, Disasters and Climate Change" defines it as an integrated ecological approach that links ecosystems to disaster and climate risk management. EbA measures frequently complement disaster risk reduction objectives. EbA integrates the use of ecosystem services into an overall strategy to help people adapt to the adverse impacts of climate change and reduce the risk of climate-induced disasters. It includes the sustainable management, conservation, and restoration of ecosystems to provide services that help people adapt to both current climate variability and climate change. According to CBD, an ecosystem-based approach uses biodiversity and ecosystem services, and generates socio-economic as well as ecological benefits. An EbA approach, thus, entails an ecosystem adaptation approach that promotes activities which address both human and biological aspects of the ecosystem. It relies on recovering and maintaining ecosystem services and the resilience of natural resources rather than on engineering and costly man-made infrastructure like sea walls, jetties, and levels. This task is relatively arduous and requires more expertise and resources in South Pacific and Caribbean small island states, and states in South Asia, which have a large number of islands. According to Jessica Mercer and Ilan Kelman, experts on EbA in developing countries, it also solicits people to meaningfully interact with essential climate change impacts and requires a fusion of national and foreign expertise in the subject matter.[4] Ecosystem services are those aspects of ecosystems that are consumed or utilized to yield human well-being.

The climate change control approach is divided into two measures — "hard" and "soft". "Hard" measures involve high capital engineering and building of climate-friendly structures, which require high budgetary input. "Soft" measures involve EbA, which is cost-effective by maintaining existing natural ecosystem structures and improving ecosystem services. EbA approaches have the potential to significantly increase the functionality of current adaptation practices and reduce the tension of the "hard" and "soft" approach by following three major ways: (1) valuing ecosystems and biodiversity in adaptation; (2) promoting development in adaptation; and (3) building long-term resilience with multiple socio-economic benefits.[5] According to the author, there can be a fourth way: to develop heat tolerant plant and animal species, e.g., developing, especially by gene manipulation or crossbreeding, heat tolerant poultry based on the African

[4]For the position of EbA in Caribbean island states, see J. Mercer, I. Kelman, B. Alfthan, and T. Kurvits, "Ecosystem-Based Adaptation to Climate Change in Caribbean Small Island Developing States: Integrating Local and External Knowledge," *Sustainability*, 4(8) (2012), 1908–1932. Also refer to J. Mercer, T. Kurvits, I. Kelman, and S. Mavrogenis, "Ecosystem-based Adaptation for Food Security in the AIMS SIDES: Integrating External and Local Knowledge," *Sustainability*, 6(9) (2014), 5556–5597.

[5]N. Huq, F. Renaud, and Z. Sebesvari, "Ecosystem-Based Adaptation (EbA) to Climate Change — Integrating Actions to Sustainable Adaptation," (2013). Retrieved 12 January 2014, from http://www.climate-impacts-2013.org/files/cwi_huq.pdf.

naked-neck chickens and Brahman breed of cattle of India.[6] Developing intensity by conserving, restoring, and creating new ecosystems of coral reefs is also a good adaptation technique in view of the rise of the sea level and is a better choice instead of building concrete walls, as they have tremendous capacity to dissipate the sea current thrust.[7] Research should also be augmented to develop marine species that could change their eating habits and could protect their shells from acidification of the seawater. It will ensure all functions that a healthy ecosystem can render to the nature and people dependent on it.[8] It will then have a dynamic relationship between and among species as well as biotic environmental interactions.[9] EbA, in practice, varies depending on the peculiar environment and natural landscape. However, it involves a wide range of ecosystem management activities to increase resilience and reduce the vulnerability of people and the environment to climate change. According to IUCN (2009), EbA practices may include the following:

- Sustainable water management, where river basins, aquifers, flood plains, and their associated vegetation are managed to provide water storage and flood regulation services;
- Disaster risk reduction, where restoration of coastal habitats such as mangroves can be a particularly effective measure against storm surges, saline intrusion, and coastal erosion;
- Sustainable management of grasslands and rangelands, to enhance pastoral livelihoods and increase resilience to drought and flooding;

---

[6]H. Evan, "Coping with Global Warming," *The Star*, 19 May 2014, p. 8; F. Ferrario *et al.*, "The Effectiveness of Coral Reefs for Coastal Hazard Risk Reduction and Adaptation," *Nature Communication*, 5 (Article 3794) (2014). DOI: 10.1038/incomms4794.

[7]R. Ingham, "Natural Buffers for Coastlines," *The Star*, 19 May 2014, p. 8.

[8]A. Colls, N. Ash, and N. Ikkala, *Ecosystem-Based Adaptation: A Natural Response to Climate Change* (Switzerland: IUCN, 2009); B. Boer and P. Clarke, *Legal Frameworks for Ecosystem-Based Adaptation to Climate Change in the Pacific Islands* (Samoa: SPREP PROC, 2012); A. A. Pérez, B. H. Fernández, and R. C. Gatti, *Building Resilience to Climate Change: Ecosystem-Based Adaptation and Lessons from the Field.* (Switzerland: IUCN, 2010); E. Pramova, B. Locatelli, M. Brockhaus, and S. Fohlmeister, "Ecosystem Services in the National Adaptation Programmes of Action," *Climate Policy*, 12(4) (2012), 393–409.

[9]Boer and Clarke (2012), *op. cit.*

- Establishment of diverse agricultural systems, where using indigenous knowledge of specific crop and livestock varieties, maintaining genetic diversity of crops and livestock, and conserving diverse agricultural landscapes secures food provision in changing local climatic conditions;
- Strategic management of shrub lands and forests to limit the frequency and size of uncontrolled forest fires; and
- Establishing and effectively managing protected area systems to ensure the continued delivery of ecosystem services that increase resilience to climate change.

Malaysia is working on the suggested measures. Additional measures with respect to transboundary haze, which has become an annual occurrence in the region, have also been adopted. EbA, when applied to indigenous peoples' rights, focuses on conserving biodiversity and the natural resources essential for the survival and livelihood of coastal and agricultural communities in Malaysia. Sustainable livelihood for rural people, who are highly dependent on natural marine and terrestrial resources in their habitat, can be achieved better and faster through well-designed EbA. Many of these resources have international and national economic value, e.g., coral reefs, forests for timber, water, medicinal plants, and habitat for animal species.[10] This has proven to be successful in maintaining the Tasek Bera wetlands, the first Malaysian Ramsar site of Malaysia, by training the inhabitants about eco-tourism.[11] A community-based adaptation program was launched by Green Coast[12] in the tsunami-affected areas in Indonesia, Sri Lanka, India, Thailand, and Malaysia.[13] Adopting a community-led approach, the Green Coast Program seeks to improve local livelihoods via four main activities: (1) Coastal ecosystem rehabilitation; (2) Building environmentally sustainable livelihoods; (3) Developing village regulations to support environmental conservation efforts; and (4) Environmental

[10]Huq, Renaud, and Sebesvari (2013), *op. cit.*; Pérez, Fernández, and Gatti, (2010), *op. cit.*
[11]A. H. Ansari and A. O. Umar, *Wetlands and Global Warming: Impacts, Adaptation and Mitigation (Developing Countries' Perspective)* (India: Serials Publications, 2012), pp. 112–161.
[12]Green Coast is a coastal rehabilitation program helping restoration and management of damaged coastal ecosystems in the Southeast Asia funded by Oxfam Novib (Netherlands).
[13]Colls, Ash, and Ikkala (2009), *op. cit.*

education campaigns. The climate of Malaysia is very much influenced by the mountainous topography and complex land–sea interactions. Malaysia's climate is a tropical climate with uniform temperatures (ranging between maximum 33°C and minimum 23°C) with relatively high humidity. It is divided into two seasons, which are southwest monsoon, northeast monsoon and two shorter periods of inter-monsoon seasons. It is geographically located just outside the "Pacific Rim of Fire" but has not recorded severe natural disasters such as earthquakes, volcanic eruptions, and typhoons. Although Malaysia is spared from the threats of severe natural disasters and calamities, it is not spared from other disasters such as floods, man-made disasters, landslides, and severe haze.[14]

Climate change and its impacts in the ASEAN region, including Malaysia, are serious issues in the region, as it is seriously causing unusual weather conditions resulting in scanty rain leading to water shortage and loss of crops resulting in food crisis.[15] The most recent floods in 2010 in Kedah and Perlis and in 2013 in Pahang were among the worst floods ever experienced by the country. The total economic loss and the financial burden on the government were enormous. Flood poised the most significant hazard affecting 4.82 million people and causing damages amounting to RM915 million yearly. Other than flooding, the country also, from time to time, experiences some man-made disasters, which cause considerable damage to properties and loss of lives.[16] Climate change can trigger national and international distributional conflicts in Malaysia, e.g., within its fisheries sector, and intensify problems already hard to manage in the region.[17]

---

[14]NRE, "Malaysia: Second National Communication to the UNFCCC," Putrajaya, Malaysia, 2007. Retrieved 12 November 2013, from http://nc2.nre.gov.my.

[15]Koh and Lovleen (2011), *op. cit.*; Othman (2011), *op. cit.*; R. Ruhizal and Phil O'Keefe, "Evolution of Disaster Planning and Housing in Malaysia: A Review," *World Applied Sciences Journal*, 21(7) (2013), 945–959.

[16]AIPA Secretariat, "3rd AIPA Caucus Report 2011," Philippines, 2012, 1–40. Retrieved 12 December 2013, from http://www.aipasecretariat.org/wp-content/uploads/2011/07/5.third_.caucus.report.pdf.

[17]W.S. Robert, "Scoping Assessment on Climate Change Adaptation in Malaysia," (October 2011). Retrieved 10 March 2015, from http://www.climateadapt.asia/upload/punlications/files/4f27b1074e7aeMalaysia.pdf.

The policy framework on climate change adaptation in Malaysia includes: National Policy on Climate Change, Coral Triangle Initiative-CTI, Southeast Asia Disaster Prevention Research Institute-SEADPRI-UKM, Malaysia Action Plan for Disaster Risk Reduction-MyDRR, Initial National Communication (INC), and Second National Communication to the UNFCCC (NC2).[18] The policy and framework are implemented in conjunction with ministerial adaptation awareness programs together by MOSTI, MNRE, NAHRIM, MKN, etc. Seven sectors have been identified in the Vulnerability and Assessment of updates contained in the NC2, which is an update of the (INC) assessment in identifying potential impacts. These sectors are: water resources, agriculture, biodiversity, forestry, coastal and marine resources, energy, and public health (NRE, 2007). Some of the sectors mainly focus on mitigation and not adaptation; therefore, this research will shed light on those sectors, which are related to ecosystem adaptation to climate change in Malaysia.

4.2. Adaptation and Mitigation: Meaning and Scope

4.2.1. *Adaptation*

Adaptation is the adjustment of the natural or human system in response to actual or expected stimuli, or its effects, to moderate harm or exploit beneficial opportunities (IPCC, 2007). It is a process of adjustment to actual or expected climate and its effect. In human systems, adaptation seeks to *moderate* or *avoid harm* or *exploit beneficial opportunities* (IPCC, WGII AR5, 2014). It has appropriately been defined by the European Commission as "anticipating the adverse effects of climate change and taking appropriate action to *prevent* or *minimize* the damage they can cause, or *taking advantage of opportunities* that may arise." For example, using scarce water resources more efficiently; adopting building codes to future climate conditions and extreme weather events; building flood defenses and raising the levels of dykes; developing drought-tolerant crops; choosing tree species and forestry practices less vulnerable to unusual heat, storms

[18] Coral Triangle Initiative, "Annex-5 Decision Document on: Climate Change Adaptation Working Group Report," 2012, 1–17; J. J. Pereira, *Effective Networking Strategies for Adaptation Governance — A Perspective from Malaysia* (Malaysia: Seadpri-UKM, 2013); Robert (2011), *op. cit.*

and fires; and setting aside land corridors to help species to migrate.[19] With the presence of constant levels of hazards, natural or otherwise, over time, adaptation will allow a system to reduce the risk associated with these hazards by reducing social vulnerability. Faced with increased hazard, natural systems may maintain current levels of risk through an enhanced Ecosystem-based Adaptation.[20]

The UNFCCC and the Kyoto protocol prompt for climate change adaptation. Adaptation formed an integral element of the Bali Action Plan of 2007 and the Copenhagen Accord of 2009, as well as the mandate of the Ad Hoc Working Group on Long-Term Cooperative Action.[21] The Third Assessment Report of the IPCC (IPCC–TAR 2001) identified various types of adaptation, which can be distinguished from each other. They include: anticipatory and reactive adaptation; private and public adaptation; and autonomous and planned adaptation. *Reactive measures* are implemented in response to current climate variability and observed impacts, whereas *anticipatory measures* are undertaken before impacts are observed to reduce exposure to future risks. The *systemic* and *proactive* approach of EbA contrasts with the interventionist and reactive nature of many existing adaptation proposals and plans that portray adaptation as a tool of the last resort in dealing with the threat of an unpredictable climate. This reactive approach supports the prioritization of "... already existing strategies" and results in "end-of-the pipe" infrastructural, land use planning, and technological interventions that are responsive to only a narrow range of outcomes and probabilities. Although until recently, efforts to reduce current impacts of climate change have primarily focused on mitigating the accumulation of greenhouse gases in the atmosphere, adaptation is recognized as fundamental.[22]

---

[19]European Commission, "Adaptation to Climate Change." Retrieved 12 March 2014, from http://ec.europa.eu/clima/policies/adaptation/index_en.htm.

[20]N. Brooks, "Vulnerability, Risk and Adaptation: A Conceptual Framework," Tyndall Centre for Climate Change Research Working Paper 38 (2003), 1–16.

[21]Koh and Lovleen (2011), *op. cit.*

[22]Klein *et al.* (2007), *op. cit.*; Koh and Lovleen (2011), *op. cit.*; Wertz-Kanounnikoff *et al.*, "Ecosystem-Based Adaptation to Climate Change: What Scope for Payments for Environmental Services?," *Climate and Development*, 3(2) (2011), 143–158.

The evolution of climate change adaptation policy and legal framework has shifted from the initial *top-down* analyzes of impacts assessment and vulnerability assessments to a *bottom-up* approach through various National Adaptation Planning Programs of Action or NAPAs.[23] Prior to NAPAs, there was clear policy direction towards mitigation under the mechanisms in the Kyoto Protocol.

Experts have not reached a consensus on a framework for adaptation to climate change.[24] They have suggested some basic principles on the basis of which a viable climate change adaptation action plan can be developed. The suggestions include the following:

- Adaptation to *short-term climate variability* and *extreme events* as a basis for reducing vulnerability to longer-term climate change.
- Adaptation policy and measures assessed in a developmental context.
- Adaptation occurs at different levels in the society, including the local level.
- Both the strategy and the process by which adaptation is implemented are equally important.

4.2.2. *Relationship between adaptation and mitigation*

The IPCC–TAR, 2001, defined climate change mitigation as *an anthropogenic intervention to reduce the sources or enhance the sinks of greenhouse gases.* A combined effort of adaptation and mitigation is necessary for an effective climate policy aimed at reducing the risks to natural and human systems. This involves a portfolio of diverse adaptation and mitigation actions at all levels of policy-making. Overtime, mitigation efforts on the international scene have proven to be inadequate to control the growing threat of climate change impact. Although mitigation encompasses all greenhouse gas emitters based on international agreement and adaptation involves participation of the government officials and local people, they are

---

[23] V. Nelson, R. Lamboll, and A. Arendse, "Climate Change Adaptation," Adaptive Capacity and Development Discussion Paper, DSA-DFID Policy Forum, 2008. For more information on NAPAs, see http://unfccc.int/adaptation/napas/items/2679.php. Also see "National Adaptation Programmes of Action," United Nations Framework on Climate Change. Retrieved 26 May 2015, from http://unfcc.int/national/ report/napa/items/2719.php.

[24] B. Lim *et al.*, *Adaptation Policy Frameworks for Climate Change: Developing Strategies, Policies and Measures* (UK: Cambridge University Press, 2005); Nelson, Lamboll, and Arendse (2008), *op. cit.*

coextensive and coexistent. IPCC 2007 identified four models for inter-relationship between adaptation and mitigation[25] that can take several forms:

- Adaptation actions that have consequences for mitigation;
- Mitigation actions that have consequences for adaptation;
- Decisions that include trade-offs or synergies between adaptation and mitigation; and
- Processes that have consequences for both adaptation and mitigation.

The working relationship between the whole scenario of emission impacts and socio-economic and environmental processes with adaptation and mitigation can be best understood from Figure 2.

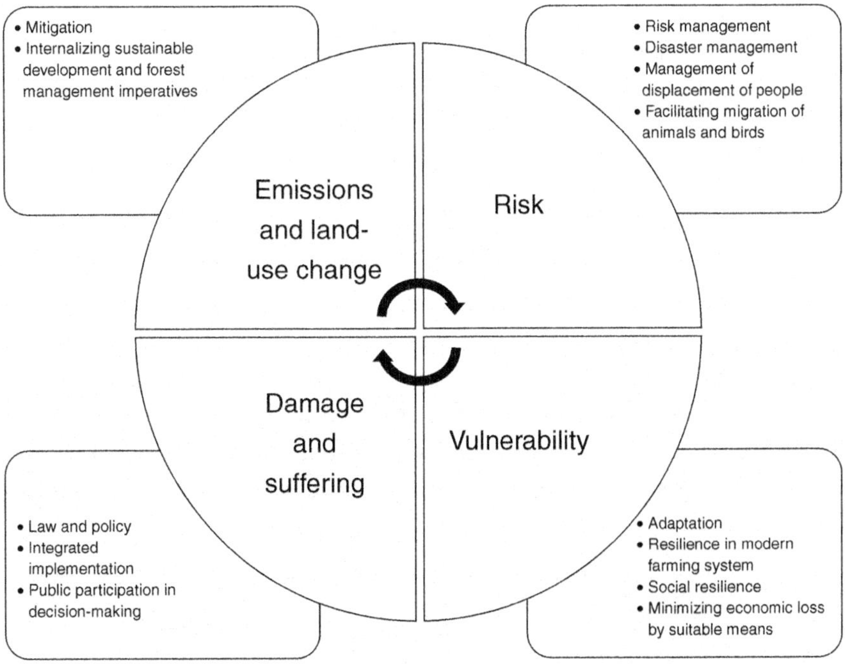

Figure 2. The Dynamics of Carbon Emission Impacts and Legal, Socio-economic and Environmental Strategies for Adaptation and Mitigation.

Note: Prepared on the basis of the IPCC WGII AR5 entitled "Climate Change 2014: Impacts, Adaptation and Vulnerability" released on 1 April 2014.

---

[25]Klein *et al.* (2007), *op. cit.*

It should be noted, that for sustainable development, the imperatives encompass adaptation strategies.[26]

4.2.3. *Adaptive capacity of the Malaysian ecosystem*

Adaptive capacity of Malaysia, like other places, is the ability of an ecosystem to (1) adjust to climate change (including climate variability and extremes), (2) moderate potential damages, (3) take advantage of opportunities, and (4) cope with the consequences (IPCC, 2007).[27] However, Malaysia's adaptive capacity is not so strong.[28] It is for the reasons highlighted in IPCC (2007) that adaptation is increasingly becoming an important factor of the development agenda and planning, especially in developing countries, including Malaysia, and least developed countries, as most of them are at high risk and in a vulnerable situation.[29]

4.3. Environmental Governance in Malaysia

4.3.1. *Sources of Malaysian climate change regulations*

In Malaysia, there exist a series of laws, which define the roles and responsibilities of various concerned agencies for specific actions to combat climate change and its impacts. Among them, at the federal level are: the Land Conservation Act; the Environmental Quality Act; Town and Country Planning Act; Irrigation and Drainage Act; and Uniform Building By-Law. The result of this is that climate change adaptation strategies and plans are embedded within the organization and functioning of governmental agencies at all levels and sectors. As suggested by adaptation experts, the federal government is working with state governments, local authorities,

[26]A.A. Miguel, K. Parviz, and N. Clara, "Strengthening Resilience of Modern Farming Systems: A Key Prerequsite for Sustainable Agricutural Production in an Era of Climate Change," TWN Briefing Paper 70, 2014. Retrieved 3 March 2014, from http://www.twn.my.

[27]Nelson, Lamboll, and Arendse (2008), *op. cit.*

[28]S. Adhikary *et al.*, "Murari Lal (India), Hideo Harasawa (Japan) and Daniel Murdiyarso (Indonesia)." Retrieved 3 March 2014, from http://www.grida.no/climate/ipcc_tar/wg2/pdf/wg2TARchap11.pdf.

[29]S. Fankhauser, *The Costs of Adaptation* (London: Wiley, 2009).

and all stakeholders at the grassroots level, based on a bottom-up approach.[30]

4.3.2. *Regional framework for Ecosystem-based Adaptation*

Southeast Asia is one of the most dynamic and fastest growing regions in the world. But with long coastlines, population, and economic activity concentrated in coastal areas, reliance on agriculture in providing livelihoods for a large segment of the population, especially those living in poverty, and dependence on natural resources and forestry to drive development, it is highly vulnerable to the harsh impact of climate change.[31]

Malaysia has taken perhaps a unique approach to climate change adaptation over and above other countries in the ASEAN region, possibly due to the economic development surpassing its many neighbors.[32] Accordingly, much of Malaysia's adaptation responses come in the form of improved ecosystem management, water resource management, and secured agricultural production — each with a backdrop of doing so to improve productivity, efficiency in resource use, and optimized economic benefit for the state and to individuals. *While there has been much focus on assessments and planned strategies to achieve these, little attention has been given to autonomous climate change adaptation in practice.* Although adaptation efforts are more enhanced at the national and local levels, there are ongoing regional efforts within the Southeast Asia region. Funding is the bane of most of the adaptation projects within the region as projects are financed through foreign aid via the Global Environment Facility (GEF), USAID, and other international donor agencies.

---

[30] C. Sterrett, "Review of Climate Change Adaptation Practices in South Asia," (2011), 1–100. Retrieved 10 March 2015, from http://www.oxfam.org/sites/www.oxfam.org/files/file_attachments/rr-climate-change-adaptation-south-asia-161111-en_3.pdf. M. Baker-Jones *et al.*, *Climate Change Adaptation Guided by the Law* (Brisbane: DLA Piper, 2013).

[31] Asian Development Bank (ADB), "The Economics of Climate Change in Southeast Asia: A Regional Review," April 2009, 1–225.

[32] G. Philip and H. Hilary, "Review of Current and Planned Adaptation Action: East and Southeast Asia Action," Adaptation Partnership, 2011. Retrieved 3 November 2013, from http://www.adaptationpartnership.org/.

Malaysia has taken the lead with the establishment of the Southeast Asia Disaster Prevention Research Institute (SEADPRI) of the National University of Malaysia (UKM) on their own initiative with the supervision of the National Security Council (NSC). The mission of the institute is as follows:[33]

- Conduct holistic research on hazards and disasters at national and regional levels;
- Enhance human capital and capacity at national and regional levels, particularly in Southeast Asia;
- Support knowledge-based decision-making on climatic, geological, and technological hazards.

The Coral Triangle Initiative on Coral Reefs, Fisheries, and Food Security (CTI-CFF), a USAID-funded program, is a multilateral partnership of six countries[34] working together to sustain extraordinary marine and coastal resources by addressing crucial issues such as food security, climate change, and marine biodiversity. It is aimed at improving the management of biologically and economically important coastal and marine resources, and associated ecosystems that support livelihoods and economies in the Coral Triangle; and assist the six CTI countries in implementing the CTI Regional and National Plans of Action with activities that focus on instituting an ecosystem approach to fisheries management, creating marine protected areas, building climate change adaptive capacity, and establishing regional platforms to promote cross-country learning and enhance sustainability.[35] The Hyogo Framework for Action (HFA) is a key instrument to facilitate the implementation of disaster risk reduction measures. Adopted during the World Conference on Disaster Reduction held in Kobe, Hyogo, Japan in 2005, HFA provides strategies to build the resilience of nations and communities to disasters, with the goal of achieving substantive reduction of disaster losses by 2015. This is an overall

[33] Southeast Asia Disaster Prevention Research Institute. Retrieved 19 November 2013, from http://www.ukm.my/seadpri/?page_id=2.

[34] Indonesia, Malaysia, Papua New Guinea, the Philippines, Solomon Islands, and Timor-Leste are known as the "CT6".

[35] Philip and Hilary (2011), *op. cit.*

framework for Disaster Risk Reduction (DRR), which can be tailored toward landslide disaster risk reduction. The upgrading of the ecosystem is central to the disaster management agenda of the HFA. Separately, world policy-makers addressing the issue of disasters have noted the need to address climate change concerns within disaster risk reduction strategies, as part of the Hyogo Framework for Action 2005–2015, which was adopted in Kobe, Japan, in 2005. In order to further solidify Malaysia's commitment towards achieving the strategic goals and priorities of the Hyogo Framework for Action (HFA), the 2011 National Disaster Awareness Day culminated with the Melaka Declaration on Disaster Risk Reduction.

4.3.3. *Legal aspects of National Adaptation Programme of Action (NAPA)*

Changes in social and economic conditions may lead to shifts in a sector's activities, change its need for particular resources or the way such resources would be utilized, and, thus, affect its vulnerability. This, in turn, may change the sector's or broader society's adaptation needs.[36] These aspects must be taken into account while designing adaptation-related plans.

4.3.4. *Government initiatives for Ecosystem-based Adaptation*

The cross-sectoral approach given to climate change in Malaysia[37] has been an impetus to varieties of initiatives across all relevant sectors towards combating climate change.[38] In terms of infrastructure, Malaysia has various Flood Mitigation Projects, the most innovative of which is the Stormwater Management and Road Tunnel (SMART) constructed in the Kuala Lumpur city center. The SMART system diverts floodwater via a holding pond, a bypass tunnel, and a storage reservoir, preventing spillover into the city during heavy downpours. In addition, local authorities have implemented the Urban Stormwater Management Manual (MSMA) and the Land Use Planning Appraisal for Risk Areas (LUPAr) in the feasibility assessment

[36] S. Chinvanno, "A Holistic Approach to Climate Change Vulnerability and Adaptation Assessment: Pilot Study in Thailand Regional Climate Change," Bangkok, 2013. Retrieved 15 May 2014, from http://www.asiapacificadapt.net.

[37] Pereira (2013), *op. cit.*

[38] Robert (2011), *op. cit.*

and execution of physical developments. Finally, the Public Works Department developed the National Slope Master Plan, which enables slope management and disaster risk reduction in incidents of landslides.[39] Other initiatives in the area of rivers and water resources management include efforts by the Department of Irrigation and Drainage (DID), Malaysia, to introduce ecosystem friendly irrigation and the Integrated Flood Forecasting and River Monitoring System (IFFRM). The responsibility for water resources planning and development in Malaysia is shared by various government agencies, such as the Ministry of Natural Resources and Environment (NRE), Ministry of Health (MOH), Public Works Department (PWD), Ministry of Housing and Local Government, and public water supply companies or privatized utility companies.[40] The Integrated Flood Forecasting and River Monitoring System (IFFRM) for the Klang Valley is inclusive of a flood modelling system in the infrastructure networks which have been completed with 88 new telemetric systems. An Integrated Flood Forecasting and Warning System for the Muda River Basin in Northern Peninsular Malaysia is also being developed simultaneously. This includes developing a radar rainfall analyzer and integrator for Malaysia (RAIM), which feeds information to a real-time flood forecasting system, which is already in place in 13 river basins throughout the country. Upon completion, this project will be extended to the Kelantan, Pahang, and Johor river basins. On another note, flood-forecasting models in the Johor, Muar, and Batu Pahat river basins are being upgraded to the real-time computerized HEC-HMS model.

In a hydrological context, the Department of Irrigation and Drainage (DID) under the Ministry of Natural Resources and Environment (NRE), is responsible for the planning, implementing, and operating of irrigation, drainage, and flood control projects throughout the country. Besides, DID took up the new and expanded task of River Basin Management and Coastal Zone, Water Resources Management and Hydrology, Special Projects, Flood Management, and Eco-Friendly Drainage.

---

[39] N.W. Chan, "Impacts of Disasters and Disasters Risk Management in Malaysia: The Case of Floods," in Y. Sawada and S. Orum (eds.), *Economic and Welfare Impacts of Disasters in East Asia and Policy Responses* (Indonesia: ERIA Research Project Report 2011-8, 2012), pp. 497–545.

[40] Othman (2011), *op. cit.*; Pereira (2013), *op. cit.*

4.3.5. *Institutions involved in climate change adaptation in Malaysia*

Climate change and Ecosystem-based Adaptation takes a center stage in the programs and policies of several governmental institutions in Malaysia.[41] Among the institutions are the following:

(1) Ministry of Natural Resources and Environment (NRE) — Environmental Management and Climate Change Division;
(2) National Hydraulic Research Institute of Malaysia (NAHRIM): Regional Water Knowledge Hub for Water and Climate Change Adaptation in Southeast Asia;
(3) Malaysian Agricultural Research and Development Institute (MARDI);
(4) Forest Research Institute Malaysia (FIRM);
(5) The Institute for Environment and Development (LESTARI); and
(6) The Malaysian Network for Research on Climate, Environment and Development (MyCLIMATE).

4.3.6. *Legal mechanisms for ecosystem restoration and resilience*

In its Second National Communications, released in April 2011, Malaysia identified seven sectors where vulnerability assessments were undertaken, including water resources, agriculture, biodiversity, forestry, coastal and marine resources, energy, and public health. Concerns related to these sectors and expressed through both of Malaysia's National Communications included:

(1) *Freshwater resources* — Concerns expressed relate to supply, floods, and erosion (MRRE, 2011). Water resources are already under particular strain in Malaysia because of increasing demand and, in particular, because of irrigation — which accounts for 83% of total water usage in the country. Increased periods of drought could further threaten water resources (MSTE, 2000).
(2) *Agriculture* — The potential impact of climate change on key crops, namely palm oil, rice, rubber, and cocoa, are a concern (MRRE, 2011). Flooding due to sea level rise threatens as much as 6% of palm oil

[41] Robert (2011), *op. cit.*

production and 4% of rubber production. Temperature rise is also a major concern because of its ability to hamper crop production (MSTE, 2000).

(3) *Forest resources* — Mangrove forests are projected to be adversely affected by the temperature rise and dry spells (MRRE, 2011).

(4) *Coastal and marine management* — Sea level rise, higher sea surface temperatures, and increasing intensity, duration, and frequency of storms are predicted to occur with negative consequences (MRRE, 2011). Shoreline erosion and loss of mangroves are also presented as significant threats (MSTE, 2000).

(5) *Energy* — Changes in weather patterns and extreme weather events are projected to have negative impacts on the energy production and development sector, while increases in ambient temperature reduce power generation and put additional stress on transportation infrastructure (MRRE, 2011).

(6) *Human health* — Climate change will lead to an increase in vector-borne diseases, such as malaria and dengue, and water-borne diseases such as diarrhea.

Malaysia launched its National Policy on Climate Change in 2010. Although previous environmental strategies were in place, this policy was the first document that contains a major focus on adaptation. Within it, the government advocates a balanced approach between mitigation and adaptation (MNRE, 2009). The document contains a number of "Strategic Thrusts" which operate as policy goals and "Key Actions" that are proposals to address climate change and meet these goals. While some of the actions proposed are quite specific, there is not any detailed description of how these actions will be undertaken, or a detailed timeline for their achievement.[42]

4.3.7. *Terrestrial ecosystem*

4.3.7.1. *Flood control*

Floods occur annually in flood-prone states of Malaysia, causing damage to property and loss of life. It is useful to distinguish "normal" from "major"

[42]P. Gass, H. Hove, and J.-E. Parry, "Review of Current and Planned Adaptation Action: East and Southeast Asia," Adaptation Partnership, 2011. Retreived 3 November 2013, from http://www.adaptationpartnership.org.

flood events. "Normal" floods are seasonal monsoon floods (November to March) whereby waters do not normally exceed the stilt height of traditional Malay houses. Thus, people living in stilt houses in the rural areas are well-adapted to normal floods. It is the major floods which are "unusual" or "extreme" events that render people helpless.[43] The recent example is the March flood in Kuantan of Pahang state.

The amount of flood control infrastructure and planning deployed by Malaysia is appreciable. As a developing country, Malaysia's flood mitigation policy can be described as commendable. Since the First Malaysia Plan (1971–1975), the country's expenditure on flood mitigation has increased substantially. From a mere US$4.56 million in this plan, it has shot up to a massive US$228.2 million for the Sixth Malaysia Plan (1991–1995), a 50-fold increase over a 20-year period. During the 10th Malaysia Plan, the budget allocated for flood management was US$1.17 billion, a 256-fold increase.

The legislative framework related to flood control is indirect as there is no flood legislation. Existing legislation is also sector-based and outdated. While there are currently some laws governing the regulation of river use (e.g., the Waters Enactment 1920, the Mining Enactment 1929, the Drainage Works Ordinance 1954, the Land Conservation Act 1960, and others) and have some bearing on flood mitigation, they are not sufficiently clear or as forceful enough as measures for flood mitigation.

The flood early warning system has been adopted by the Drainage and Irrigation Department of Malaysia, viz. the Linear Transfer Function Model (LTFM) at Pahang River and the Tank Model at Kelantan River. The two flood forecasting models have been developed and use a system which consists of dissemination systems such as automatic warning sirens, the Short Messaging System (SMS), telephone, fax, and the Internet. Other flood forecasting mechanisms include automated telemetric rain gauges and river level recorders.

Relocation and rehabilitation of flood victims has been the focus of the Malaysian government with the establishment of the National Flood Disaster Relief and Preparedness Committee (NFDRPC). Its main task is to ensure that assistance and aid are provided to flood victims in an orderly

[43]Chan (2012), *op. cit.*; Othman (2011), *op. cit.*; Robert (2011), *op. cit.*

and effective manner from the national level downwards. Its approach is largely reactive to flood disasters. The NFDRPC is activated via a National Flood Disaster Management Mechanism (NFDMM). Members of the NFDRPC include government departments/agencies and social organizations, which provide shelter, rescue, and food supplies in case of a disaster. The NFDRPC meets at least once a year, normally before the onset of the northeast monsoon, in order to focus on the need to get ready before the monsoon arrives (bringing floods with it). Its purpose is to ensure that its machinery will run smoothly. At the national level, the NSC is the secretariat for the NFDRPC, which comprises members from the Ministries of Information, Finance, National Unity and Social Development, and Transport; the Federal Chief Secretary; the Federal Police Department; and the Federal Armed Forces. The NFDRPC coordinates all relief operations from the Malaysian Control Centre in Kuala Lumpur. The NFDMM is basically a mechanism responding to disasters, as its name suggests. As such, its approach towards disaster management/reduction is largely reactive.[44]

As noted in Malaysia's Second National Communication to the UNFCCC (NC2), there is the need to review the flood and erosion management plan to accommodate flexible ecosystem-based mechanisms as follows (NRE, 2007):

- Review flood management plans and assess integrity of existing structures particularly where failure could result in loss of life.
- Review design standards for flood risk management in all new infrastructure including water control structures, transportation structures, and electrical, water, and waste amenities to incorporate climate change factors;
- Complementation of structural approaches with non-structural approaches such as improved rainfall and flood forecasting, disaster warning systems, and flood hazard mapping as part of a coordinated disaster prevention and management plan;
- Drainage system improvements to regulate water table depth and prevent floods;

[44] Chan (2012), *op. cit.*

- Further establishment of irrigation facilities (particularly during crop establishment); and
- Cultivar breeding programs to develop new varieties with high Water Use Efficiency (WUE) traits and drought tolerance.

The absence of a robust Ecosystem-based Adaptation framework for the management of flood-prone areas is taking its toll on the environment in Malaysia. Flood legislations are largely focused on disaster risk reduction and risk management. EbA practices if entrenched in future flood legislations can help to reduce the effect of erosion on livelihoods and save huge relocation and rehabilitation costs that usually follow every major flood disaster.

4.3.7.2. *Forest management*

The forest ecosystem plays a major role in the global carbon cycle, being the carbon sink of the world. Forests are important for water regulation, both on the ground and under the ground, and flood management in river basins. They are also a source of timber products, an income for the rural population, as well as home to local ethnic peoples. Malaysian forests are very rich in plant and animal species and are extremely complex ecosystems. As of today, approximately 56% of the total land area of Malaysia is still forested. Forests include permanent reserve forests (PRFs), state land forests, national parks, as well as wildlife and bird sanctuaries. The remaining comprise of areas of agricultural crops, rubber plantations, oil palm plantations, urbanization, or other uses. In 2007, 18.30 million ha, or approximately 55% of the total land area of Malaysia was still forested. Malaysia's commitment to maintain 50% of its land covered with forests has been reiterated by its Prime Minsters from time to time.

For a number of countries, including Malaysia, the management of forests is a central aspect of Ecosystem-based Adaptation approaches. With adequate relevant legislation to incorporate ecosystem-based approaches, the aim of the Malaysian forest policy is to retain and enhance forest cover and other vegetation to the greatest extent possible.

To maintain forest ecosystems and their carbon storage capacity and to maintain its forest cover, the Malaysian government introduced a legislation that:

- Requires the completion of national forest inventories, which is almost complete now. It is supported with a regional barcode of forests and eco-labeling in order to curb illegal logging;
- Requires mandatory EIAs for converting forests for non-forest use beyond prescribed hectares;
- Prohibits clearing forests on steep mountains;
- Requires the preparation of sustainable harvest plans;
- Sets high penalties for breaches of certain provisions of the National Forestry Act 1984, especially for illegal logging which carries a maximum RM500,000 fine and/or 20 years of imprisonment with one year mandatory imprisonment;
- Creates a legal obligation to ensure reforestation of logged areas; and
- Provides for the use of environmental rehabilitation bonds.

Malaysia has a sizeable population of indigenous people who reside in the forest ecosystem. Providing an opportunity for indigenous people (IP) from Myanmar, Cambodia, Indonesia, Malaysia, the Philippines, Thailand, and Vietnam was the focus of REDD initiatives on Southeast Asia Indigenous Peoples Regional Consultation on REDD, 9–11 November 2008, Baguio City, Philippines. The regional initiative was to discuss the impact of REDD strategies on the rights of the indigenous people (IP). Among the highlights of the forum were:

(1) Review of REDD human rights framework;
(2) Issues of land and resource control for indigenous people (IP) and forest governance; and
(3) Effective participation of IPs in consultation, capacity building, and raising awareness on REDD.[45]

REDD has further been enhanced by REDD+, which has an emission mitigation solution and which subsumes REDD, with further novel initiatives, e.g., Forest Carbon Partnership Facility and Forest Investment

[45]ADB (2009), *op. cit.*

Program of the World Bank. But for its success, the REDD+ must be supported by better incentives, especially financial and technological.

4.3.7.3. *Agricultural ecosystem*

The projection by NAHRIM (2006) signifies paddy agricultural yield in Malaysia as a result of positive or negative variation of both rainfall and temperature, and which is expected to continue by the year 2060. Negative impact of climate change on agriculture might be worrisome in Southeast Asian countries due to heavy reliance on and the central role of the agro-allied sectors in terms of national income and employment.[46] Some agricultural factors which affect the adaptability of farmers are presence of injurious insects, temperature variation, soil fertility loss, cost of input, shortage of rainfall, and excessive rainfall, among others.[47]

Research on the adaptation options provided by ecological agriculture must take into account context and location specificities such as soil types, crop types, and management practices which are capable of helping to adapt to the effects of climate change.[48] Other factors are: understanding localized research and extension for the development and sharing of "best practices"; "fair trade" to promote ecological agriculture and small rural farm holders; knowledge management arrangements for the sharing of information and experiences, transfer of and training in good practices that constitute adaptation in ecological agriculture, including through extension services. In these areas *policy, law, and good governance* play a vital role (emphasis added). Law can also play a role in the consolidating of lands, granting lands to landless people, encouraging cooperative and community farming, providing facilities to farmers to hold roadside farmers' markets, which is being practised by Malaysia, and providing soft loans to farmers via village banks supported by the central bank, which have been well-adopted in India and Bangladesh, and can be emulated by other countries. Likewise, conservation of forests and forest resources, and all kinds of flora and fauna, can best be managed with participation of forest inhabitants. South Africa

---

[46] Koh and Lovleen (2011), *op. cit.*

[47] M.M. Alam *et al.*, "Agricultural Vulnerability and Adaptation to Climatic Changes in Malaysia: Review on Paddy Sector," *Current World Environment*, 8(1) (2013), 1–12.

[48] Robert (2011), *op. cit.*

is the best example of this. All these help adaptation in view of global warming, along with alleviating poverty. In view of climate change and its adverse impact on food production, food security is one of the most crucial issues relating to adaptation in the ASEAN countries, especially in Malaysia, Indonesia, and the Philippines, as a result of increasing reliance on agriculture, fisheries, and forestry for livelihood and employment.[49] This is stated under the ASEAN Multi-Sectoral Framework on Climate Change: Agriculture, Fisheries and Forestry towards Food Security (AFCC Framework), which represents a comprehensive and strategic approach to address the impacts of climate change.[50] Its overall goal is to contribute to food security through sustainable, efficient, and effective use of land, forest, water, and aquatic resources by minimizing the risks and impacts of and contributions to climate change. This is further supplemented with the existing ASEAN Integrated Food Security (AIFS) Framework and the Strategic Plan of Action on Food Security in the ASEAN Region (SPA-FS) 2009–2013.[51] Climate vulnerability in Malaysia has impacted agriculture due to the fluctuation in amount of rainfall. Farmers' adaptation practices are based on their own ideas, *which are helpful* (emphasis added) but are unlikely to stem the trend of reduced crop production.

4.3.8. *Freshwater ecosystem*

Freshwater ecosystems in many Southeast Asian and island countries are under threat due to the rise in temperatures. Rising sea levels, floods, together with storm surges, are likely to result in salination of freshwater lenses. Over time, population growth and sea level rise are likely to lead to an increased clearing of water catchments, as settlement and agricultural activities extend further inland.[52]

To maintain water quality, it will be necessary to introduce effective mechanisms for the protection and management of water catchments,

[49]Koh and Lovleen (2011), *op. cit.*

[50]ASEAN, "ASEAN Multi-Sectoral Framework on Climate Change: Agriculture and Forestry towards Food Security," (2011). Retrieved 19 November 2013, from http://aseanfoodsecurity.asean.org/wp-content/uploads/2011/08/afcc.pdf.

[51]The ASEAN Integrated Food Security (AIFS) Framework and the Strategic Plan of Action on Food Security in the ASEAN Region (SPA-FS) 2009–2013.

[52]Boer and Clarke (2012), *op. cit.*; Sterrett (2011), *op. cit.*

including restrictions on clearing and burning in drinking water supply catchments and riparian zones. Relevant legislation concerning the protection of freshwater ecosystems will require review and amendment; where no such provisions currently exist, various jurisdictions will require preparation of new legislation.

4.3.8.1. *Coastal and marine ecosystem*

Due to Malaysia's wet equatorial climate regime with frequent heavy rainstorms of high rainfall intensities, landslide disasters are common. In recent decades, landslide disasters in the Klang Valley Region and elsewhere have caused significant loss of life, property and infrastructure damage, environmental destruction, and anxiety.[53]

Furthermore, NAHRIM offers specialist consultancy services in project planning and impact assessment in all facets of hydraulic engineering encompassing river and coastal engineering, water resources, and water environment. Its main clients include both the public and private sectors in the water-related industry.[54] Thus, NAHRIM was highlighted on the urgent need to develop vulnerability and adaptation process and procedures, as well as to conduct further research and data collection for a quantified and qualified development of options and decision-making.

4.3.8.2. *Fisheries management and marine protected areas*

Protection of coastal waters and its resources is very important for the maintenance and enhancement of the marine ecosystem. Marine Protected Areas — Marine Parks created under the Fisheries Act 1985 (read with the Establishment of Marine Parks Malaysia [amendment] Order, 2012) and managed under this Act with the Exclusive Economic Zone 1994 by the Department of Marine Parks — have the major role of maintaining and restoring the marine ecosystems including coral reefs. This has been recognized in the Pacific Islands and the ASEAN region.[55] It has

---

[53]P. Thanapackiam, O.S. Khairalmaini, and A.G. Fauza, "Slope Failure Hazards: A Spatial and Temporal Study of the Klang Valley Region, Peninsular Malaysia," *Disaster Advances*, 5(3), 43–51.

[54]Othman (2011), *op. cit.*

[55]Boer and Clarke (2012), *op. cit.*

been identified that global climate change is the greatest threat to the marine ecosystem including the world's coral reefs along with ocean warming, increasing ocean acidification, more intense tropical storms, plus overfishing, pollution, and sedimentation.

The regional program to conserve marine ecosystem and coral reefs was launched under the CTI-CFF initiative in the Coral Triangle. The goals of the initiative, which spans along countries are:

(1) Priority seascapes designated and effectively managed;
(2) Ecosystem approach to management of fisheries (EAFM) and other marine resources fully applied;
(3) Marine protected areas (MPAs), which are known as marine parks, established and effectively managed;
(4) Climate change adaptation (CCA) measures; and
(5) Improving the status of threatened species.

Malaysia has recognized the need for integrated planning and management of both land and sea to control and minimize adverse impacts on the marine environment and fisheries management (visit the official website of the Department of Marine Parks, 2012, created under the Ministry of Natural Resources and Environment). In Peninsular Malaysia marine parks were created under the Marine Parks Order 1994. In Sarawak, they were created under the National Park and Reserve Ordinance 1998. In Sabah, they were brought into existence under the Parks Enactment, 1984 and Wildlife Conservation Enactment, 1997. There is a three-tier enforcement of these laws, i.e., at the level of the federal government, state government, and the local authority. Efforts are being taken to ensure sustainable development and management of coastal areas especially to cope with impacts of climate variability including sea level rise. The implementation of the Integrated Shoreline Management Plans (ISMP) by local authorities has already begun in selected coastal areas of the country.[56] The principal goal in the establishment of MPAs, which have wider restrictive measures, in Peninsular Malaysia is "to protect, conserve and manage in perpetuity, representative marine ecosystems of significance in

[56]Robert (2011), *op. cit.*

order that they remain undamaged for future generations and to inculcate public understanding, appreciation, and enjoyment of Malaysia's marine heritage".

In 2004, the responsibility of Marine Park management was transferred from the Department of Forest Management to the Marine Parks Department Malaysia (MPDM) under the Ministry of Natural Resources and Environment (MNRE). The main reason for this arrangement was to promote sustainable resource management and enhance tourism in the marine parks. The Department of Forest Management and the MPDM fall under the federal government's jurisdiction, while the land in the island belongs to the state government, which makes decisions on all land-based development activities. Coordination between state and federal governments is important for the development of adaptive capabilities of the ecosystem in the marine environment.

4.4. Environmental Assessment and Strategic Planning

4.4.1. *Environmental impact assessment*

Environmental impact assessment (EIA) and social impact assessment (SIA) are important tools for assessing and managing probable impacts of development on ecosystem resilience. They have to go hand in hand for minimizing the negative impacts of climate change on the environment and people. It is important in this context to give input of ecosystem resilience into the environment impact assessment made under the Environmental Quality Act, 1974, the Environmental Quality (Prescribed Activities) (Environment Impact Assessment) Order, 1987, and the EIA Guidelines, 1987. In order to provide an accurate view of the landscape in the long term, which is the timescale of climate change, in supporting adaptation planning, additional vulnerability and adaptation assessment is undertaken taking into consideration socio-economic factors as well as interactions amongst sectors. In Malaysia, EIA is made mandatory on activities which are deemed to have adverse effects on the environment under the Environmental Quality Act, 1974, to be undertaken by the Department of Environment. Mitigating measures, if necessary, are required prior to the EIA in respect to the proposed project approved. EIA approvals are generally granted with certain necessary conditions to be complied with by the proponent. It is suggested that the adaptive capacity of the ecosystem of any given project area should be a necessary consideration in project implementation. Robust EIA and SIA together with adaptation assessment mechanisms can be used to prevent development around wetlands, by declaring more wetlands as Natural Protected Areas as well as extending existing ones to cover future inadaptable lands, establishing wetland-friendly land use around current wetlands through the different zoning schemes, and requiring buffer areas as part of the mitigation measures in EIA processes. It is to be noted here that in Malaysia, both EIA and SIA are assessed together.[57]

4.4.2. *Strategic planning and disaster management*

Adaptation to climate change and disaster risk reduction share common features — they are not sectors in themselves but must be implemented through the policies of other sectors, in particular, those of agriculture, water resources, health, land use, environment, finance, and planning.[58] There are also linkages with other policies, most notably poverty eradication and planning for sustainable development, education and science.

The law of climate change adaptation in Malaysia cannot be viewed in complete isolation from disaster risk reduction or vice versa. As experience with both disaster risk reduction and climate change adaptation grows, there is increasing recognition in the ASEAN and Pacific region that these two fields share a common focus, i.e., reducing the vulnerability of communities and contributing to sustainable development. Given the high magnitude of climate-related risks in the ASEAN region, disaster risk reduction and climate change adaptation have been one of the major policy goals of the Malaysian government. In Malaysia, the National Security Council (NSC) is the principal policy-making and coordinating body for disaster management. The NSC coordinates and plans all activities related to disaster preparedness, prevention, response/relief operations, and recovery/rehabilitation.

"The National Security Council Directive No. 20 (NSC No. 20): The Policy and Mechanism for National Disaster and Relief Management" is

---

[57] Boer and Clarke (2012), *op. cit.*

[58] "International Strategy for Disaster Reduction, Adaptation to Climate Change by Reducing Disaster Risks: Country Practices and Lessons," (2009), 1–12.

the main guideline for disaster management in Malaysia. The Directive prescribes the mechanism for the management of disasters, including the responsibilities and functions of related agencies under an integrated emergency management system. This is achieved through the establishment of the Disaster Management and Relief Committee at three different levels (federal, state, and district levels), pending the severity of the disaster. The Directive is supported by other standard operating procedures which outline the mechanism as well as roles and responsibilities of various agencies for specific disasters, i.e., flood, open burning, forest fire, haze, industrial disasters, etc.[59]

4.4.3. *Protection of the rights of indigenous people*

Adaptation is at the core of the livelihood of the local and indigenous people globally, as it tends to affect their sustenance and reliance on the ecosystem for their basic needs. Through the International Indigenous Peoples' Forum on Climate Change (IIPFCC), indigenous peoples seek recognition of their contribution to climate change adaptation issues, and respect for their right to actively engage and participate in decision-making pertaining to climate change policy.[60] Demands on climate finance, as stated through global collective statements and declarations, relate to ensuring that indigenous peoples do not become more vulnerable in the execution and implementation of climate change projects, and ensuring their full and effective participation in global and national efforts to mitigate and adapt to climate change.

Malaysia's population of 28.3 million[61] comprises an indigenous population of about 67.4%, of which Malays are the majority. The immigrant populations comprise the Chinese (24.6%) and Indians (7.3%). In Peninsular Malaysia, the aboriginal people (Orang Asli) form about

[59]AIPA Secretariat (2012), *op. cit.*

[60]M. Francesco and R. Jen, *Indigenous Peoples and the Green Climate Fund: A Technical Briefing for Indigenous Peoples, Policymakers and Support* (Malaysia: Forest Peoples Programme and JOAS, 2012).

[61]Malaysian Department of Statistics (DOS), "Population Distribution and Basic Demographic Characteristic Report 2010." Retrieved 5 June 2014, from http://www.statistics.gov.my/portal/ index.php?option=com_content&id=1215.

0.5% of the population[62] whereas the indigenous groups in Sarawak comprise above 70%, and in Sabah, above 60% of the respective states' populations.[63]

Malaysia is a signatory to the UN Declaration on the Rights of Indigenous Peoples (UNDRIP), which requires states to consult and cooperate in good faith with indigenous peoples through their own representative institutions in order to obtain their free, prior, and informed consent (FPIC) before adopting and implementing legislative and administrative measures that may affect them. Also in 1994, Malaysia ratified the Convention on Biological Diversity (CBD), whereof Article 8(j) requires states to "respect, preserve and maintain knowledge, innovations and practices of indigenous and local communities embodying traditional lifestyles relevant for the conservation and sustainable use of biological diversity". The rights of aborigines are protected under the Aboriginal Peoples Act, 1954, which provides for the declaration of any area predominantly or exclusively inhabited by aborigines to be an *aboriginal area* or *aboriginal reserve*.[64]

4.4.4. *Financing mechanism for ecosystem adaptation*

4.4.4.1. *Legislated funding mechanism*

This section concerns examination of appropriate funding mechanisms to promote incorporation of Ecosystem-based Adaptation mechanisms into the relevant legislation. These climate finance mechanisms are needed to help vulnerable communities to adapt to the adverse effects of a changing climate. Climate funding is a crucial issue for countries in South Asia, both in terms of how international sources of finance are governed, as well as the capacity of national institutions to make the best use of climate finance when it is available.[65]

In Malaysia, finance and incentives for climate change adaptation are a key policy and governance action element of the National Policy on

---

[62]Colin Nicholas puts the figure at 147,412 individuals in 2003. See N. Colin, Centre for Orang Asli Concerns. Retrieved 2 November 2009, from http://www.coac.org.my.

[63]Malaysian DOS, State/District Data Bank (2005).

[64]Aboriginal Peoples Act 1954, Section 6.

[65]Fankhauser (2009), *op. cit.*

Climate Change of the Ministry of Natural Resources and Environment, NRE, 2011. Malaysia, like other developing countries, requires funding for adaptation from various developed countries under the UNFCCC Adaptation Fund and funding sources — the Global Environment Facility (GEF); the World Bank; the Asian Development Bank; the European Communities; the EC-ASEAN Energy Facility (EAEF); the DANIDA and Japan International Cooperation Agency (JICA); and the New Energy and Industrial Technology Development Organization of the Ministry of Energy, Trade and Industry, Japan.[66]

The National Adaptation Programme of Action (NAPA) has been prepared under the auspices of the 1992 UN Framework Convention on Climate Change, and financed through the Global Environment Facility. In 2007, the Secretariat of the UNFCCC commissioned five sector studies to get a better handle of investment needs for adaptation both globally and in developing countries (UNFCCC, 2007).

Malaysia, since joining the GEF, has received GEF grants totaling US$63 million that leveraged US$179 million in co-financing resources for 18 national projects, and 24 regional and global projects. These include eight projects in biodiversity, eight in climate change, and two in multifocal areas.[67]

4.4.4.2. *Ecosystem-based mechanism at local level*

Community-based management of ecosystem combines traditional and local knowledge, scientifically sound monitoring methods, and responsive decision-making processes to provide a strong foundation for improving resilience of ecosystems and responding to new challenges, including the impacts of climate change.[68]

Despite the robust climate change adaptation initiatives at the regional and national levels, it is of utmost priority to further identify public safety vulnerabilities and delivery processes at the local level. In Malaysia,

---

[66] Othman (2011), *op. cit.*

[67] Global Environment Facility, "Malaysia and the GEF," (2013). Retrieved 14 January 2014, from http://www.thegef.org/gef/sites/thegef.org/files/publication/Malaysia%20-%20Fact%20Sheet%20-%20Feb2013_EN.pdf.

[68] Boer and Clarke (2012), *op. cit.*

local participation has been delivered through localized information related to climate hazard risks and associated vulnerability mapping; research on the development (required parameters) and delivery of vulnerability assessments (including facilitation of knowledge) by local level researchers/facilitators; and an understanding of local government (top-down) processes and procedures for cross-sector and vertical collaboration for effective risk reduction responses.

In addition, raising awareness and sensitization efforts for local government actors to community level participants will be required to ensure that assessments are contextualized, understood to include ecosystem services, and eventually moved towards the acceptance of a standard operational procedure (SOP).[69]

4.4.5. *Implementation, compliance, and enforcement*

Implementation, compliance, and enforcement of environmental legislations are distinct aspects to be understood differently.[70] In climate change, adaptation concerns policy formulation functions channeled through international instruments such as the UNFCCC, United Nations Convention on Biological Diversity (CBD), and related decisions reached at the COPs.

In Malaysia, climate change policy concerns are directed through various sectors such as: energy, irrigation, natural resources, land-use planning, and solid waste, among others, and facilitated by the National Steering Committee on Climate Change (NSCCC) under the leadership of the Secretary General of the Ministry of Natural Resources and Environment. The implementation of climate change policy including adaptation is saddled with the National Policy on Climate Change and the National Green Technology Policy, which is chaired by the Prime Minister of Malaysia.[71] The Adaptation Working Committee is one of the several committees of the council. One of the functions of the council is to ensure implementation and compliance with international agreements on climate change, to which Malaysia has become a party.

---

[69]Robert (2011), *op. cit.*

[70]Boer and Clarke (2012), *op. cit.*

[71]Robert (2011), *op. cit.*

Each sector or ministry enforces compliance with climate change and adaptation policies determined by the council. Relevant government environmental departments are charged with enforcement of regulations aimed at achieving the overall objective of reducing the impact of climate change and adaptation in the country. However, there is a pressing need for capacity-building programs to ensure adequate focus on adaptation to climate change through the instrumentality of the natural ecosystem, if need be, with the help of prevalent traditional adaptation techniques.

4.5. Conclusion

In view of the adverse impacts of climate change on people and resources for their survival, Malaysia has used law and policy to augment mitigation and adaptation supported with instilling the element of the necessity for preventing wastage with respect to consumption of natural resources, especially water. Among them, boosting conservation of energy, maximizing the use of energy efficient equipments, minimizing use of wood by the village households, conserving forests in general and wetlands specifically, enlarging the tapping of REDD+, shifting to genetically modified crops which can survive in adverse weather conditions, learning from customary adaptation techniques, minimizing use of air conditioners, making people used to water rationing in crisis situations, maximizing the existing public transport, increasing the use of PV cells for fulfilling energy needs, harnessing rainwater for domestic use and irrigation, and increasing constructive peoples' participation in coastal management are notable. However, due to the lack of enough technological and financial resources to encourage adaptation, the pace of adaptation is not encouraging. Still, there is plenty of scope to learn from other countries. It is well-accepted that mitigation and disaster management have to go along with adaptation. The track record Malaysia has with respect to disaster management, especially landslides, is quite good. The country has undertaken the task to review all landslide prone areas and to take necessary preventive measures. As far as mitigation is concern, the country is doing its best.

The following are suggestions for consideration, based on the review of the current scientific literature on adaptation strategies, Malaysian Second National Communication to the UNFCCC, National Adaptation Plans, and adaptation made by relevant departments:

(1) There is a need for a detailed review of existing environmental laws and policies both at national and state levels to include natural resources management towards adaptation and ecosystem resilience to the impacts of climate change. There must be close cooperation between state and federal authorities on this matter. They should work together to make best use of available resources. They should work together to increase resources and building capacity.
(2) There is a need to provide robust adaptation mechanisms for consideration in the environmental impact assessment and social impact assessment to include specific provision on Ecosystem-based Adaptation assessment at all levels of government including the community levels.
(3) The Adaptation Working Committee under the National Policy on Climate Change should work to promote community-level integration in adaptation planning with national laws. This will strengthen and secure the adaptive capacity of the ecosystem.
(4) Malaysian universities and research institutes should engage more with the regional research networking efforts among universities in Asia, for example, the Asian University Network of Environment and Disaster Management (AUEDM). This will enable cross-fertilization of ideas in the area of Ecosystem-based Adaptation in the region. It is important to note that networking is most effective when its purpose is clear, its members are committed, and its mechanism and activities are sustainable.
(5) Help of NGOs and social activists in the adaptation processes should be enhanced.
(6) A regulatory approach to tourism activities and ecosystem deterioration in the designated marine protected areas in Malaysia to protect and conserve biodiversity is warranted.
(7) Climate adaptation funding should be enhanced through private initiatives to support the inadequate finance mechanisms sourced from international and multilateral institutions.
(8) Customary mechanisms for implementation and enforcement of adaptation measures should be encouraged in specific legislations at local

and community levels to include Ecosystem-based Adaptation practices. They can be further enhanced, as coupling adaptive improvements in infrastructures with efforts to improve ecosystem resilience will also make the communities resilient.

(9) There is a need for both top-down and bottom-up assessments, conducted from time to time so that necessary changes are made in order to make risks minimal.

These suggestions are in line with the observations made by the Working Group II of the Fifth Assessment Report on Climate Change 2014: Impacts, Adaptation and Vulnerability in Chapters 14, 15, and 16.[72]

[72] See IPCC Intergovernmental Panel on Climate Change, "Climate Change 2014: Impacts, Adaptation and Vulnerability," (2014). Retrieved 10 October 2014, from http://www.ipcc.ch/report/ar5/wg2.

Chapter 5

Policy and Legal Responses to Climate Change Adaptation in China: New Developments, New Challenges

Wang Xi and Gao Qi†*

5.1. Introduction

As a contribution to the IPCC's Fifth Assessment Report, the Working Group I of IPCC further raised the probability of human influence in global warming from 90–95% to 95–100% (extremely likely).[1] Compared to the traditional focus on climate change mitigation, IPCC observed that adaptation has emerged as a central area in climate change research, planning, and relevant implementation.[2] In order to better respond to climate-related risks, iterative risk management is highlighted as "a useful framework for decision-making in complex situations characterized by large potential consequences, persistent uncertainties, long time frames, potential for learning, and multiple climatic and non-climatic influences changing over time".[3] In many ways, adaptation to climate change requires

*Professor of Law, Director of Environmental and Resources Law Institute (ERLI), Shanghai Jiao Tong University. A member of the National Academic Committee on Environmental and Resources Law, National Key Institution in Human and Social Science Research, the Education Ministry of the PRC; Vice Chairman of the Chinese Society of Environment and Resources Law; Counselor of the Municipal Government of Shanghai.
†Lecturer, Environmental and Resources Law Institute, Shanghai Jiao Tong University.

[1] IPCC, *Climate Change 2013: The Physical Science Basis (Summary for Policymakers)* (Switzerland: IPCC WGI AR5 SPM, 2013), p. 17.

[2] IPCC, *Climate Change 2014: Impacts, Adaptation, and Vulnerability (Technical Summary)* (Switzerland: IPCC WGII AR5 TS, 2014), p. 2.

[3] IPCC, *Climate Change 2014: Impacts, Adaptation, and Vulnerability (Summary for Policymakers)* (Switzerland: IPCC WGII AR5 SPM, 2014), p. 9.

transformation of environmental law and policy.[4] Meanwhile, successful adaptation builds on the effectiveness of established environmental law regimes in a multitude of decision contexts.

Tremendous pressure has been put on China to respond to climate change adaptation needs and priorities at different levels. Being the world's biggest CO_2 emitter[5] and the largest developing country, China is very vulnerable to climate change itself. According to the IPCC, "warming in the period 1979 onwards was strongest over China in winter".[6] Existing research has confirmed that China is challenged by increasing extreme weather events, vegetation changes, insufficient water supply, and negative impacts on yields correlated with high temperature and climate change-disease prevalence.[7] The needs for climate change adaptation are particularly pressing with regard to agriculture, animal husbandry, forestry, water resources, coastal and urban areas, biodiversity, and human health.[8]

Climate change adaptation was initially mentioned in China in the 1994 China's Agenda 21. China is also the first developing country that has adopted a National Climate Change Programme.[9] Two white papers titled "China's Policies and Actions for Addressing Climate Change" have been issued in 2008 and 2011 respectively, which include adaptation measures and tasks.[10] Annual reports are also issued to assess the relevant progress

[4]See J.B. Ruhl, "Climate Change Adaptation and the Structural Transformation of Environmental Law," *Environmental Law*, 40 (2010), 363.

[5]International Energy Agency, "CO_2 Emissions from Fuel Combustion: Highlights," 2011 Edition, p. 9.

[6]IPCC, *Climate Change 2014: Impacts, Adaptation, and Vulnerability* (Switzerland: IPCC WGII AR5, 2014), Chapter 24, p. 4.

[7]See generally *ibid. China's National Strategy for Adaptation to Climate Change* (China: National Development and Reform Commission, 2013) (in Chinese).

[8]*China's Policies and Actions for Addressing Climate Change* (China: National Development and Reform Commission, 2008) (in Chinese), p. 1; *China's National Strategy for Adaptation to Climate Change*, supra note 7, p. 3.

[9]*China's National Climate Change Programme* (National Development and Reform Commission, 2007).

[10]See *China's Policies and Actions for Addressing Climate Change*, supra note 8; *China's Policies and Actions for Addressing Climate Change* (China: National Development and Reform Commission, 2011) (in Chinese).

per annum.[11] The 12th Five-Year Plan (2011–2015) specifically states that climate change factors should be taken into adequate consideration during the distribution of productivity, and the planning and construction of infrastructure and major projects. In addition, the capacity to adapt to climate change should be strengthened and the adaptability of certain key sectors and areas should be improved.[12] Over the years, a series of sectoral policies have been adopted to address climate change adaptation, particularly with respect to agriculture, forestry, water resources, marine environment, public health, and urban and rural construction.[13] During the 2013 Warsaw Climate Change Conference, China released the National Strategy for Adaptation to Climate Change, which for the first time considers adaptation as a national strategy.[14] It sets targets for adaptation, identifies key tasks, determines regional arrangements, and makes safeguard measures. The strategy is valid until 2020 and can be adjusted and revised during implementation. Here, learning by doing is considered as an important approach for adaptive management.

At the moment, there is no national legislation on climate change in general or specifically on climate change adaptation. The National Development and Reform Commission (NDRC), together with other relevant departments, have set up a leading group for drafting laws on addressing climate change.[15] Some local regulations have been adopted to cope with climate change. For example, provincial governments of Qinghai and Shanxi have issued the Measures on Addressing Climate Change that are applicable in their respective jurisdictions. Adaptation is specifically addressed in both regulations.[16] But whether specific legislation on climate

[11] See *2013 Report on China's Policies and Actions for Addressing Climate Change* (China: National Development and Reform Commission, 2013).

[12] *12th Five-Year Plan of the People's Republic of China* (China: Fourth Session of the 11th National People's Congress, 2011) (in Chinese).

[13] *China's National Strategy for Adaptation to Climate Change*, supra note 7, p. 4.

[14] See *ibid.*

[15] *2013 Report on China's Policies and Actions for Addressing Climate Change*, supra note 11, p. 9.

[16] Measures on Addressing Climate Change in Qinghai Province 2010, Arts. 6–13 (in Chinese); Measures on Addressing Climate Change in Shanxi Province 2011, Arts. 26–38 (in Chinese). It should be noted that the hierarchy of local rules is the lowest in China, same as the department rules issued by the Ministries and Commissions of the State Council.

change is indeed necessary is still doubtable. Part 3 of this chapter will review the above regulations to further explore the necessity for adopting new legislation on climate change in China.

Climate change adaptation in China is challenged to various extents. Integrating existing or future climate change legislation and policies with the established environmental law in China is one of the fundamental legal challenges both in theory and in practice. Fragmented legislation and policy development at national and local levels and across different administrative sectors further increases difficulty for effective adaptation. Climate change adaptation not only requires transformation and renovation of traditional environmental law, but also demands better design and implementation of many existing environmental mechanisms and the general improvement of an enabling environment. For example, information transparency, public participation, and environmental impact assessment (EIA) still need to be significantly enhanced even in a climate change context. With a problematic record on the effectiveness of China's environmental law in tackling environmental threats, the transformation of environmental law for climate change adaptation purposes alone is not enough to secure effective implementation. Moreover, the administrative setting for climate change adaptation has strongly influenced the formulation of relevant legislation and policy and their implementation. As will be discussed in detail later, the current administrative structure for climate change adaptation can cause coordination problems. Some key aspects of reform may be overlooked or undermined due to insufficient governmental cooperation or involvement. In addition, scientific uncertainties have negative impacts on the will of society to take preventive action, and a lack of adaptive capacity in China also creates barriers for effective implementation of relevant legislation and policies.

The following discussion will review the current administrative structure for climate change adaptation and the relevant legislation and policies that have been put in place to respond to adaptation needs. The status of implementation will be taken into consideration. Both progress and problems will be identified and analyzed. Recommendations will be made to improve regulatory and institutional frameworks to promote more effective, responsive, and coordinated actions across different regions, different levels and sectors of governments, and more importantly, among the government, business, and civil society. In particular, how China's environmental legislation and policy shall respond to adaptation challenges and targets will be highlighted.

5.1.1. *Administrative structure for climate change adaptation*

In face of increasing international and domestic pressure on climate change, the Chinese government has created a management system to address relevant issues. In 2007, the National Leading Group to Address Climate Change was established as a mechanism for deliberation and coordination. It is headed by the Premier and includes 20 ministers as its members. It is responsible for developing major strategies, guidelines, and measures for addressing climate change and coordinating relevant implementation. Under the leadership of this loose cooperative structure, the NDRC is designated to conduct specific coordination and supervision activities. A subordinate department was then established under the NDRC to address climate change issues. Accordingly, other relevant government sectors have further established specific agencies or working mechanisms to address climate change within their respective jurisdictions. In 2010, a liaison office was set up within the leading group in order to coordinate inter-ministry coordination. At the local level, all provincial governments have established their own leading groups and specific agencies to address climate change. But only some municipal governments have done so.

Several features can be concluded from the above structure. First of all, more emphasis has been put on mitigation instead of adaptation. Among the major tasks listed for the leading group, climate change adaptation is not specifically mentioned.[17] In contrast, energy conservation and emission reduction is emphasized. While mitigation remains to be crucial and urgent, climate change adaption should attract more attention from the administration and more administrative resources should be redirected accordingly.

Climate change adaptation requires comprehensive coordination and cooperation among different government sectors. Based on the departments involved in drafting and implementing the 2013 National Strategy for

---

[17]National Development and Reform Commission, "National Leading Group to Address Climate Change (in Chinese)." Retrieved 26 May 2014, from http://qhs.ndrc.gov.cn/ldxz/.

Adaptation to Climate Change, nine government sectors are involved in this area, namely, the NDRC, Ministry of Finance, Ministry of Housing and Urban-Rural Development, Ministry of Transport, Ministry of Water Resources, Ministry of Agriculture, Forestry Bureau, Meteorological Bureau, and State Ocean Administration.[18] The National Strategy requires that cooperation on specific adaptation issues among different government sectors should make better use of other existing advisory and coordinating organs under the State Council,[19] such as the National Patriotic Health Campaign Committee, National Committee for Disaster Reduction, Flood Control and Drought Relief Headquarters, and Leading Group on Poverty Alleviation and Development.

Notably, the Ministry of Environmental Protection (MEP) has not been involved. The MEP is set to supervise and manage national environmental protection administration in an integrated manner.[20] With the NDRC taking a leading role in climate change adaptation, the regime for climate change adaptation is alienated from the traditional environmental governance regime. This administrative setting may create institutional obstacles for the transformation of environmental law in the age of climate change. For example, all EIA reports in China shall be reviewed by environmental protection authorities to decide whether the proposed specific activities can be approved or not. A lack of participation by the environmental protection authorities will restrict the prospect of transforming environmental impact assessment (EIA) in order to integrate adaptation considerations during decision-making processes.

Similar challenges exist for transboundary cooperation among different administrative areas. The national strategy encourages neighboring areas, regions located in the same river basin, and areas that share similar climate conditions to establish communication and coordination mechanisms.[21] This requirement is aimed at solving the problem of lack of communication and cooperation among various relevant administrative authorities.

---

[18] National Development and Reform Commission, "Notice on the Issue of National Strategy for Adaptation to Climate Change (in Chinese)." Retrieved 27 May 2014, from http://www.gov.cn/zwgk/2013-12/09/content_2544880.htm.

[19] *China's National Strategy for Adaptation to Climate Change*, supra note 7, p. 28.

[20] Environmental Protection Law of PRC (2014), Art. 10 (in Chinese).

[21] *Ibid.*, p. 28.

For example, water governance in China is highly fragmented by administrative boundaries despite the theoretical prevalence of integrated river basin management. The awkward design of river basin organizations indeed creates further conflicts between river basin management bureaucracies and jurisdictional agencies.[22] Combined with the strong local protectionism that exists in water resources governance in China, their real influence on water resources management and utilization is indeed restricted in the field.[23] Climate change adaptation concerns will likely add another level of uncertainty and controversy to the ongoing disputes surrounding transboundary river basin management.

Another administrative factor that hinders the development and application of adaptive measures on the ground is insufficient participation by governments at the local levels. It is argued that "adaptation policy must operate at all scales in an interconnected network of decision making".[24] At present, only a few municipal governments have established leading groups and specific agencies to address climate change. Climate change adaptation has been mainly promoted by top-down efforts, while bottom-up initiatives are still rare in China. While many aspects of adaptation policy are inherently national or regional in scope, it should be acknowledged that national or regional adaptation efforts can only be realized or tested via local practice. Increasing participation of local governments is crucial for future reforms to generate productive communication and cooperation across all government levels.

As stated by the IPCC, "decision support is most effective when it is sensitive to the context and the diversity of decision types, decision processes, and constituencies".[25] In addition to insufficient participation by local governments, the current climate change decision-making process in China falls short of effective public participation. In this sense, climate change discourse adds to the growing momentum for more

[22]See Qi Gao, *A Procedural Framework for Transboundary Water Management in the Mekong River Basin* (Netherlands: Brill, 2014), pp. 33–35.

[23]Ji Dou, "A Legal Analysis of the Institutional Obstacles on Market Allocation of Water Resources," (16 September 2003). Retrieved 27 May 2014, from http://www.riel.whu.edu.cn/article.asp?id=25668.

[24]Ruhl, supra note 4, p. 428.

[25]IPCC, supra note 3, p. 23.

information transparency and better public participation in environmental decision-making in China, which is partly responded to by the most recent revision of the Environmental Protection Law in China.[26]

In general, China has established a basic administrative framework to address climate change. But adaptation concerns should enjoy more administrative resources and potentials for establishing specific agencies to promote adaptation can be explored. The current framework builds on China's governance regime on environmental protection, and economic and social development. The introduction of adaptation discourse is expected to further bridge the gaps among and within the government, business, and civil society. But it still encounters, if not aggravates, deep-rooted institutional obstacles, competing bureaucratic interests, problematic implementation of environmental law, and a lack of effective public participation. Climate change adaptation is still distant from local governments and general public as well. While it is generally commendable for the progress so far, there is plenty of room for the current administrative structure and governance regime to improve in the future.

5.2. Policy and Legal Responses to Climate Change Adaptation in China

As briefly discussed before, China has not yet adopted a nationwide legislation on climate change, although the process of drafting one is currently under way[27] and some local regulations are issued. In contrast, most responses to climate change adaptation remain at a policy level, represented by the National Strategy for Adaptation to Climate Change. The following analysis will first review the local regulations on climate change in order to provide valuable experiences and lessons for the drafting of law on climate change. Attention will then be paid to the

[26] See "Environmental Protection Law of PRC 2014," Arts. 53–58 (in Chinese). The revised Environmental Protection Law not only embraces public participation as one of the environmental protection principles in China, but also clearly recognizes the public's right to know and participate. However, it fails to provide sufficient framework requirements on how to secure the right to participate. Instead, it authorizes and requires the government to improve the process of public participation.

[27] NetEase, "State Council Information Office: Legislation Outline Is Ready for Law on Climate Change (in Chinese)," (5 November 2013). Retrieved 28 May 2014, from http://money.163.com/13/1105/14/9CU3MI5700252G50.html.

major policies guiding climate change adaptation in China, bearing in mind the question on how environmental law should be transformed to respond to the adaptation needs. Three approaches will be considered: (1) enhancing the implementation of existing legislation; (2) construing the current environmental law to reflect adaptation needs; and (3) revising the existing legislation or adopting new laws to incorporate adaptive mechanisms and adaption concerns.

5.2.1. *Local regulations on addressing climate change*

As two examples of learning by doing, provincial governments of Qinghai and Shanxi have promulgated the Measures on Addressing Climate Change in 2010 and 2011 respectively. Adaptation is specifically addressed in both regulations.[28] Notably, the whole Qinghai Province is located on the Qinghai–Tibet Plateau, which has been recognized as most sensitive to climate change.[29] In contrast, Shanxi Province has known for its economic dependence on coal mining industries. This difference in background leads to varied priority choices during the drafting of regulations in both provinces. The Qinghai version focuses more on adaptation while the Shanxi version highlights mitigation.[30]

Governments at or above the county level are held responsible by both regulations for addressing climate change within their respective jurisdictions.[31] They shall urge relevant government agencies, business, and other kinds of organizations to achieve adaptation and mitigation targets and to implement detailed measures on addressing climate change. Governments shall deal with major climate change issues in a cooperative and organized manner. Chapter 2 of the Qinghai version and Chapter 3

[28]"Measures on Addressing Climate Change in Qinghai Province 2010," Arts. 6–13 (in Chinese); "Measures on Addressing Climate Change in Shanxi Province 2011," Arts. 26–38 (in Chinese). It should be noted that the hierarchy of local rules is the lowest in China, same as the department rules issued by the ministries and commissions of the State Council.

[29]Ma Yong, "Icebreaking Progress towards Legislating on Addressing Climate Change (in Chinese)," (4 December 2010). Retrieved 28 May 2014, from http://news.sina.com.cn/c/sd/2010-12-04/020921580039.shtml.

[30]This preference can be discerned from the different chapter sequence and the number of articles on adaption compared to mitigation.

[31]"Measures on Addressing Climate Change in Qinghai Province 2010," Art. 4 (in Chinese); "Measures on Addressing Climate Change in Shanxi Province 2011," Art. 4 (in Chinese).

of the Shanxi version specifically address climate change adaptation. But most articles resemble policy statements rather than legal norms. The latter usually contains a matter of fact and the corresponding legal effects. Most provisions on climate change adaptation require the government to enhance the *implementation* (emphasis added) of existing legislation and policies, such as strengthening the management of permits on water usage, improving monitoring of extreme weather events, reinforcing infrastructure for agriculture and forestry, rehabilitating the mining area, and enhancing the preservation of typical ecoregions.[32] Some provisions are rather general in stating that governments should promote sustainable development.[33] Several articles specifically designate governments at or above the county level or certain government sectors as the responsible party for implementing the policies,[34] but many other provisions have failed to clarify this.

Climate change adaptation requires financial support from both public finance and other funding sources. The Qinghai version encourages the society to fund climate change mitigation and adaptation, but only requires local public finance to earmark funding for energy conservation, which is more relevant to mitigation than adaptation.[35] The Shanxi version requires governments at or above the county level to establish relatively stable government funding channels to support the construction of projects addressing climate change and relevant scientific research and renovation.[36] Climate change adaptation is not specifically mentioned. It further calls for the attraction of venture capital into climate change related business and research. Enterprises shall be guided to increase investment in climate change-related technologies.[37]

---

[32] See "Measures on Addressing Climate Change in Qinghai Province 2010," Arts. 8, 11, and 12 (in Chinese); "Measures on Addressing Climate Change in Shanxi Province 2011," Arts. 27, 29, 30–37 (in Chinese).

[33] "Measures on Addressing Climate Change in Qinghai Province 2010," Art. 6 (in Chinese); "Measures on Addressing Climate Change in Shanxi Province 2011," Art. 26 (in Chinese).

[34] "Measures on Addressing Climate Change in Qinghai Province 2010," Arts. 6, 8, and 10–12 (in Chinese); "Measures on Addressing Climate Change in Shanxi Province 2011," Arts. 26 and 37 (in Chinese).

[35] "Measures on Addressing Climate Change in Qinghai Province 2010," Art. 20 (in Chinese).

[36] "Measures on Addressing Climate Change in Shanxi Province 2011," Art. 47 (in Chinese).

[37] *Ibid.*, Art. 48.

Some interesting legal progress can be identified in both regulations. For example, the Qinghai version explicitly recognizes public participation as a principle for addressing climate change.[38] The Shanxi version calls for the establishment of an incentive mechanism to promote public participation and supervision.[39] It also requires the government to improve channels and mechanisms for information disclosure regarding climate change and to enhance transparency of relevant decision-making.[40] Both regulations require the government to seek public opinions during the plan making process.[41] A series of plans are specifically mentioned, including comprehensive plans on land utilization, urban and rural plans, plans on environmental protection, construction plans for the preservation of ecosystems, plans on water resources, and plans on soil and water conservation. Notably, some of these are considered as comprehensive plans and are not required to consult with the public under the current EIA Law and the SEA Regulation in China. Moreover, while public participation during the SEA process only focuses on the SEA report, issues subject to public comments under both regulations are not limited to environmental impacts.

Both regulations require the decision-making process of plans to fully consider the impacts of climate change on social and economic development.[42] This general requirement is different from the SEA procedure that considers potential environmental impacts of proposed plans. But how to fully consider climate change impacts in practice remains challenging. Climate feasibility study is probably the only established mechanism to consider risks of climate change and meteorological disasters. Both regulations require climate feasibility study for a mix of plans and specific projects.[43]

---

[38]"Measures on Addressing Climate Change in Qinghai Province 2010," Art. 3 (in Chinese).

[39]"Measures on Addressing Climate Change in Shanxi Province 2011," Art. 54 (in Chinese).

[40]*Ibid.*

[41]"Measures on Addressing Climate Change in Qinghai Province 2010," Art. 21 (in Chinese); "Measures on Addressing Climate Change in Shanxi Province 2011," Art. 46 (in Chinese).

[42]*Ibid.*

[43]"Measures on Addressing Climate Change in Qinghai Province 2010," Art. 13 (in Chinese); "Measures on Addressing Climate Change in Shanxi Province 2011," Art. 38 (in Chinese).

Before the adoption of these regulations, the China Meteorological Administration already issued the Management Measures on Climate Feasibility Study in 2008. The study aims to avoid or alleviate the impacts of climate change and meteorological disasters on plans and specific projects after their implementation or to avoid or alleviate such impacts on the local climate.[44] Different from the EIA procedure, it is administrated by the China Meteorological Administration.

One of the only two restrictions imposed by the regulations is that governments at or above the county level should report to governments at an upper level on their implementation of duties to address climate change.[45] The other one is that the performance evaluation of government officials and the person in charge of state enterprises should consider their performance on energy conservation and emission reduction as a major factor.[46]

Both regulations have put in place some general requirements on public participation and some can be considered as an extended application of previous legislation on public participation. But there is no available remedy to secure its implementation. In addition, the regulations apply climate feasibility study in order to evaluate the impact of climate change on social and economic development.

5.2.2. *Major policy responses to climate change adaptation and transformation of environmental law*

5.2.2.1. *National strategy for adaptation to climate change*

5.2.2.1.1. Overview

The 2013 National Strategy for Adaptation to Climate Change is China's pivotal policy on climate change adaptation.[47] It mainly targets climate change sensitive sectors (e.g., agriculture and water resources), regions, and population. Proactive adaptation is highlighted and adaptive measures should be reasonably tailored in accordance with unique circumstances.

---

[44]"Management Measures on Climate Feasibility Study 2008," Art. 1 (in Chinese).

[45]"Measures on Addressing Climate Change in Qinghai Province 2010," Art. 26 (in Chinese); "Measures on Addressing Climate Change in Shanxi Province 2011," Art. 57 (in Chinese).

[46]*Ibid.*

[47]Information in this section regarding the strategy is extracted from: *China's National Strategy for Adaptation to Climate Change*, supra note 7.

The strategy also calls for an integrated consideration of climate change adaptation and mitigation, local, regional, and national adaptation needs, and short-term and long-term adaptation targets and measures. Collaboration and cooperation across different government sectors, administrative boundaries, and among all levels of government should be enhanced. Another guiding principle of the strategy is extensive participation by civil society and diversified international cooperation, particularly South–South cooperation.

The strategy then lists a series of key adaptation tasks for infrastructure, agriculture, water resources, coastal areas, forest and other ecosystems, human health, tourism, and other relevant industries. For infrastructure, the strategy calls for the establishment of climate change risk evaluation and information exchange mechanisms. In light of variations of climate conditions, the technical standards for construction, operation, and maintenance of infrastructure should be revised. The safety of major hydrological projects should be re-examined. The monitoring, early warning, and emergency response systems for natural disasters should be established or enhanced to secure proper operation of major infrastructures.

With regard to agriculture, early warning systems and measures to prevent or mitigate agricultural disasters are highlighted. Crop varieties with high photosynthentic efficiency or those resistant to high or low temperature and drought should be cultivated. Other adaptation options are also mentioned, such as early or later planting, choosing the varieties with longer growing duration, adjusting the northern boundary of certain varieties, and redistribution of crop varieties. Pilot projects on insurance of agricultural disasters should be increased and the insurance coverage should be expanded.

Major adaptation measures regarding water resources management focus on prevention of water and soil erosion; maintaining ecological flow and water levels; aqua-ecosystem protection; monitoring; improving utilization and allocation of water resources through water storage projects, diversion, and irrigation projects; tightening water resources management via strict total quantity control, quota management, and water permit systems; and enhancing the ability of flood control and drought relief.

The strategy further states that China will establish a climate change impact assessment system with a coverage of all coastal areas and islands. Risks of marine disasters will be assessed and relevant areas will be regionalized accordingly. All kinds of marine-related plans shall fully consider climate change factors. Marine conservation areas will be established and the ecosystems of typical islands and coastal areas and waters should be restored. More coast protection forests should be cultivated. Marine disasters and the fluctuation of sea level should be more widely and better monitored. The mechanisms for early warning and emergency response shall be enhanced as well.

For forest and other ecosystems, climate change impacts shall be monitored and evaluated. Plans on forestation should be adjusted to cope with fluctuation of temperature and precipitation. Water and soil preservation should be enhanced for grasslands. The ability to prevent and control bushfire, grassland fire, rats, and pests should be strengthened and specific targets are set to be achieved by 2020. Wildlife habitats and biodiversity should be conserved. The protection of ecosystems should be improved and degraded ecosystem should be restored. Ecomigration can be considered to reduce pressure on ecologically fragile areas. Targets are set for wildlife protection, wetland conservation, and restoration of desertificated land.

With respect to human health, public health service should be improved and relevant standards should be updated and revised. Knowledge to cope with extreme weather and information on climate change should be provided for the public, particularly the sensitive population. Emergency response plans should be made or revised to cope with high or low temperatures, extreme precipitation, and haze. Climate change impacts on sensitive population shall be assessed.

With regard to tourism and relevant industries, labour protection and emergency rescue should be improved in the context of increasing extreme weather events. The government should strengthen the protection of scenic attractions, culture, and natural heritages from climate change impacts. The tourism industry should also utilize positive impacts of global warming and promote more diversified tourism services and products.

Different regional priorities are then taken into consideration. Based on the National Plan for Major Function-Oriented Zone,[48] the strategy

[48]The 2011 version is China's first national territory development plan, which classifies Chinese territory into four types of major function-oriented zones. Regional policies

identified three kinds of key adaptation zones: Urbanized Regions, Major Agricultural Areas, and Typical Eco-Regions. Specific adaptation targets for those zones are set accordingly.[49]

The above adaptation targets and measures are less likely, if not impossible, to be achieved or implemented without safeguard arrangements. The strategy addresses this issue mainly from four aspects: institutional arrangements, funding and financial policies, technical support, and international cooperation.

The strategy urges the establishment of a legal framework and a comprehensive adaptability evaluation index system for climate change adaptation. All adaptation tasks should be incorporated into national economic and social development plans and should be considered as an important content for mid- or long-term development strategies and plans. In accordance with the national strategy, governments at all local levels should develop their respective climate change adaptation strategies as well. Administrative structure should also be improved. The strategy calls for the establishment and enhancement of administrative coordination and cooperation mechanisms, which have been discussed in great length in Section 5.2.

A blueprint will not become a reality without necessary funding. The strategy attempts to resolve this problem via both public financing and financial markets. It is stated that public finance should increase spending on adaptation capacity building, major technological innovation, and secure funding for the achievement of adaptation targets in key sectors and regions. It also urges the clarification of fiscal responsibility among central and local governments. Efforts should be made to relieve the financial burden of less developed regions on adaptation. Tax incentives should be provided to attract funding from the business sector. Financial markets should explore innovative approaches to share climate change risks and provide more

---

and government performance evaluation should be adjusted or redirected towards the development direction set by this plan. For example, areas designated as an optimized development zone are characterized by high density of land use and decreasing carrying capacity of resources and environment. According to this plan, such areas should alter the economic development mode which traditionally based on extensive resource consumption and land occupation. See *National Plan for Major Function-oriented Zone* (China: State Council, 2011) (in Chinese).

[49] See *China's National Strategy for Adaptation to Climate Change*, supra note 7, pp. 20–28.

funding for adaptation. Several options are mentioned in particular, such as catastrophe bond and insurance products relating to agriculture and forestry. International financial support is another source of funding and the management and spending of which should be improved.

Science and technology development regarding climate change adaptation is another notable weakness identified by the strategy. Science and technologies regarding climate change monitoring and forecast, impact assessment, and adaptation should be improved. Relevant research and application should be encouraged and supported. Databases should be created and inventories on adaptation technologies should be issued on a regular basis. Multidisciplinary, transboundary, and cross-sectoral cooperation network on adaptation technology should be established.

International cooperation on climate change adaptation is encouraged and supported, not only among governments, but also among business sectors and NGOs. It also reaffirms China's position on urging developed countries to provide assistance to developing countries with regard to funding, technology, and capacity building. Cooperation and communication among developing countries are mentioned as well.

5.2.2.1.2. Comments

As the very first national strategy on climate change adaptation in China, the document is generally commendable. Compared to mitigation, climate change adaptation is less controversial. In this sense, adaptation agenda certainly enjoys stronger political will. In majority cases, climate change adaptation coincides with other purposes of environmental protection or natural preservation. Many measures or mechanisms listed in the strategy have been put in place for years in China, represented by adaptation strategy regarding water management. In this respect, strengthened or stricter enforcement of laws and policies is emphasized.

Numerous mechanisms or tools are required to incorporate climate change concerns, such as standard setting, monitoring, early warning, emergency response system, planning, and performance evaluation.

A number of specific measures are listed to enhance adaptive capacity, such as an enhancement of seawalls and other infrastructure, a change of crop or tree varieties, and altering planting time. Explicit targets are set in some aspects of adaptation, which could provide more coercion for implementation, especially if it can be incorporated into the future performance evaluation system. As a national strategy, a great deal of detailed measures are probably unrealistic and unnecessary, but it is suggested that more sectoral, regional, and local adaptation policies should be promulgated to add more concrete, operable, and practical contents to this national and cross-sectoral adaptation strategy.

Compared to the business sector and the public, it is the government that should take the main responsibility for climate change adaptation. This can be discerned from the content and focus of the strategy as well. Governments are required to play key roles as regulators, facilitators, providers, or even insurers. In order to achieve the ambitious targets in practice, a comprehensive and responsive administrative structure is essential. The strategy recognizes administrative collaboration and cooperation as one of its guiding principles and lists several measures to improve the existing administrative structure. These measures, if implemented, may partly address some of the problems identified in Section 5.2. But many profound institutional barriers remain and will have negative influences on the implementation of the strategy. As a country that is used to considering the government as a superior regulator, a changing understanding of the government's role in the age of climate change should be a fundamental shift of view angled at relevant administrative reform. Such administrative transformation is still in an early stage in China.

Among all kinds of stakeholders, the climate change sensitive population is considered as one of the key focus groups. Some specific measures are listed regarding their right to know and climate change impacts on their health, but no further arrangement is made to improve their capacity to actively adapt to climate change. Command-and-control mechanisms alone are insufficient to build resilience to climate change. A more participatory approach should be adopted to encourage and secure active public participation, which can contribute to front-end policy and decision-making and supervision of the government's implementation of relevant laws and policies.

Economic and market tools are emphasized in the text, but uncertainties still abound regarding funding and market building. In addition to the necessity to explore innovative approaches to improve government funding and expand market regarding climate change, the stagnation of many long overdue reforms on government finance and market economy should not be overlooked, such as a lack of supervision and debate on government budget and spending, obstacles of inter-governmental fiscal transfer, difficulty to clarify on the funding responsibility among central and local governments, and what the government should or should not do to promote the development of climate change-related economy, to name but a few.

The release of the strategy is certainly one major step ahead towards better climate change adaptation in China. But policy alone is not enough to promote and secure effective, responsive, and coordinated adaptive actions. Further down this road, legislation may be necessary to crystallize experiences gained from learning and to establish a more stable institutional framework for continuous climate change adaptation.

5.2.2.2. *Transformation of environmental law for dealing with climate change adaptation*

Having discussed climate change adaptation at the policy level, this section will attempt to provide some initial answers to the following key questions: (1) where does climate change adaptation put pressure on existing legislation? and (2) how can environmental law respond to such challenges?

To begin with, climate change shakes the scientific foundation of traditional environmental protection and natural preservation. Environmental law has been based on assumptions of ecological stationarity and targets are set accordingly. According to Craig, neither those assumptions nor those goals "fit a world of continual, unpredictable, and nonlinear transformations of complex ecosystems".[50] This scientific divergence seems to be largely ignored by current climate change-related discussion in China. In many respects, the regulations and policies analyzed before attempt to arbitrarily fit climate change considerations into the existing regulatory regime. For example, non-stationarity has important implications for studies of species

[50]Robin Kundis Craig, "'Stationarity Is Dead'—Long Live Transformation: Five Principles for Climate Change Adaptation Law," *Harvard Environmental Law Review*, 34(1) (2010), 9–75.

diversity patterns and their scaling,[51] but this is not reflected in the current climate change adaptation targets relating to biodiversity preservation.

A similar problem exists in the climate feasibility study, which is an evaluation process run by the meteorological administration and independent from EIA. The study is commendable for considering not only the impacts of plans and projects on local climate, but also investigating the impacts of climate change and climate disaster on proposed plans and projects. But the feasibility study is mainly conducted based on existing databases or historical records.[52] Despite the partial recognition of non-stationarity, the current evaluation approaches and standards seem unable to effectively assess the uncertainty and complexity associated with climate change. To some degree, this can be attributed to the fact that the feasibility study does not or is unable to differentiate climate change with climate variability. According to the United Nations Framework Convention on Climate Change (UNFCCC), "'Climate change' means a change of climate which is attributed directly or indirectly to human activity that alters the composition of the global atmosphere and which is in addition to natural climate variability[53] observed over comparable time periods".[54]

As discussed before, the climate change feasibility study also fails to integrate concerns on local climate with other environmental impacts. This is not difficult to understand, since it is managed by the meteorological administration, and the EIA and SEA procedures are administrated by environmental protection authorities like the MEP. Compared to EIA and SEA, climate feasibility study is subject to different screening standards. According to the regulation, the following plans and construction projects that are closely related to climate conditions should go through climate

[51] See Giles M. Foody, "Spatial Nonstationarity and Scale-Dependency in the Relationship between Species Richness and Environmental Determinants for the Sub-Saharan Endemic Avifauna," *Global Ecology and Biogeography*, 13(4), 315–320.

[52] "Management Measures on Climate Feasibility Study 2008," Arts. 8 and 9 (in Chinese).

[53] Climate variability merely indicates a year-to-year fluctuation in the climate record. Xiangbai He, "Integrating Climate Change Factors within China's Environmental Impact Assessment Legislation: New Challenges and Developments," *Law Environment and Development Journal*, 9(1) (2013), 50, 55.

[54] United Nations Framework Convention on Climate Change, 1992, 1771 UNTS 107, Art. 1(2).

feasibility study, including: (1) urban and rural plans, and key area or regional development and construction plans; (2) major infrastructure, and public or major projects; (3) major projects on regional economic development or restructuring of regional agriculture; (4) major projects on solar energy and wind energy; and (5) others that are required by law.[55] Obviously, the range of application for climate feasibility study is much more restricted than for EIA and SEA.[56] In addition, penalties for violation of the regulation are less severe compared to that of the EIA and SEA procedures. Information transparency and public participation is not mentioned in the climate feasibility study, while the EIA process has been considered as a relatively advanced example in securing the public's procedural rights.

Another major institutional implication of ecological non-stationarity is the necessity to increase the significance of post-project/plan monitoring and analysis. This is a compulsory procedure during the EIA and SEA processes in China, but it is not required by the current climate feasibility study. If the EIA and SEA procedures were to be applied to assess climate change impacts, back-end monitoring and analysis should receive more emphasis than it currently does. A more dynamic, learning-oriented decision adjustment process is therefore highlighted, in contrast to the traditional front-end focused EIA and SEA procedures.[57]

In China, climate change has been compounded by numerous other environmental problems. Many of them are equally, if not more, urgent and severe than climate change. There could be potential conflicts among different environmental protection needs. For example, the construction of irrigation infrastructure could be beneficial to climate change adaptation, but it may also result in a negative impact on ecosystems. On the other hand, mutual benefits widely exist in many climate change adaptation measures and other environmental protection efforts. This is best illustrated by the aforementioned national strategy. Significant co-benefits, synergies, and

---

[55]"Management Measures on Climate Feasibility Study 2008," Art. 4 (in Chinese).

[56]See "Environmental Impact Assessment Law of PRC 2002," Arts. 7, 8 and 16 (in Chinese).

[57]Ruhl, supra note 4, p. 420; Daniel A. Farber, "Adaptation Planning and Climate Impact Assessments: Learning from NEPA's Flaws," *The Environment Law Reporter*, 39(7) (2009), 10605, 10610.

trade-offs that exist between climate change adaptation responses and other environmental protection agendas call for more integrated consideration of climate change and other environmental protection goals. In part, such comprehensive consideration can be achieved through transformed EIA and SEA procedures.

Although the EIA and SEA procedures in China have been criticized for many deficiencies,[58] they are still considered as the key environmental law mechanisms in China and these relevant regulations are more mature and tested than the climate feasibility study. Overall speaking, there are more scientific and institutional merits to dynamically assess climate change impacts as part of EIA or SEA procedures rather than to introduce a brand new instrument. Pilot projects and research should be encouraged to explore feasible and reliable ways to reshape EIA and SEA procedures to incorporate climate change concerns. The existing deficiencies of EIA and SEA procedures itself should be addressed as well. The biggest obstacle against the mainstreaming of climate change impact assessment into EIA and SEA procedures lies in the current administrative structure for climate change, which to a large extent excludes closer and deeper involvement of MEP and other environmental protection authorities at lower levels.

The capability of environmental law in directing China towards more climate-resilient pathways is further challenged by the rigid command-and-control mechanisms that are most frequently used to protect environment in China. Past experience has repeatedly demonstrated that a command-and-control approach without sufficient external supervision can easily result in rent seeking and government failure. Internal supervision, such as government performance evaluation and administrative disciplinary procedures, can only have partial or temporary effects. The aforementioned regulations and policies highlight the importance of better law enforcement, but only make very limited progress on external supervision by civil society and legislative or judicial branches. It is argued that climate change adaptation discourse can play a bigger role in promoting information transparency, public participation, and stricter regulation on government finance and spending.

[58] See Gao, supra note 22, pp. 234–278.

Climate change adaptation is a process of continuous learning and adjustment. Different priorities, choices, trade-offs among competing environmental goals, and levels of tolerance should be decided deliberatively on a country or local basis. While the law may not always secure a climate resilience result, it can require decisions to be the outcome of a process which promotes climate change adaptability. Through comprehensive deliberation and communication, the forming and development of substantive contents of climate change adaptation can be cultivated. Scientific uncertainty relating to climate change and multiple values of modern society can easily lead to non-cooperative dilemmas. But procedural justice and values can help promote cooperative behaviors and dispute resolution. The recently revised Environmental Protection Law in China has made some notable progress in terms of the public's procedural rights to know, to participate, and to access judicial remedies.[59] But more detailed arrangements should be made to gain normative contents in the specific context of climate change adaptation. Notably, the major deficiencies of public participation in environmental decision-making in China have been analyzed in Section 5.2, which should not be overlooked in climate change adaptation research.

In addition to the need to enhance external supervision on the government, the exploration of economic instruments to foster adaptation is another approach recognized by existing regulations and policies. Due to the initial stage of progress in this direction, current legislation and policies only include some general provisions or arrangements. While it may be too early to crystallize pilot experiences into legal documents, academics should pay more attention to advanced legal research in this area and policies can be used as a more flexible form of regulation. Due to the background of unfinished economic reform in China, many administrative, civil, and commercial law issues should be addressed to provide a better legal and institutional environment for healthier expansion and operation of climate change-related market and economic instruments. In particular, economic methodology on the analysis of law should be emphasized.

Nevertheless, it is expected that climate change adaptation in China will continue to rely on command-and-control approaches. Other than

[59] See "Environmental Protection Law 2014," Arts. 53–8 (in Chinese).

improved government transparency and external supervision, government cooperation across different administrative hierarchies, sectors, and regions should be highlighted as well. Deficiencies of Chinese legislation in this respect are best illustrated by trans-administrative boundary cooperation on water resources management. Problems have been discussed regarding the current administrative structure on climate change adaptation. As an initial start, the national strategy only partially recognized and addressed these problems. Without timely adjustments along the process, the administrative structural barriers will undermine the implementation of the strategy in practice.

5.3. Conclusions

This chapter begins by discussing the tremendous pressure on China to enhance climate resilience. Specific analysis is then made regarding the administrative structure of climate change and existing regulations and policies on climate change adaptation. While China has adopted a national strategy on climate change adaptation in 2013, the majority of China's adaptation responses are still in the planning phases. Evaluations have been made mainly based on the contents of the strategy rather than its implementation in reality. These initial conclusions still need to be tested or verified in practice.

With regard to specific legislation on climate change or adaptation in particular, it is suggested that while there are merits to formulate an overarching legislation on climate change, it is not necessarily a priority issue of concern at the moment. Experiences in Qinghai and Shanxi Provinces have illustrated the immaturity of current legal research and legislative techniques on many relevant issues in China. It is suggested that China should not rush into legislating on climate change without the strong support of more solid research and abundant experience in the field.

As a country facing compounded environmental problems, steering environmental law and policy towards stronger climate resilience while exploring resolutions to other pressing environmental concerns at the same time is a profound challenge facing developing countries like China. While climate change adaptation is now a prevalent discourse worldwide and it is necessary to learn from international experiences, China's approach to enhanced adaptability should always be sensitive and oriented to its unique legal and institutional circumstances. Based on the analysis above, it can be concluded that both traditional and innovative reforms of environmental law are urgently needed in China. Along with the adaptation process, effective cooperation among different social sectors is necessary. Continuous feedbacks and adjustments should be emphasized as opposed to mere front-end approaches. With down-to-earth self-evaluation and proper priority choices, it is not impossible to turn climate change test into a testimony in China.

Theme III

Key Future Impacts and Vulnerabilities

Chapter 6

The Laws, Policies, and Institutions Relating to Climate Change in Thailand: Balancing between "Mitigation" and "Adaptation"

Kanongnij Sribuaiam

6.1. Introduction

Thailand signed the United Nations Framework Convention on Climate Change (UNFCCC) on 12 June 1992 and ratified the Convention on 28 December 1994.[1] Its date of entry into force is 28 March 1995. Thailand signed the Kyoto Protocol on 2 February 1999 and ratified the Protocol on 28 August 2002.[2] Its date of entry into force is 16 February 2005. In Thailand, dialogues, debates, policies, institutions, and laws on climate change issues, both mitigation and adaptation dimension aspects, are at an incipient stage. Policies move more actively than laws on climate change issues. More emphasis, however, has been laid on "mitigation". Although national policies mention both "mitigation" and "adaptation", Thai laws provide no provision on climate change adaptation and thus are in need of reform by striking a balance between "mitigation" and "adaptation".

6.2. Policies on Climate Change Mitigation and Adaptation

Following the direction of the UNFCCC, Thailand developed two documents on climate change at the national level: first, the 2008–2012 National Strategy on Climate Change; second, the 2013–2050 Draft of Thailand

[1] Status of Ratification of the UNFCCC. Retrieved 10 March 2015, from http://unfccc.int/essential_background/convention/status_of_ratification/items/2631txt.php.

[2] Status of Ratification of the Kyoto Protocol, *UNFCC*. Retrieved 4 February 2015, from http://unfccc.int/kyoto_protocol/status_of_ratification/items/2613.php.

Climate Change Master Plan.[3] The common character of the two is that they both address *adaptation* and *mitigation* as the tools for enhancing competitiveness, achieving both sufficiency economy and sustainable development, and finally, making Thailand a low-carbon society.

The *2008–2012 National Strategy on Climate Change*, the first document formulating a national strategy on climate change, which was initiated and approved by the Cabinet in 2008, provides a comprehensive guideline for national responses to climate change challenges. According to this document, there are six strategies, enumerated as follows:[4]

- Strategy 1: Build capacity to *adapt* and reduce vulnerabilities to climate change impacts.
- Strategy 2: Promote GHG *mitigation* activities based on sustainable development.
- Strategy 3: Support R&D to better understand climate change, its impacts, and *adaptation and mitigation* options.
- Strategy 4: Create awareness and participation for problem solving on climate change.
- Strategy 5: Build capacity of relevant personnel and institutions and establish a framework of coordination and integration.
- Strategy 6: Support international cooperation to achieve the common goal of climate change *mitigation* and sustainable development.

The present national master plan on climate change is the *2013–2050 Draft of Thailand Climate Change Master Plan*. It is a framework of integrated policies and action plans relating to climate change to support climate change preparedness initiatives so that they are in line with Thailand's economic and socio-cultural contexts as well as sufficiency economy philosophy. There are five principles under this master plan:[5]

---

[3]Thailand's Climate Policy, "European External Action Service." Retrieved 4 February 2015, from http://eeas.europa.eu/. . ./thailand/. . ./thailande_eu. . ./onep_climate_policy_en.pdf.

[4]Tubtim Limsoontorn, "National Strategy on Climate Change Management: Modeling and Data Application." Retrieved to 10 March 2015, from http://file:///D:/my%20documents/NUS/%E0%B8%82%E0%B8%AD%E0%B8%87%E0%B9%84%E0%B8%97%E0%B8%A2/National%20Strategy%20on%20Climate%20Change.pdf.

[5]Nirawan Pipitsombat, "Thailand Climate Policy: Perspectives beyond 2012." Retrieved 10 March 2015, from http://eeas.europa.eu/delegations/thailand/documents/thailande_eu_coop/environment_energy/onep_climate_policy_en.pdf.

- Thailand places great emphasis on preparedness, solution, coping, and *adaptation* to impacts of climate change, especially for the sectors which are highly at risk and exposed to biological, physical, social, economic, and livelihood vulnerability, such as small-scale agriculture and traditional fishing.
- Thailand voluntarily has a part in *reducing* the greenhouse gases (GHGs) with common but differentiated responsibilities in operation according to UNFCCC's principles.
- Thailand uses the crisis and effects of climate change as an opportunity to support and sustainably develop the country in every aspect and according to the sufficiency way of life.
- Adhere to the concept of public sector stewardship in inducing change, as well as promoting partnership between the public, private, and civil society sectors and participation of a strong community together with the use of Polluter Pays Principle (PPP).
- Divide the operational framework into three periods: the short-, medium-, and long-term and arrange for follow-up and evaluation of the operation as well as the improvement of the master plan every 10 years in order to improve the overall efficiency of the master plan and operation.

The draft master plan states the vision and mission in dealing with climate change; the *Vision* being that "Thailand [should have] a framework and guideline on climate change preparedness, *adaptation*, appropriate and efficient application of as well as enhancing competitiveness and development toward sufficiency economy and low carbon society". The *Missions* are:

- To equip all related parties with preparedness and *adaptation* skills to cope with the effects of climate change.
- To encourage all related parties to take part in *reducing* GHG emissions on the basis of sustainable development and co-benefits in the context of Thai society based on the principle of sufficiency economy philosophy and way of life.
- To develop the body of knowledge and mechanisms in motivating integrated planning in effectively solving the problem of Thailand's climate change with concrete effectiveness at every level.

- To work with the global community in solving the issue of climate change without producing negative impacts on the country's economic, social, and environmental progress (development).
- To promote the adjustment of the economic and social structures to gain immunity impediment against interference of the nation's development an competitiveness in transitioning Thailand into a low-carbon society.

The objectives of the draft master plan are:

- To serve as a framework and mechanism for solving the nation's climate change problem both in *adapting* to climate change and *reducing* GHG emissions in every sector.
- To motivate every sector and level to be able to create operational plans/implementation plans for climate change properly, appropriately, efficiently, and with concrete effectiveness.
- To promote Thailand's sustainable development in a way that is in line with the international endeavor in solving the climate change problem.
- To transition Thailand into a sustainable low-carbon society in line with sufficiency economy philosophy.
- To continue Thailand's economic, social, and environmental developments in accordance with sufficiency economy philosophy and to *cut* GHG emissions by 2050, without impeding the country's gross domestic product (GDP) or reducing its growth of developmental capability and competitiveness.

To implement the master plan, three key *strategies* which determine the principles and operational guidelines for coping with the effects of climate change were introduced:

- Strategy 1: *Adaptation* for coping with the negative effects of climate change.
- Strategy 2: *Mitigation* of GHG emissions and increase of greenhouse gas sinks.
- Strategy 3: *Strengthening the capacity* of human resources and institutions and managing the risks from the effects of climate change and other cross-cutting issues.

The objectives of the first strategy on *adaptation* are:

- To raise awareness of the issue and effects of climate change so that all parties are capable of coping with and mitigating climate change impacts in the short-, medium-, and long-term.
- To encourage all climate change-related parties at every level to work together in an integrated manner to prepare a work plan and action plan for *adaptation* in an appropriate and corresponding way, and in the same direction.

The approaches of the first strategy are:

- Creating knowledge and awareness for climate change and the negative effects of climate change to the public in every sector.
- Creating financial mechanisms to support the implementation of *adaptation* for coping with the negative effects of climate change.
- Developing databases of climate change to date and making it easier to use.

The objectives of the second strategy on *mitigation* are:

- To promote participation from every sector and level in *reducing* GHG emissions and increasing GHG sinks.
- To promote efficient use of the nation's energy and resources both in terms of *reducing* GHG emissions and increasing GHG sinks.
- To create bodies of knowledge on environmentally-friendly manufacturing processes, consumption, and services as well as creating the potential of each sector in *reducing* GHG emissions as well as seeking alternatives for the alleviation of its effects.
- To promote capability of energy conservation, alternative energy, and green energy, and to reduce energy import from foreign countries with focus on public participation.
- To develop operation projects to *reduce* GHG emissions-related to the forestry sector (REDD).
- To promote and support the preparation of the carbon footprint for all products and services.

The approaches of the second strategy are:

- Developing a system for imposing carbon tax and other environmental taxes with the use of the polluter pays principle (PPP) in a way that is fair.
- Promoting and supporting development of cities and communities toward an efficient and sustainable low-carbon society.

The objectives of the third strategy on *capacity strengthening and risk management* are:

- To build the capacity of human resources, organizations, and the nation to manage risks from the effects of climate change.
- To create mechanisms that will drive the master plan towards systematic, concrete, and effective implementation.
- To integrate cooperation in determining the nation's positions in international negotiations to generate maximum benefits for the country.

Approaches of the third strategy are:

- Adjusting and enforcing rules and laws to support the implementation of the climate change policy in all sectors.
- Setting plans to promote international cooperation and capacity building in the negotiations of the climate change with sustainable development and co-benefits.

6.3. Institutions and Actors Responding to Climate Change Mitigation and Adaptation

Under the Ministry of Natural Resources and Environment (MONRE), Office of Natural Resources and Environmental Policy and Planning (ONEP), the *Climate Change Management and Coordination Office (CCMCO)* is the *national focal point* and responsible agency for the implementation of the United Nations Framework Convention on Climate Change (UNFCCC) and the Kyoto Protocol.

To realize the challenges of climate change issues, in 2007, the Cabinet approved two legal documents relevant to climate change management: The first is the *Order of the Prime Minister on Climate Change B.E.2550 (2007)*. The second is the *Royal Decree on Thailand Greenhouse Gas Management*

Organization (Public Organization) Establishment B.E.2550 (2007).[6] This creates four major institutions which are: (1) the National Committee on Climate Change Policy; (2) the Climate Change Management and Coordination Office (CCMCO); (3) the Thailand Greenhouse Gas Management Organization (TGO); and (4) the Thailand Greenhouse Gas Management Board.

The National Committee on Climate Change Policy is a high level committee having the Prime Minister as a Chairperson and the Minister of Environment as a Vice Chairperson, with Permanent Secretaries from relevant Ministries, and experts on various disciplines relevant to different dimensions of climate change issues. The Committee's authorities and responsibilities are:[7]

- To formulate policies and strategies on the prevention and solution of climate change problems in Thailand, and absorption and emission of GHGs, including research and development relating to climate change.
- To formulate policies, directions, rules, and mechanisms for the international operation concerning climate change management in accordance with socio-economic conditions and national interest including relevant international agreements.
- To propose the amendment or improvement of laws necessary or useful for the implementation according to Thailand's commitment to the UNFCCC Convention or the Kyoto Protocols, or the operations for promoting and supporting the principles and objectives of the UNFCCC Convention or the Kyoto Protocols, in accordance with socio-economic conditions and national interest including relevant international agreements.
- To set directions and negotiation positions on the UNFCCC Convention or the Kyoto Protocols, in accordance with socio-economic conditions and national interest including relevant international agreements.
- To consider and support the budget allocation for state organizations and the private sector concerning appropriate operation on climate change.

[6] "Talking Points on the UNFCCC and the Kyoto Protocol." Retrieved 4 February 2015, from http://www.mfa.go.th/fealac/images/OtherDocumentDoc23.doc.
[7] "Order of the Prime Minister on Climate Change B.E.2550," 2007, Art. 8.

- To set measures for enhancing cooperation and coordination among state organizations and the private sector concerning the operation on climate change.
- To propose appointing the Thailand Greenhouse Gas Management Board to the Cabinet.
- To appoint a sub-committee or working committee for implementation according to this Order of the Prime Minister on Climate Change, or other assignments from the Committee.
- To conduct any other activities in achieving this Order, other laws or other assignments from the Cabinet or the Prime Minister.

The Climate Change Management and Coordination Office (CCMCO) was established under the ONEP by the Order of the Prime Minister on Climate Change to serve as a secretariat of the National Committee on Climate Change Policy, with authorities and responsibilities as follows:[8]

- To propose policies, strategies, and plans on national climate change management, and absorption and emission of GHGs, including research and development relating to climate change.
- To propose directions, rules, and mechanisms for the international operation concerning climate change management, including directions and negotiation position on the UNFCCC Convention or the Kyoto Protocols, in accordance with the socio-economic conditions and national interest.
- To propose measures for enhancing cooperation and coordination among state organizations and the private sector concerning the operation on climate change.
- To monitor and coordinate the operation with state organizations and the private sector concerning the operation on climate change.
- To conduct any other activities according to assignments from the Committee.

Thailand's Greenhouse Gas Management Organization (TGO) is a public organization attached to MONRE as an implementing agency on GHG emission reduction and other relevant activities. The TGO also serves as a Designated National Authority of CDM (DNA CDM) for Thailand.

[8]"Order of the Prime Minister on Climate Change B.E.2550," 2007, Art. 10.

The organization has the following objectives:[9]

- To analyze, screen, and make opinions on the Clean Development Mechanism project approval including to monitor and evaluate the approved CDM projects.
- To promote the development of CDM projects and trading markets for the quantity of approved GHG.
- To serve as an information center concerning the GHG management situation.
- To establish an information base concerning the approved CDM projects and the selling of approved GHGs, according to policies formulated by the National Committee on Climate Change Policy and the Thailand Greenhouse Gas Management Board.
- To promote and develop capacity including advising state organizations and the private sector concerning GHG management.
- To disseminate and publicize GHG management.
- To promote and support climate change management.

Thailand's Greenhouse Gas Management Board was established under the Royal Decree on Thailand Greenhouse Gas Management Organization (Public Organization) Establishment B.E.2550, 2007, having one Chairperson with high expertise and experiences on management appointed by the Cabinet, together with four high level officers from relevant Ministries, and experts on various disciplines relevant to different dimensions of climate change issues.

The Board's authorities and responsibilities are:[10]

- To formulate direction, goals, and policies in administering the TGO.
- To set rules and methods in considering and approving the CDM projects.
- To approve the CDM projects for the sake of selling GHG quantities.
- To set fees, expenses, or charges for the service provided by the TGO.
- To monitor and evaluate the approved CDM projects including the quantities of GHG selling.

---

[9]"Royal Decree on Thailand Greenhouse Gas Management Organization (Public Organization) Establishment B.E.2550," 2007, Art. 7.

[10]"Royal Decree on Thailand Greenhouse Gas Management Organization (Public Organization) Establishment B.E.2550," 2007, Art. 20.

- To approve plans (investment plans, financial plans), projects, and the annual budget of the TGO.
- To set up and dissolve branch offices in necessary and appropriate cases in order to achieve the TGO's goal, as well as to formulate administrative procedures of such branch offices.
- To recruit, appoint, evaluate the operation, and dismiss the TGO's Director.
- To make rules, orders, regulations, or announcements concerning general administration of the TGO, coordination between the TGO and its branch offices, the division of the TGO's work and scope of each division's responsibility, and human resources management, including other routine works of the TGO.
- To conduct any other necessary activities for achieving the TGO's goals and its authorities and responsibilities, or according to the assignment from the Cabinet and the National Committee on Climate Change Policy.

6.4. Law Reforms for Balancing between "Mitigation" and "Adaptation"

Although Thailand is highly vulnerable to climate change particularly on agricultural, forestry, coastal, and health issues, there is no specific law on climate change "adaptation" responding to such issues with vulnerability as natural disaster and food security. By contrast, there is more progress on laws on "mitigation" measures. One example is the most advanced legislation on mitigation, *the 2007 Royal Decree on Establishment of TGO* for facilitating various mechanisms: GHG inventory, Clean Development Mechanism project (CDM), CDM project approval, carbon market, carbon footprint, etc. Other examples are mitigation legislation in the energy and industry sector which can be seen through promotion of more efficient use of energy in the *Energy Conservation and Promotion Act;* or mitigation in the forestry sector which can be seen via forest legislations, logging ban regulations, or promotion of community forest, etc.

6.4.1. *Efficient use of energy: The Energy Conservation Promotion Act B.E. 2535*

The Energy Conservation Promotion Act B.E. 2535 was established with three principles: (1) To promote energy efficient production and consumption; (2) To promote the production of energy efficient equipment and machinery; and (3) To promote the concrete energy conservation by establishing the Energy Conservation Fund to provide financial support to people who have to implement energy conservation as required by law. Target groups of the Act are the designated factory; designated building; and manufacturer or supplier of energy efficient equipment, machinery, and material used for energy conservation. Certain buildings and factories are designated according to the type of energy consumption as follows: (1) Installed power meter of $\geq$1,000 kW; (2) Total transformer capacity $\geq$1,175 kVA; or (3) Annual energy consumption from 1 January–31 December $\geq$20 million megajoules.

The Act also stipulates that owners of the designated building and designated factory shall have duties to conserve energy as follows:

(1) Implement the energy conservation to be complied with specified standards.
(2) Appoint at least one Person Responsible for Energy (PRE) to work full-time at the designated building or designated factory.
(3) Submit data of energy generation, consumption, and conservation to the Department of Alternative Energy Development and Efficiency (DEDE), Ministry of Energy.
(4) Record data of energy consumption, data of installation, or modification of equipment or machinery affecting energy consumption and conservation.
(5) Establish an energy conservation target and plan and submit it to the DEDE.
(6) Audit and analyze the implementation of the energy conservation target and plan.

The procedure of the Act's implementation composes of three elements:

- *Person Responsible for Energy (PRE)* — for assisting the owner of the designated building and designated factory in implementing energy conservation by performing the duties to: (1) periodically maintain and inspect the efficiency of equipment and machinery that use energy; (2) verify an energy consumption and conservation report to be submitted to DEDE; (3) assist the owner of a designated building or a designated factory in establishing energy conservation target and plan; and (4) verify the result of the energy conservation audit and analysis.

- *Energy Consumption Report* — for the benefit of both the owner of the designated building and designated factory and DEDE in analyzing and planning energy conservation measures. The production data, as well as energy consumption and conservation data, shall be reported to DEDE.
- *Establishment of Energy Conservation Target and Plan* — by using the result of the audit and analysis of energy consumption to be complied with the energy conservation standard. The energy conservation target and plan shall be submitted to DEDE every three years.

According to the Act, the Energy Conservation Promotion Fund is established under a supervisor of the Energy Conservation Promotion Fund Committee, to promote and support the designated building and designated factory to implement energy conservation as required by law, as well as to support others who want to conserve energy.

6.4.2. *Mitigation in forestry sector: Acts, cabinet resolution, and draft law*

The 1913 Forest Protection Act[11] sets out principles for the long-term exploitation of forests benefiting the state. Besides this Act, there are several forest legislations including the 1960 Wildlife Protection and Preservation Act,[12] the 1961 National Park Act,[13] and the 1964 National Forest Reserve Act.[14] In January 1989, the national "logging ban" in natural forests was officially declared by the Royal Thai Government (RTG) due to various pressures, particularly the devastating floods in southern Thailand. The goal of the logging ban was to protect the remaining natural forests and

[11] "Forest Act," B.E. 2484 (A.D. 1941). Retrieved 4 February 2015, from http://www.thailawforum.com/database1/forest-act.html.

[12] "Act for the Conservation and Protection of Wildlife," B.E. 2535, 1992. Retrieved 4 February 2015, from http://www.fao.org/fishery/shared/faolextrans.jsp?xp_FAOLEX=LEX-FAOC004724&xp_faoLexLang=E&xp_lang=en.

[13] "National Park Act." Retrieved 4 February 2015, from http://www.thailawforum.com/database1/national-park-act.html.

[14] "National Reserved Forest Act." Retrieved 4 February 2015, from http://www2.austlii.edu.au/~graham/AsianLII/Thai_Translation/National%20Reserve%20Forest%20Act.pdf.

to enforce rules and punishment of encroachers in the protected forests. Since then, conservation measures have played a major role in Thailand's forestry policy. The core legally designated conservation areas, declared by Cabinet Resolution, are national parks, wildlife sanctuaries, forest parks, non-hunting areas, watershed Class 1, and mangrove conserved forests. In 1990, the Community Forest Act was drafted. In 1992, the Forest Plantation Act was enacted. In 1996, the RTG also revoked logging licenses in mangrove forests to halt their destruction which was especially excessive during the 1980s and early 1990s.

Compared to "mitigation", "adaptation" is relatively new for Thai laws. Attempts to adopt climate change adaptation initiatives can be found in the making of policies, plans, strategies, and capacity building, but not in making laws and regulations. Environmental law cannot accommodate climate change adaptation policies without *reforming* its infrastructure for responding more actively to green philosophy and the sustainable development concept, and at the same time, keeping up with scientific uncertainty of climate change impacts by using precautionary principles. In addition, environmental law focusing purely on natural resources management and pollution control is insufficient unless components of foreseeable disruptive climate events are identified to be managed and interconnected with other aspects of environmental law through the exercise of due care. Finally, both "mitigation" and "adaptation" should be well-integrated in environmental law or other appropriate new law.

6.5. Change of Understanding for Welcoming New Paradigm

In *reforming* Thai environmental laws or other relevant laws for introducing climate change adaptation to the Thai legal system, it needs to start at the basic foundation on cultural change and a paradigm shift of all stakeholders who will be actors in the adaptation policy: government at all levels, private sector, civil society, both individuals and communities, academia, media, etc. Certain dialogues need to be initiated among these stakeholders at the early stage particularly on some issues regarding the nature of climate change, climate change management, climate change mitigation, climate change adaptation, differences between mitigation and adaptation, interconnection between mitigation and adaptation, and the necessity to have both mitigation and adaptation within the same legal structure for successful climate change management.

The following are examples of the understanding that needs to be promoted:

> Climate change problem is not only a pollution problem. Climate change adaptation is essential because only climate change mitigation cannot avoid impacts of climate change and thus does not form part of the whole picture of climate change management, regardless of the most stringent efforts or the success of the mitigation.[15] The difference between mitigation and adaptation is that "mitigation" is directed at shaping human behavior to reduce the causes of climate change, while "adaptation" is directed at building abilities of the nature system to respond to climate change through adjustments.[16] The goal of climate change adaptation under the new paradigm embraces both reducing "harm", as well as, securing "benefits" from climate change.[17] Climate change adaptation will not be used to replace climate change mitigation; in other words, although mitigation and adaptation are different and sometimes competing objectives, mitigation still needs adaptation and vice versa.[18]

6.6. New Paradigm for New Legal Structure

Natural resources and pollution laws can be established on two different sets of values on different types of change: "restoration paradigm" or recovery approach predominantly used in pollution control management and pollution regulation, thus struggling to recover from the effects of climate change; and "preservation paradigm" or preserving approach reasonably applied for natural resources management and law which accept the realities of ecosystem change, thus trying to adapt for change. The restoration paradigm establishes discontinuity with the realities of ecosystem change deriving from climate change impacts. While the restoration paradigm

[15] Robin Kundis Craig, "'Stationarity Is Dead' — Long Live Transformation: Five Principles for Climate Change Adaptation Law," *Harvard Environmental Law Review*, 34(1) (2010), 14–15.

[16] *Ibid.*, 21.

[17] J.B. Ruhl, "Climate Change Adaptation and the Structural Transformation of Environmental Law," *Environmental Law Review*, 40 (2010), 376, 383.

[18] *Ibid.*, 369.

is the basis of mitigation law, the preservation paradigm is the basis of adaptation law. The former is not considered as an appropriate tool for dealing directly with climate change adaptation. The latter is regarded as a new paradigm referring to an essential factor for making a new law for climate change adaptation. However, the adaptation law will not replace the mitigation law. Both need to coexist within the larger scope of climate change management law.[19]

Based on different values, mitigation law, while recognizing legal stationarity and adaptation law while responding to ecological dynamism and legal transformation, must use different types of tools. Unfortunately, the acknowledgement of this difference in Thai laws is indiscernible. Moreover, understanding of most policy-makers, decision-makers, regulators, and courts about the "realities of ecosystem change" is relatively limited.

Following are examples of the understanding that needs to be promoted:

> Ecological change can be both controllable and uncontrollable by humans over a particular human timescale, and thus, in some cases, it is irreversible. Ecological change may not be predictable especially when there are external factors. To elaborate new adaptation laws under the new paradigm, the "adaptive capacity" of humans, society, species, and the ecosystem needs to be increased for continuing in the new human society, natural world, and socio-ecological systems. Making law reforms by introducing adaptation law for answering the continuity between the human and the nature under climate change pressure is therefore much more complicated than making mitigation law. Such challenges as poor understanding, unpredictability, non-linear changes, adaptive management, and complex systems will create an uncomfortable feeling among people involved in the law reform process.

6.7. Legal Structure for New Adaptation Law

To respond to the complexity of the new paradigm, all possible measures and tools reflecting the new paradigm should be used with good integration: political will in initiating the climate change adaptation and the adaptation law, policies, plans, strategies, decision-making process, executive orders, judicial reviews, economic instruments, awareness building and education

[19]Craig, supra note 7, pp. 32–35.

through research and development and information base support, best practices and voluntary promotion, including law amendments in responding to the new culture of thinking or the new paradigm. Such law amendments must interact with other fields of law and policy such as national interest, national security, human rights, public health, international trade, etc.

"Adaptive capacity" should be put as the *goal* of climate change adaptation law. According to the International Panel on Climate Change/IPCC, adaptive capacity means:[20]

> The ability of systems, institutions, humans, and other organisms to adjust to potential damage, to take advantage of opportunities, or to respond to consequences.

While IPCC defines "adaptation" as:[21]

> The process of adjustment to actual or expected climate and its effects. In human systems, adaptation seeks to moderate harm or exploit beneficial opportunities. In natural systems, human intervention may facilitate adjustment to expected climate and its effects.

The *scope* of the climate change adaptation law should be well-designed by differentiating between non-climate change stressors, such as pollution, habitat destruction, and loss of biodiversity, and climate change stressors because existing laws already cope with the non-climate change stressors. While the existing laws, such as natural resources laws, pollution control laws, and environmental law, need to be recognized and coordinated, the climate change adaptation law must fill the gaps and incorporate, by ensuring that other laws will not undermine the adaptive capacity goal. Law reformers need to consider the broad range of relevant laws, make coordination among them, and set priority in case there are some overlapping areas of different management regimes.[22]

The *character* of the climate change adaption law should be in the form of policy law or framework law. Policy law is mostly used for dealing with

---

[20]"IPCC WGII AR5 Glossary," p. 2. Retrieved 15 March 2015, from http://ipcc-wg2.gov/AR5/images/uploads/WGIIAR5-Glossary_FGD.pdf.

[21]*Ibid.*, p. 1.

[22]Craig, supra note 7, pp. 43–44, 60.

complicated issues, embracing both law and policy to be closely linked and work more efficiently together. Framework law normally provides general terms and overall principles, sets clear objectives and guidelines, lays down a framework, and leaves more details to the government, other public bodies, or other administrative authorities to spell out regulations, as a form of delegation of legislative powers. While the policy law endorses the complex character of the climate change adaptation law, the framework law facilitates the flexibility character of the climate change adaptation law. The two can be made out of choice or be combined depending on the legal system, administrative structure, as well as social, economic, political, and socio-ecological conditions of each society.

The provisions of the climate change adaptation law may be elaborated in more details to achieve the goal of such law. Examples include:

- Provisions on information, monitoring, and research for increasing knowledge, building adaptive capacity, coping with uncertainty, avoiding political or popular resistance, and initiating innovative measures for climate change adaptation.[23]
- Provisions on duties in producing reports on adaptation planning and adaptation results at the national level on a regular basis such as every two or five years.[24]
- Provisions on requirements for climate change adaptation statements at the project level together with or separate from the environmental impacts assessment.[25]

6.8. Conclusion

In balancing between "climate change mitigation law" and "climate change adaptation law" in the Thai legal system, it requires law reform because the existing laws — natural resources laws, pollution laws, and environmental laws — cannot respond to uncontrollable impacts from climate change. Using mitigation law to deal with the uncontrollable climate change impacts

[23] *Ibid.*, 42–43.
[24] Daniel A. Farber, *A Legal Framework for Climate Adaptation Assessment* (Washington, DC: Resources for the Future, 2009), Issue Brief 09–14, p. 7.
[25] *Ibid.*, p. 1.

is not possible since it is based on the restoration paradigm. A new paradigm on preservation is thus introduced through the introduction of the climate change adaptation law. This knowledge is necessary before the law reform actually starts. With the support of the new paradigm understanding, new adaptation law can be systematically constructed. However, clear goals, scope, character, and content are also essential components.

In achieving the law reform, several missions must be conducted. *Firstly,* a clear goal must be set that "adaptation" is necessary to be incorporated in the present legal system, for which "adaptive capacity" is the key element in fulfilling adaptation, and that "adaptation" will not replace "mitigation". *Secondly*, a clear scope must be drawn for identifying differences between "adaptation" and "mitigation", as well as for linking interdependence between the two. Also, the linkage must be made between the "climate change law" and the "non-climate change law". *Thirdly*, a clear character of the new law must be designed either as "policy law" for capturing its complex character, or as "framework law" for responding to its flexible character, or both. *Lastly,* clear content must be identified and spelled out for covering all issues and aspects of the new law.

References

Craig, Robin Kundis. "'Stationarity Is Dead' — Long Live Transformation: Five Principles for Climate Change Adaptation Law," *Harvard Environmental Law Review*, 34(1) (2010), 9–75.

Farber, Daniel A. *A Legal Framework for Climate Adaptation Assessment* (Washington, DC: Resources for the Future, 2009), Issue Brief 09–14, pp. 1–11.

"IPCC WGII AR5 Glossary," p. 2. Retrieved 12 March 2015, from http://ipcc-wg2.gov/AR5/images/uploads/WGIIAR5-Glossary_FGD.pdf.

Nirawan Pipitsombat. "Thailand Climate Policy: Perspectives beyond 2012." Retrieved 12 March 2015, from http://eeas.europa.eu/delegations/thailand/documents/thailande_eu_coop/environment_energy/onep_climate_policy_en.pdf.

"Order of the Prime Minister on Climate Change B.E.2550," (2007).

"Royal Decree on Thailand Greenhouse Gas Management Organization (Public Organization) Establishment B.E.2550," (2007).

Ruhl, J.B. "Climate Change Adaptation and the Structural Transformation of Environmental Law." *Environmental Law Review*, 40 (2010), 343.

Tubtim Limsoontorn. "National Strategy on Climate Change Management: Modeling and Data Application." Retrieved 12 March 2015, from http://file:///D:/my%20documents/NUS/%E0%B8%82%E0%B8%AD%E0%B8%87%E0%B9%84%E0%B8%97%E0%B8%A2/National%20Strategy%20on%20Climate% 20Change.pdf.

Chapter 7

The Warsaw International Mechanism: Exploring the Structures and Functions to Address Loss and Damage Associated with Climate Change Impacts

*M. Hafijul Islam Khan**

7.1. Introduction

The loss and damage resulting from adverse impacts of climate change is now a reality. In the absence of inadequate efforts of mitigation and adaptation, the adverse impacts of climate change are causing harm to the lives and livelihoods of millions of people all over the world and inflicting significant economic and noneconomic losses and damages particularly in developing countries.[1] Therefore, climate change-induced loss and damage have imposed additional challenges for the global community, to institute an appropriate mechanisms to deal with such losses and damages related to life and livelihood, property, environment and ecology, values, culture, heritage, and territory. But conceptualization and identification of the nature of losses and damages associated with climate change impacts is also a challenge, and the prerequisite for ascertaining appropriate approaches to address such losses and damages.

Some of the recent research studies provide the empirical data, which sustain the evidence on losses and damages resulting from climate change impacts,[2] and also attempt to define and identify the nature of loss and

*An environmental lawyer and Executive Director, Centre for Climate Justice-Bangladesh. The author would like to acknowledge and thank Ms. Sharaban Tahura Zaman, Research Assistant, Centre for Climate Justice-Bangladesh, for assisting him to review the relevant decisions of UNFCCC on Loss and Damage.

[1]Khan (2013).

[2]Warner and Van der Geest (2013).

damage associated with climate change impacts. One of the working papers provided a definition of loss and damage as, "*the negative effects of climate variability and climate change that people have not been able to cope with or adapt to*".[3] In particular, "damage" on the one hand, which can be put on a level with tort, describes harming climate change impacts afflicting a person or entity possible to repair or rebuild. "Loss", on the other hand, can be understood as harming climate change impacts not possible to repair or rebuild.[4] So, loss and damage in nature can be permanent and temporary, while damage can be recovered with corrective measures and loss needs to be addressed with remedial measures.

Moreover, there are three types of losses and damages identified as avoided, unavoided, and unavoidable. Avoided loss and damage is used to characterize the impacts of climate change that are avoided by mitigation and adaptation. Unavoided loss and damage could have been avoided, but have not been avoided because of inadequate mitigation and adaptation efforts. Lastly, there is some loss and damage that is unavoidable no matter how ambitious mitigation and adaptation efforts are.[5] So, mitigation and adaptation as to proactive response, can avoid the potential loss and damage, whereas remedial approaches are needed for unavoidable losses and damages with regard to reactive responses. So, loss and damage resulting from climate change impacts go beyond adaptation, which cannot be dealt with adaptation measures and needs a separate approach. In addition, in terms of risk management, some of the approaches including risk retention and risk transfer are identified within the proactive responses to loss and damage.

The losses and damages resulting from the adverse impacts of climate change include those from sudden onset events and slow onset processes of climate change. While slow onset events like extreme events are difficult to attribute to climate change, the risks of some climate-related events, such as heat waves, extreme precipitation, and coastal flooding are already moderate and are expected to increase as temperatures rise.[6] A sudden

---

[3]Warner *et al.* (2012).

[4]Schäfer and Kreft (2014).

[5]Verheyen (2012).

[6]IPCC (2014).

onset event may be a single, discrete event that occurs in a matter of days or even hours, whereas slow onset events evolve gradually from incremental changes occurring over many years or from an increased frequency or intensity of recurring events.[7] For instance, high temperatures and sea level rise will lead to uninhabitable and unproductive territories. As the global temperature increases, there will be a reduction in moisture of soil, rendering land arid for cultivation in entire regions, groundwater aquifers in coastal areas will become too saline to consume, and so on. Consequently, these will result in permanent and irreversible loss and damage on assets and livelihoods in various regions, predominantly in developing countries.[8]

Loss and damage emanating from climate change impacts can be economic in nature, such as loss of income or damage to property and assets, and non-economic, which include the cultural, social, and mental impacts of climate change, as well as the loss of biodiversity and ecosystem services, amongst others.[9] According to a technical paper on non-economic losses as prepared in accordance with Decision 3/CP.18, para, 10(b), non-economic losses and damages are classified as those "not commonly traded in markets" as opposed to economic impacts for which a market price can be attributed (UNFCCC, 2013a). As such, these types of losses and damages are a subset from the total damage cause by environmental shocks and climate change.

Thus, the nature and the contexts of loss and damage associated with climate change impacts require appropriate assessment tools and methodologies for conceptualization and identification of such losses and damages, which is also the prerequisite for ascertaining appropriate approaches to address losses and damages. Understanding of nature of losses and damages and identification of appropriate approaches need to be employed with effective institutional arrangements under appropriate legal and policy mandates. The Loss and Damage Work Programme was established under the United Nations Framework Convention on Climate Change (UNFCCC, also known as "the Convention") in 2010,[10] which

[7]Siegle (2012).
[8]Balogun (2013).
[9]Morissey and Oliver-Smith (2013).
[10]Decision 1/CP.16, para. 26.

explored the different tools and methodologies for assessment of loss and damage and approaches and mechanisms to address loss and damage. The outcomes of the activities of the two year Loss and Damage Work Programme influenced the Parties of the Convention to establish the Warsaw International Mechanism (WIM) for Loss and Damage in 2013.

The WIM for Loss and Damage was established at the 19th Conference of the Parties (COP19) of the Convention in 2013 to deal with loss and damage associated with climate change in a comprehensive, integrated, and coherent manner under the guidance of the Convention.[11] Consequently, it also was mandated to complement, draw upon the work of and involve, as appropriate, existing bodies and expert groups under the Convention as well as outside the Convention at all levels (national, regional, and international levels). Nevertheless, further development of the comprehensive structures, functions and appropriate mechanisms of the WIM is a major challenge for the Parties of the Convention, taking into account the demands for broad, system-changing solutions to the climate crisis and to address the loss and damage associated with climate change.

This chapter intends to provide related policy analysis on the establishment of the Warsaw International Mechanism (WIM) and to explore the potential functions, structures, and mechanisms of the WIM to deal with loss and damage resulting from climate change with a view to influence the ongoing decision-making process of the Convention. Initially, this chapter provides a contextual evolution of the decision for establishing the WIM for loss and damage and attempts to identify the issues and challenges involved with assessment and approaches to address loss and damage. The second part of the chapter attempts to identify the functions of WIM with a view of exploring the potential structures of WIM. The concluding section provides some insights on structures and functions of the WIM and proposes some policy guidance for establishing national, regional, and international mechanisms to deal with loss and damage associated with the adverse impacts of climate change.

[11] Decision 2/CP.19.

7.2. Brief Overview of UNFCCC Process and Establishment of WIM

The First Assessment Report of the Intergovernmental Panel on Climate Change (IPCC) was released in 1990 (IPCC, 1990), which influenced the global political community to take the initiative to adopt a policy framework for climate change. Consequentially, the Intergovernmental Negotiating Committee (INC) was formed in 1991[12] and the committee worked for 15 months on negotiating the UN Framework Convention on Climate Change ("the Convention"), which was adopted in 1992.[13] During the negotiation period of the Intergovernmental Negotiating Committee on adopting the UNFCCC, Vanuatu on behalf of the Alliance of Small Island States (AOSIS) proposed for the establishment of an International Climate Fund to counter the adverse consequences of climate change, and a separate International Insurance Pool to provide financial insurance against the consequences of sea level rise.[14] However, the insurance mechanism proposed by Vanuatu, was not incorporated into the Convention but concentrated on Greenhouse Gas (GHG) emissions reduction and also on adaptation to climate change.[15]

The Convention, adopted in 1992, recognized the largest historical and current global emissions of GHGs by industrialized countries and obliges developed countries to take measures related to mitigation and adaptation and also to assist the developing countries through technical and financial resources, particularly those are vulnerable to the adverse effects of climate

[12]The Intergovernmental Negotiating Committee was formed through UN Reolution 45/212, 1990.

[13]The United Nations Framework Convention on Climate Change (UNFCCC) was adopted in 1992 and entered into force in 1994.

[14]INC (1991), Vanuatu: Draft annex relating to Article 23 (Insurance) for inclusion in the revised single text on elements relating to mechanisms (A/AC.237/WG.II/Misc.13) submitted by the Co-Chairmen of Working Group II. Intergovernmental Negotiating Committee for a Framework Convention on Climate Change, Working Group II, Fourth Session, Geneva, 9–20 December 1991. Retrieved 26 March 2015, from http://unfccc.int/resource/docs/a/wg2crp08.pdf, accessed 10.06.14.

[15]Article 2, UNFCCC, 1992.

change.[16] Furthermore, the Parties of the Convention adopted the Kyoto Protocol in 1997 with legally binding commitments by developed countries to reduce GHGs emissions, and also to finance adaptation measures.[17] However, in absence of adequate efforts of mitigation by the developed countries and the influence of the Fourth Assessment Report of IPCC, 2007, which stated that mitigation efforts were insufficient to avoid all of the impacts of climate change, Parties of the Convention started to negotiate adaptation framework through adopting the Bali Action Plan (BAP).[18] In 2007, under the BAP, Parties of the Convention agreed at the Thirteenth Conference of the Parties (COP13) to enhance action on adaptation, and the concept of loss and damage associated with climate change was addressed within the adaptation framework. Two years later, at COP16 in Cancun, a Work Programme was established to consider approaches to address loss and damage from the impacts of climate change in developing countries that are particularly vulnerable to such adverse effects.[19] Thereafter, loss and damage associated with climate change impacts has been framed as a "third pillar" of negotiations, paralleling mitigation and adaptation, even within the adaptation framework.

The decision of COP17 in 2011 provided further guidance to advance the work program through identifying three thematic areas include assessing the risk of loss and damage, identifying a range of approaches to address loss and damage and determining the role the Convention in enhancing the implementation of approaches to address loss and damage.[20] It was also agreed to pursue a set of activities for each of the thematic area. Moreover, the same decision appreciated the need to explore a range of possible approaches and potential mechanisms, including an international mechanism to address loss and damage and to make recommendations to COP18 for its consideration in making further decisions.[21] The inclusion of an "international mechanism" in the COP17 decision marked an important

[16] Article 4, UNFCCC, 1992.

[17] The Kyoto Protocol was adopted in 1997, at COP3.

[18] Decision 1/CP.13.

[19] Decision 1/CP.16, para 25, 26.

[20] Decision 7/CP.17.

[21] Para 5, Decision 7/CP.17.

window of opportunity for the development of an institutional structure for addressing loss and damage. Consequently, Parties of the Convention, at COP18, acknowledged the necessity of strengthening institutional arrangements at the national, regional, and international levels and also decided to establish institutional arrangements, such as an international mechanisms, to address loss and damage in developing countries that are particularly vulnerable to the adverse effects of climate change. The said decision also mandated the establishment of such institutional arrangements and its functions and modalities at COP19.[22] So, in accordance with the mandate of paragraph 9, of the Doha Decision (COP18),[23] Parties of the Convention established the Warsaw International Mechanism (WIM) to address loss and damage associated with climate change impacts, including extreme events and slow onset events, in developing countries that are particularly vulnerable to the adverse effects of climate change, at COP19 held in 2013 in Warsaw.[24]

7.3. Contextual Evolution of the Warsaw International Mechanism for Loss and Damage: Issues and Challenges

The precedent section provided a brief background of the decision related to the establishment of the WIM for Loss and Damage in the context of the climate change regulatory regime. But to frame the appropriate structures and functions of the WIM it is necessary to identify the issues and challenges involved with addressing loss and damage. This section attempts to examine the decisions of the Convention related to loss and damage critically with a view of identifying the issues and challenges for the WIM in addressing loss and damage, which ultimately will sketch the potential structures and functions of the WIM.

[22]Decision 3/CP18, paras. 7, 8 and 9.

[23]Decision, 3/CP18, para. 9: *Decides* to establish, at its nineteenth session, institutional arrangements, such as an international mechanism, including functions and modalities, elaborated in accordance with the role of the Convention as defined in paragraph 5 above, to address loss and damage associated with the impacts of climate change in developing countries that are particularly vulnerable to the adverse effects of climate change.

[24]Decision 2/CP.19, para. 1.

7.3.1. *The Bali Action Plan and loss and damage*

Parties of the Convention adopted the Bali Action Plan (BAP) at COP13, held in 2007,[25] and agreed to enhance action on adaptation. The issue of loss and damage associated with climate change was addressed within the adaptation framework under the BAP and called for *"[r]isk management and risk reduction strategies, including risk sharing and transfer mechanisms such as insurance"* as well as *"[d]isaster reduction strategies and means to address loss and damage associated with climate change impacts in developing countries that are particularly vulnerable to the adverse effects of climate change"* (emphasis added).[26] It is worth mentioning that this call was made based on Article 4.8 of the Convention, which referred to the insurance as a tool to "meet the specific needs and concerns of developing country parties arising from the adverse effects of climate change".[27] Following the BAP, an Ad Hoc Working Group on Long-Term Cooperative Action (AWG–LCA), was formed and the issue of loss and damage was advanced further under adaptation negotiations. At COP14 in Poznan, AOSIS, proposed for a Multi-Window Mechanism to Address Loss and Damage with the following components:[28]

- A *risk management and prevention component* to promote risk assessment and risk management tools and strategies at all levels, with a view to facilitating and supporting the implementation of risk reduction and risk management measures;
- An *insurance component* to address climate-related extreme weather events, and risks to crop production, food security, and livelihood; and
- A *rehabilitation and compensation component* to address progressive negative impacts that result in loss and damage.

However, the aforesaid proposal for a Multi-Window Mechanism was not considered at COP14, and not even at the next COP15 held

[25]Decision 1/CP.13: At COP13 held in December 2007, by its decision 1/CP.13, decided to launch a comprehensive process to enable the full, effective, and sustained implementation of the Convention through long-term cooperative action, up to and beyond 2012.

[26]Decision 1/CP 13, para. 1 c.

[27]Article 4.8, UNFCCC, 1992.

[28]AOSIS (2008).

in Copenhagen in 2009. However, COP16 decided to establish a Work Programme on Loss and Damage, which led to the adoption of the WIM for loss and damage.

7.3.2. *Work programme on Loss and Damage*

The Decision of COP16 (known as the Cancun Agreements) recognized the need to strengthen international cooperation and expertise in order to understand and reduce loss and damage associated with the adverse effects of climate change. The decision of COP16 considered the loss and damage associated with adverse impacts related to extreme weather and slow onset events. A footnote clarifies what events fall under slow onset events: "sea level rise, increasing temperatures, ocean acidification, glacial retreat and related impacts, salinization, land and forest degradation, loss of biodiversity and desertification".[29] Moreover, the decision of COP16 launched a work program to "consider including through workshops and expert meetings, as appropriate, approaches to address loss and damage associated with climate change impacts in developing countries that are particularly vulnerable to the adverse effects of climate change".[30] COP16 then mandated the Subsidiary Body for Implementation (SBI) to: (1) agree on the activities of the work program; and (2) make recommendations on loss and damage for COP18 to decide further in future. Consequently, Parties of the Convention agreed on three thematic areas for the implementation of the work program at SBI 34, held in June 2011 as follows:

- Thematic area 1: Assessing the risk of loss and damage associated with the adverse effects of climate change and the current knowledge on the same;
- Thematic area 2: A range of approaches to address loss and damage associated with the adverse effects of climate change, including impacts related to extreme weather events and slow onset events, taking into consideration experience at all levels;

[29]Decision 1/CP. 16, para. 25.

[30]Para. 26, Decision 1/CP. 16.

- Thematic area 3: The role of the Convention in enhancing the implementation of approaches to address loss and damage associated with the adverse effects of climate change.

At COP17 held in 2011, Parties of the Convention decided to work further on agreed thematic areas in SBI 34 and requested the SBI to continue the implementation of the work program. The COP17 decision also provided the required guidance to advance the work program through a set of activities related to each of the thematic areas.[31] Decision 7/CP.17 also appreciated the need to explore a range of possible approaches and potential mechanisms, including an international mechanism, to address loss and damage, with a view to making recommendations on loss and damage to consider at COP18 for further decision.[32]

The activities of the work program on loss and damage provided an avenue to identify the issues and challenges related to assessing the risks of loss and damage, and exploring a range of approaches to address loss and damage. It also provided guidance on the potential role of the Convention in addressing loss and damage through different activities undertaken in the whole year of 2012. The activities include expert meetings and technical papers of the work program and produced a significant amount of knowledge, information, and technical products, which influenced the Parties of the Convention to adopt the Doha Decision on loss and damage at COP18.[33] Parties of the Convention at COP18 recognized the need for comprehensive, inclusive, and strategic responses to loss and damage and agreed that the Convention has a role in promoting the implementation of approaches to address loss and damage as follows:

(1) Enhancing knowledge and understanding of comprehensive risk management approaches to address loss and damage associated with the adverse effects of climate change, including slow onset impacts;
(2) Strengthening dialogue, coordination, coherence, and synergies among relevant stakeholders;

[31] Decision 7/CP.17, para. 1, 8–7.
[32] Para. 5, Decision 7/CP.17.
[33] Decision 3/CP 18.

(3) Enhancing action and support, including finance, technology, and capacity building to address loss and damage associated with the adverse effects of climate change.[34]

The Doha Decision acknowledged the need for further work to advance the understanding of and expertise on loss and damage and implied the country actions to enhance action on addressing loss and damage by assessing the risk of loss and damage and enhancing access to sharing and use of related data, identifying options, designing and implementing comprehensive country driven risk management strategies and approaches as we systematically observe and collect data on climate change impacts, and involving vulnerable communities and populations as well as stakeholders.[35] Moreover, the Doha Decision acknowledged the need for enhancing support including finance, technology, and capacity building for undertaking the relevant actions by the developing countries.[36]

Most importantly, the Doha Decision acknowledged the necessity of strengthening institutional arrangements at national, regional, and international levels to address loss and damage associated with the adverse effects of climate change.[37] The Decision also agreed to establish institutional arrangements, such as an international mechanism, to address loss and damage in developing countries that are particularly vulnerable to the adverse effects of climate change. The said Decision mandates the establishment of such institutional arrangements and their functions and modalities in accordance with the UNFCCC's role as defined in paragraph 5 of the same decision at COP19.[38] It is worth mentioning that Doha Decision appreciated the progress of activities of the work program and acknowledged the importance of the continuation to address the loss and damage associated with the adverse effects of climate change.[39] During the course of 2013, further technical works and a mapping of existing institutional arrangements to address loss and damage were carried out

[34]Para. 5, Decision 3/CP 18.

[35]Paras. 6 and 7, Decision 3/CP 18.

[36]Para. 9, Decision 3/CP 18.

[37]Para. 7, Decision 3/CP 18.

[38]Para. 9, Decision 3/CP 18.

[39]Preamble, Decision 3/CP 18.

under the continuation of the work program, which influenced the Parties of the Convention to decide at COP19 on loss and damage. At COP19, the WIM for Loss and Damage associated with climate change impacts was established.

7.4. Structure and Functions of the WIM

In accordance with the Mandate of paragraph 9 of the Doha Decision (COP18), Parties of the Convention established the WIM at its 19th session held in Warsaw, 2013 (COP19) to address loss and damage associated with impacts of climate change, including extreme events and slow onset events, in developing countries that are particularly vulnerable to the adverse effects of climate change.[40] However, this mechanism, established under the Cancun Adaptation Framework, is subject to review including its structure, mandate, and effectiveness by COP22 and based on the outcomes of such further review an appropriate decision will be taken by the Parties of the Convention.[41]

An Executive Committee of the WIM is suggested to be established in order to guide the implementation of functions indentified in paragraph 5 of the same decision under the guidance of and accountable to COP.[42] But the Executive Committee is requested to report annually to the COP through the Subsidiary Body of Scientific Technological Advice (SABSTA) and the Subsidiary Body for Implementation (SBI), and also requested to make recommendations as appropriate.[43] COP19 decided to form an interim Executive Committee which will consist of two representatives from each of the following bodies under the Convention, ensuring a balanced representation between developed and developing country Parties: the Adaptation Committee, the Least Developed Countries Expert Group, the Standing Committee on Finance, the Technology Executive Committee, and the Consultative Group of Experts on National Communications from Parties not included in Annex I to the Convention.[44] The same decision

[40]Decision 2/CP.19, para. 1.
[41]Decision 2/CP 19, paras. 1 and 15.
[42]Para. 2, Decision 2/CP 19.
[43]Para. 3, Decision 2/CP 19.
[44]Para. 4, Decision 2/CP 19.

requested the Executive Secretary of UNFCCC to convey the initial meeting of the executive committee by March 2014, in consultation with the COP President.[45]

The interim Executive Committee will develop a two-year workplan to implement its functions outlined in paragraph 5 of the same decision taking into account the issues outlined in decision 3/CP.18, paragraphs 6 and 7, for consideration at the 41st session of SBSTA and SBI to be held in December 2014.[46] Moreover, the SBSTA and SBI are requested to consider the composition of, and procedures for, the Executive Committee, and to make recommendations thereon for adoption by the Conference of the Parties at its 20th session (COP20), with a view to finalizing the organization and governance of the executive committee.[47]

7.4.1. *Interim executive committee of WIM*

The initial meeting of the interim Executive Committee was held from 25 to 28 March 2014 in Bonn, Germany and initiated their considerations for developing the two-year workplan for Executive Committee of WIM for Loss and Damage in accordance with the mandate of decision 2/CP.19. In addition to the 10 members of the interim Executive Committee, 47 observers from Parties and relevant organizations and stakeholders, engaged in work related to addressing loss and damage, attended the initial meeting. The initial meeting of the interim Executive Committee concluded with clustering the diverse range of elements into possible action areas together with specific activities of the two-year workplan and agreed to resume at a later date to complete the remaining work of the workplan.

Later, in June, 2014, during the SBs 40, held in Bonn, Germany, the members of the interim Executive Committee shared information on the process to fulfill its mandate of developing the two-year workplan and called for inputs for the Executive Committee's draft workplan. In response to the said call of the interim Executive Committee, Parties of the Convention and other related organizations provided their inputs and

---

[45]Para. 8, Decision 2/CP 19.

[46]Para. 9, Decision 2/CP 19.

[47]Para. 10, Decision 2/CP 19.

the interim Executive Committee analyzed these inputs with a view of consolidating them with the outcomes of the first part of the initial meeting. The second meeting of the interim Executive Committee held on 17 and 18 September, in Bonn, Germany, was opened by introducing the draft initial two-year workplan for further consultations with the observers with a view of finalizing the workplan. The second meeting of the interim Executive Committee was resumed through adopting the revised final version of the workplan.[48]

7.4.2. *Permanent executive committee of WIM*

In accordance to the decision of COP19, SBSTA and SBI jointly recommended the COP on approval of the two-year workplan developed by the interim Executive Committee and three different proposals on the composition of the permanent Executive Committee. At the end, the Parties of the Convention approved the initial two-year workplan of the Executive Committee of the WIM for Loss and Damage associated with Climate Change Impacts[49] and decided on the composition of the Executive Committee of 20 members — 10 members from Parties included in Annex I to the Convention (Annex I Parties) and 10 members from Parties not included in Annex I to the Convention (non-Annex I Parties).[50] Eight of the non-Annex I Party members are stipulated in the decision: two each from the African, Asia-Pacific, and the Latin American and Caribbean States, and one each from Small Island Developing States (SIDS) and Least Developed Country Parties (LDCs). The two remaining non-Annex 1 slots need to be decided further by non-Annex I Party members.

It is worth mentioning that the Parties of the Convention decided that the Executive Committee shall be composed of the aforementioned regional representations taking into account the goal of gender balance pursuant to decision 23/CP.18. The same decision also encouraged Parties to nominate to the Executive Committee a member who has the expertise

[48]See FCCC/SB/2014/4, "Report of the Executive Committee of the Warsaw International Mechanism for Loss and Damage Associated with Climate Change Impacts."

[49]Decision 2/CP 20, para. 1.

[50]Decision 2/CP 20, para. 5(b).

with the diversity of experience and knowledge relevant to loss and damage associated with climate change impacts.[51] It was further decided that members of the Executive Committee will serve for a term of two years and shall be eligible to serve a maximum of two consecutive terms of office. Half of the members shall be elected initially for a term of three years and half of the members shall be elected for a term of two years and thereafter, the COP shall elect members for a term of two years. The members shall remain in office until their successors have been elected.[52] It was further decided in COP20 that decisions of the Executive Committee shall be taken by consensus and that the members of the Executive Committee shall elect annually Co-Chairs from among its members to serve for a term of one year, with one being a member from an Annex I Party and the other being a member from a non-Annex I Party. It is also decided that, if one or both Co-Chairs are absent from a particular meeting, any other member designated by the Executive Committee shall temporarily serve as the Co-Chair or Chair of that meeting and if a Co-Chair is unable to complete the term of office, the Executive Committee shall elect a replacement to complete that term of office.[53]

Moreover, it is decided that the Executive Committee shall meet at least twice per year, while retaining its flexibility to adjust the number of meetings, as appropriate. It is also decided that the Executive Committee shall convene its first meeting as soon as it is practical, following the election of its members commencing at COP20, but no later than March 2015, and at its first meeting the Executive Committee shall adopt its rules of procedure and begin implementing its workplan approved by COP20.[54] Most importantly, it is decided that the Executive Committee may establish expert groups, subcommittees, panels, thematic advisory groups, or task-focused ad hoc working groups to help the Executive Committee to execute its work in guiding the implementation of the WIM.[55] This provision

[51] Para. 6, Decision 2/CP 20.

[52] Para. 7, Decision 2/CP 20.

[53] Paras. 9, 10 and 11, Decision 2/CP 20.

[54] Paras. 12 and 13, Decision 2/CP 20.

[55] Para. 8, Decision 2/CP 20.

provides further scope for enhancement of the structure of the Executive Committee of the WIM for Loss and Damage associated with climate change impacts.

7.4.3. *Functions of WIM*

The function of the WIM is to address loss and damage associated with impacts of climate change, including extreme events and slow onset events, in developing countries that are particularly vulnerable to the adverse effects of climate change, at COP19.[56] The Interim Executive Committee of WIM, established at COP19 as to the first function, developed a two-year workplan and the Permanent Executive Committee, established at COP20, has the responsibility of implementing this workplan. Parties of the Convention further mandated the Executive Committee to establish expert groups, subcommittees, panels, thematic advisory groups, or task-focused ad hoc working groups as appropriate to help the Executive Committee to execute its work. Thus, it needs to analyze the workplan to understand the function of the WIM by the next two years so that it can provide insights on further development of structuring the WIM.

In accordance with decisions 3/CP.18 and 2/CP.19, the Interim Executive Committee of the WIM developed the initial two-year workplan, where the specific activities and timeline are identified based on the nine action areas. The action areas include:[57]

Action area 1:

Enhance the understanding of how loss and damage associated with the adverse effects of climate change affect particularly vulnerable developing countries, segments of the population that are already vulnerable owing to geography, socio-economic status, livelihoods, gender, age, indigenous or minority status or disability, and the ecosystems that they depend on, and of how the implementation of approaches to address loss and damage can benefit them.

---

[56] Decision 2/CP.19, para. 1.

[57] See FCCC/SB/2014/4, "Report of the Executive Committee of the Warsaw International Mechanism for Loss and Damage Associated with Climate Change Impacts."

Action area 2:

Enhance the understanding of, and promote, comprehensive risk management approaches (assessment, reduction, transfer, retention), including social protection instruments and transformational approaches, in building long-term resilience of countries, vulnerable populations and communities.

Action area 3:

Enhance data on and knowledge of the risks of slow onset events and their impacts, and identify ways forward on approaches to address slow onset events associated with the adverse effects of climate change with specific focus on potential impacts, within countries and regions.

Action area 4:

Enhance data on and knowledge of non-economic losses associated with the adverse effects of climate change and identify ways forward for reducing the risk of and addressing non-economic losses with specific focus on potential impacts within regions.

Action area 5:

Enhance the understanding of the capacity and coordination needs with regard to preparing for, responding to, and building resilience against loss and damage associated with extreme and slow onset events, including through recovery and rehabilitation.

Action area 6:

Enhance the understanding of and expertise on how the impacts of climate change are affecting patterns of migration, displacement, and human mobility; and the application of such understanding and expertise.

Action area 7:

Encourage comprehensive risk management by the diffusion of information related to financial instruments and tools that address the risks of loss and damage associated with the adverse effects of climate change to facilitate finance in loss and damage situations in accordance with the policies of each developing country and region, taking into account the necessary national efforts to establish enabling environments. These financial instruments and tools may include: comprehensive risk management capacity with risk pooling and transfer; catastrophe risk insurance; contingency finance; climate-themed bonds and their certification; catastrophe bonds; and financing approaches to making development climate resilient, among other innovative financial instruments and tools.

Action area 8:

Complement, draw upon the work of, and involve, as appropriate, existing bodies and expert groups under the Convention, as well as relevant organizations and expert bodies outside the Convention at all levels, as the Executive Committee executes the above-mentioned elements of the workplan.

Action area 9:

Develop a five-year rolling workplan for consideration at COP22 building on the results of this two-year workplan to continue guiding the implementation of the functions of the Warsaw International Mechanism.

Thus, the workplan identified nine action areas with 30 activities to deliver by the next two years. Some of the action areas suggested establishing a technical expert panel or group and the Executive Committee also got the mandate from the COP20 decision to establish such an expert panel or group to deliver its functions effectively. Also, the action area number nine mandates the Executive Committee to develop a five-year rolling workplan for consideration at COP22, building on the results of this two-year workplan to continue guiding the implementation of the functions of the WIM. Since COP20 approved this workplan, it has become the COP20 decision to develop a five-year rolling workplan for the Executive Committee. Moreover, challenges remains at COP22 to review the WIM including its structure, mandate, and effectiveness, and based on the outcomes of such further review an appropriate decision will be taken at COP22. Another important challenge related to loss and damage is to incorporate the issue in the Agreement to be adopted at COP21 in 2015. The COP20 decision related to the current negotiating text of the 2015 agreement referred to the WIM in its preamble and provided several options to address loss and damage in the 2015 Agreement.[58] Hence, there are many dimensions that need to be considered in addressing loss and damage in the UNFCCC processes.

7.5. Potential Functions and Structures of WIM for Loss and Damage

The first and foremost challenge of WIM for Loss and Damage is to conceptualize and to identify the nature of losses and damages associated with climate change impacts, including extreme events and slow onset events. The identification of the nature of losses and damages, including economic and non-economic, avoidable and unavoidable, repairable or irreversible permanent losses and damages is required for ascertaining appropriate response measures to address such losses and damages. Moreover, appropriate assessment tools and methodologies are needed to sort out the nature of loss and damage and the appropriate approaches need to be employed with effective institutional arrangements under appropriate legal and policy mandates.

However, the international climate regime, which began in 1992 with the adoption of the UNFCCC, is still struggling to set up governance mechanisms for mitigation and adaptation with respect to climate change. Even if adequate mitigation measures are taken now, given the levels of GHGs that have already been released into the atmosphere, some climate change impacts and the associated loss and damage are inevitable. While adaptation measures can reduce loss and damage resulting from climate change to some extent, there will be a certain unavoidable degree of loss and damage, for which a separate framework is needed (Khan, 2013). Nonetheless, Parties of the Convention, particularly states from developing countries negotiated enthusiastically for the structure and functions of WIM, which still remains within adaptation framework. But, further negotiation efforts of the Parties and the identification of appropriate functions and structure of the WIM will determine the future institutional frameworks for loss and damage resulting from climate change. Thus, at

[58] Decision 2/CP.20.

this stage, it is important to identify the appropriate functions and structure of the WIM.

The WIM was mandated to promote the implementation of approaches to address loss and damage associated with the adverse effects of climate change by enhancing knowledge and undertaking comprehensive risk management approaches to address loss and damage, strengthening dialogue, coordination among relevant stakeholders, enhancing action, and support that included finance, technology, and capacity building to address loss and damage. It was decided that the WIM shall fulfill its role under the Convention of promoting the implementation of approaches to address loss and damage associated with the adverse effects of climate change, pursuant to decision 3/CP.18, in a comprehensive, integrated, and coherent manner.[59] The WIM also mandated to complement, draw upon the work of, and involve, as appropriate, existing bodies and expert groups under the Convention as well as on that of relevant organizations and expert bodies outside the Convention, at all levels (national, regional, and international). Regarding activities of the WIM, it is decided that, in exercising the functions outlined in paragraph 5 of the same decision, the WIM will, inter alia: (1) facilitate support of actions to address loss and damage; (2) improve coordination of the relevant work of existing bodies under the Convention; (3) convene meetings of relevant experts and stakeholders; (4) promote the development of, and compile, analyze, synthesize, and review information; (5) provide technical guidance and support; and (6) make recommendations, as appropriate, on how to enhance engagement, actions, and coherence under and outside.[60]

The interim Executive Committee of WIM developed a two-year workplan, which identified nine action areas with 30 activities to be delivered by the next two years. Delivering some of the activities require establishing a technical expert panel or group under the Executive Committee. In addition, the Executive Committee is mandated to develop a five-year rolling workplan for consideration at COP22 based on the outcomes of this two-year workplan developed by the interim Executive Committee in order to continue the functions of the WIM. The structure and the functions of the

[59]Decision 2/CP.19, para. 5.

[60]Para. 7, Decision 2/CP.19.

WIM will be reviewed at COP22 and further appropriate decisions on the WIM will be taken in COP22. Moreover, the issue of loss and damage must be a separate stream going beyond the adaptation framework and the 2015 agreement can take this holistic approach. The draft negotiation text for the 2015 agreement with the COP20 decision provides different options for addressing loss and damage and the upcoming 2015 agreement can incorporate the basic elements of the structures and functions of the WIM to deal with the loss and damage. Thus, taking into account the different contexts, the Parties of the Convention need to consider the following issues for the structure and functions of the WIM:

- Developing appropriate tools and methodologies to assess the loss and damage and ascertaining appropriate approaches including proactive and reactive responses to address loss and damage.
- Developing appropriate structures and functions of WIM.
- Developing compensation and liability regime for residual loss and damage.
- Developing a financial entity to address and to redress loss and damage.
- Establishing autonomous international mechanism for loss and damage beyond adaptation, ensuring clear linkages with regional and national mechanisms for loss and damage and the coordination with relevant other institutions.
- Incorporation of loss and damage as the separate stream in the 2015 agreement and designing appropriate mechanisms.

7.6. Concluding Remarks

The losses and damages related to life and livelihood, property, environment and ecology, values, culture, heritage, and territory result from climate change impacts, including from sudden onset events and slow onset processes. Careful, proactive policy can minimize the risks of potential loss and damage and can maximize the extent to which community resilience copes with climatic hazards. Mitigation is primary in this respect, with a secondary focus being placed on adequate adaptation measures that can prevent and reduce the loss and damage related to climate change. The reactive legal response with compensation and remedial approaches is also needed to redress the unavoidable climate change-induced loss and damage.

The WIM for Loss and Damage is established under the adaptation framework. But limits of adaptation are now scientifically evident and empirical data on loss and damage provides the strong argument for establishing an autonomous approach beyond adaptation to address loss and damage resulting from climate change impacts. The first and foremost challenge for WIM is to take an autonomous approach beyond adaptation to address loss and damage resulting from climate change impacts. However, the immediate challenge for the WIM is to implement a two-year workplan developed by the interim Executive Committee of WIM and to develop a rolling five-year workplan for the WIM. It is worth mentioning that activities identified in the initial two-year workplan also suggested establishing a technical expert panel or group, which led the Parties of the Convention to mandate the Executive Committee of WIM to establish such an expert panel or group to deliver its functions effectively.

Thus, the Executive Committee can establish a technical facility to implement the activities related to understanding the nature and the contexts of loss and damage associated with climate change impacts through developing appropriate assessment tools and ascertaining appropriate approaches to address loss and damages. Understanding the nature of losses and damages and identifying of appropriate approaches to address loss and damage will create a path for developing an autonomous international mechanism with micro-level national mechanisms so that a bottom-up approach can assess and redress the loss and damage associated with adverse impacts of climate change. On the other hand, the Executive Committee of WIM can establish a financial facility to develop and manage the dedicated funding sources of the WIM.[61]

Nevertheless, successful implementation of the two-year workplan by the permanent Executive Committee of the WIM with its two wings of technical and financial facilities will produce new knowledge and scientific bases to establish an autonomous approach with liability and compensation regime to address loss and damage resulting from climate change impacts. It is off course; the parties of the Conventions from LDCs, SIDSs, and other vulnerable states such as African countries will have to work together to

[61] See "Submission of Nauru on Behalf of The Alliance of Small Island States," 12 November 2013. Retreived 26 March 2015, from http://aosis.org/wp-content/uploads/2013/11/UNFCCC-Submission-Loss-and-Damage-Warsaw.pdf.

frame the effective structures and functions of the WIM by COP22 through the review process agreed at COP19. Another important challenge for loss and damage is to incorporate the issue in the Agreement to be adopted at COP21 in 2015. The COP20 decision related to current negotiating text of the 2015 Agreement referred to the WIM in its preamble and provided several options including liability and compensation to address loss and damage in the 2015 Agreement. Hence, the 2015 Agreement provides an avenue to address loss and damage with compensation and liability approaches considering the inadequacy of mitigation and adaptation efforts to climate change.

References

Al Faruque, Abdullah and M. Hafijul Islam Khan. "Loss and Damage Associated with Climate Change: The Legal and Institutional Context in Bangladesh." *Germanwatch*, (2012). Retrieved 1 June 2014, from http://www.lossanddamage. net/download/7114.pdf.

AOSIS. "Proposal to AWG-LWG: Multi-Window Mechanism to Address Loss and Damage from Climate Change Impacts." Retrieved 10 June 2014, from http://unfccc.int/files/kyoto_protocol/application/pdf/aosisinsurance061208.pdf.

Balogun, Kehinde. "Managing Loss and Damage from Slow Onset Events: Applicability of Risk Transfer Tools Including Insurance." *Germanwatch*, (2013). Retrieved 26 March 2015, from www.lossanddamage.net/download/727/.pdf.

INC. Vanuatu: Draft annex relating to Article 23 (Insurance) for inclusion in the revised single text on elements relating to mechanisms (A/AC.237/WG.II/Misc.13) submitted by the Co-Chairmen of Working Group II. Intergovernmental Negotiating Committee for a Framework Convention on Climate Change, Working Group II, Fourth Session, Geneva, 9–20 December 1991. Retrieved 2 June 2014, from http://unfccc.int/resource/docs/a/wg2crp08.pdf.

IPCC. "Climate Change 2007: Synthesis Report. Contribution of Working Groups I, II and III to the Fourth Assessment Report of the Intergovernmental Panel on Climate Change." Retrieved 13 June 2014, from, http://www.ipcc.ch/publications_and_data/publications_ipcc_fourth_assessment_report_synthesis_report.htm.

IPCC. "Climate Change 2014: Impacts, Adaptation and Vulnerability. IPCC Working Group II Contribution to AR5, Summary for Policymakers, Intergovernmental Panel for Climate Change," (2014). Retrieved 7 June 2014, from http://www.ipcc.ch.

IPCC. "Climate Change 2013: The Physical Science Basis. Summary for Policy-Makers. Contribution of Working Group I to the Fifth Assessment Report of the Intergovernmental Panel on Climate Change," (2013). Retrieved 1 June 2014, from https://www.ipcc.ch/report/ar5/wg1/.

Khan, Hafijul I. "Legal and Policy Responses to Loss and Damage Associated with Climate." In O. C. Ruppel, C. Roschmann and K. Ruppel-Schlichting (eds.), *Legal Responses and Global Responsibility* (Germany: Nomos Publishers, 2013).

Kreft, S. "Loss and Damage: Unpacking the Doha Decision 3/CP. 18. A Detailed Look at the Doha Decision to Address Loss and Damage Associated with the Adverse Impact of Climate Change." Retrieved 5 June 2014, from http://www.lossanddamage.net/download/7011.pdf.

Schäfer, Laura and Sönke Kreft. "Loss and Damage: Roadmap to Relevance for the Warsaw International Mechanism." *Germanwatch*, (2014). Retrieved 6 June 2014, from http://www.lossanddamage.net/download/7284.pdf.

Schwarte, Christoph with Ruth Byrne. "International Climate Change Litigation and the Negotiation Process." Working Paper. Foundation for International Environmental Law and Development, London, 2010.

Siegele, L. "Loss and Damage: The Theme of Slow Onset Impact." *Germanwatch*, (2012). Retrieved 25 March 2015, from http://germanwatch.orglen/download/6674.pdf.

UNFCCC. "A Briefing Note: Initial Meeting of the Executive Committee of the Warsaw International Mechanism for Loss and Damage Associated with Climate Change Impacts." Bonn, Germany, March 2014, pp. 25–28.

UNFCCC. "A Literature Review on the Topics in the Context of Thematic Area 2 of the Work Programme on Loss and Damage: A Range of Approaches to Address Loss and Damage Associated with the Adverse Effects of Climate Change." United Nations Framework Convention on Climate Change (UNFCCC), FCCC/SBI/2012/INF.14.

Verheyen, R. "Tackling Loss & Damage — A New Role for the Climate Regime?," (2012). Retrieved 2 June 2014, from http://www.lossanddamage.net/download/6877.pdf.

Verheyen, R. and Peter Roderick. "Beyond Adaptation: The Legal Duty to Pay Compensation for Climate Change Damage." WWF-UK Climate Change Programme Discussion Paper. Surrey, World Wildlife Fund — United Kingdom, 2008.

Warner, K. and S. A. Zakieldeen, "Loss and Damage Due to Climate Change. An Overview of the UNFCCC Negotiations," (2012). Retrieved 2 June 2014, from http://www.climateinsurance.org/upload/pdf/20120220_Loss_and_Damage_in_UNFCCC_warner_zakieldeen.pdf.

Warner, K. "The Warsaw International Mechanism: A Legitimate Policy Space for Loss and Damage Widens and Deepens." (2013). Retrieved 12 June 2014, from http://www.ehs.unu.edu/file/get/11501.pdf.

Warner, K. and K. van der Geest, "Loss and Damage from Climate Change: Local-Level Evidence from Nine Vulnerable Countries." *International Journal of Global Warming*, 5(4), 1–20.

Warner, K., K. van der Geest, and S. Kreft. "Pushed to the Limits: Evidence of Climate Change-Related Loss and Damage When People Face Constraints and Limits to Adaptation." Report no. 11, United Nations University Institute of Environment and Human Security (UNU-EHS), Bonn. Retrieved 10 April 2014, from http://www.ehs.unu.edu/.

Chapter 8

Geoengineering: An ASEAN Position

*Jolene Lin**

For a collection dedicated to exploring the challenges and successes of climate change adaptation, a paper on geoengineering may appear out of place. Such a view is not misguided because geoengineering is not conventionally recognized as a form of adaptation to the impacts of climate change. It is often viewed as a third option to complement mitigation and adaptation strategies. The main reasons behind the choice to focus on geoengineering in this chapter are: (1) Geoengineering is a controversial issue which raises significant governance and ethical issues; (2) The Association of Southeast Asian Nations (ASEAN) is not likely to become a significant player in the geoengineering research scene but will be affected by any large-scale research activity that has transboundary effects; (3) As such, this chapter seeks to use the opportunity to stimulate discussion about geoengineering from an ASEAN perspective.

This chapter focuses primarily on geoengineering research rather than deployment because the technical and policy debate about geoengineering is currently at the early stage at which the broad questions are whether research into geoengineering ought to be encouraged and how such research

---

*Associate Professor, Faculty of Law, The University of Hong Kong, email address: jolene@hku.hk. I am indebted to Richard L. Ottinger and Bill Pentland for making available to me their unpublished manuscript "Geoengineering: Too Hot to Handle? Proposing Limited Research" and encouraging me to use it as a platform for my research into this topic. Parts I and II of this paper rely heavily on this manuscript.

should be regulated.[1] The first part provides some background about geoengineering and surveys the two main categories of geoengineering proposals: solar radiation management (SRM) and carbon dioxide removal (CDR) methods. The second part surveys the adverse consequences that could result from deploying these methods. The third part examines existing efforts to regulate geoengineering research. The fourth part argues that even if ASEAN does not wish or does not have the capacity to become a significant player in geoengineering research, it is salient that ASEAN adopts an informed position on geoengineering because research in the form of field tests (and possible eventual deployment of geoengineering techniques) can have transboundary and global environmental and health impacts.

8.1. Background and Survey of Geoengineering Techniques

The term "geoengineering" does not have a standard definition. This chapter adopts the widely used definition of climate change geoengineering proposed by the United Kingdom's Royal Society: "the deliberate large-scale manipulation of the planetary environment to counteract anthropogenic climate change".[2] Geoengineering has become of interest because of the disappointingly slow pace of international negotiations to reduce greenhouse gas (GHG) emissions and growing concerns about climate change risks.[3] Efforts to reduce GHG emissions constituted the first global response. Subsequently, adaptation became a second legitimate category of action. Proposals to intervene on a massive scale in global physical, chemical, and biological systems to counteract climate change are now being discussed as an additional category of action to supplement mitigation and adaptation efforts or as a "Plan B" to "buy time" in which to reduce GHG emissions and to counterbalance the effects of climate change.

---

[1] See generally David R. Morrow *et al.*, "Towards Ethical Norms and Institutions for Climate Engineering Research," *Environmental Research Letters*, 4 (4) (2009), for distinction between research and deployment based on environmental impacts, timeline, and the "intentions of those carrying out the activity".

[2] The Royal Society, "Geoengineering the Climate: Science, Governance and Uncertainty," September 2009, Retrieved 25 March 2015, from http://www.royalsociety.org, p. 1.

[3] Steve Rayner *et al.*, "The Oxford Principles," *Climatic Change*, 121 (2013), 501.

In Ambassador Richard E. Benedick's words, "[c]limate remediation [or geoengineering] could therefore be considered as a possible supplementary action in the event that climate mitigation and climate adaptation measures prove insufficient".[4]

One approach to geoengineering, endorsed by the United Kingdom's Royal Society, classifies geoengineering techniques by the mechanism they manipulate to moderate or reverse the increase in global mean temperature.[5] In this schematic, there are two groups of geoengineering applications: solar radiation management (SRM) and carbon dioxide removal (CDR).[6] A related approach, which was formulated by the Asilomar Scientific Organizing Committee, categorizes geoengineering techniques by the policy rationale for deploying them.[7] The categories are thus "intervention technologies" and "remediation technologies".[8] Generally speaking, intervention technologies include SRM and remediation technologies include CDR. The risk profile of SRM proposals are orders of magnitude more substantial compared to CDR. As a result, the regulatory framework required for CDR and SRM should not be the same. For these reasons, the analysis presented in this paper focuses primarily on SRM.

SRM refers to methods of making the Earth's atmosphere more reflective, reducing the amount of solar energy that reaches the Earth and increases the warming (thereby resulting in climate change). SRM methods operate directly on the radiative fluxes involved in the Earth's energy balance, and so take effect relatively rapidly.[9] SRM proposes to increase the planetary albedo by reflecting a larger share of the sun's incoming shortwave radiation than would otherwise occur. Albedo is the proportion

---

[4]Richard Elliot Benedick, "Considerations on Governance for Climate Remediation Technologies: Lessons from the 'Ozone Hole'," *Stanford Journal of Law, Science & Policy*, IV (2011), 7.

[5]The Royal Society (2009), *op. cit.*, 57 (finding that the "term "geoengineering" now includes such a broad spectrum of methods that general statements can be very misleading).

[6]*Ibid.*, 1.

[7]The Asilomar International Conference on Climate Intervention Technologies was held in California, 22–26 March 2010 and organized by the Climate Response Fund.

[8]"The Asilomar Conference Recommendations on Principles for Research into Climate Engineering Techniques: Conference Report," November 2010, p. 13. Retrieved 27 March 2015, from http://www.climate.org/PDF/AsilomarConferenceReport.pdf.

[9]The Royal Society (2009), *op. cit.*, 47.

of radiation reflected back into space.[10] The proposed reductions in global solar radiation absorption are usually 1% to 2%.[11] There are four main classes of SRM engineering methods for cooling Earth:

- *Stratospheric or mesospheric aerosols.* It has been proposed that small particles placed high in the atmosphere would potentially exert a net cooling effect on the Earth's climate by scattering or reflecting sunlight back into space.
- *Whitening marine clouds.* It has been proposed that low clouds in some oceanic regions could be whitened with a fine spray of seawater. If done on a large enough scale, this could potentially cool the planet.
- *Satellites in space.* It has been proposed that vast satellites could be constructed in space to deflect sunlight away from the Earth.
- *Whitening the surface.* It has been proposed that whitening roofs, crops, or the ocean surface would reflect more sunlight to space, thereby exerting a cooling influence on planetary temperatures.[12]

8.1.1. *Stratospheric aerosols*

Volcanic eruptions pump massive amounts of volcanic ash into the atmosphere, which scatters back into space some of the sunlight that would otherwise have warmed the surface. As more sunlight is scattered, the planet cools. Injecting submicron-sized particles into the stratosphere may yield cooling effects similar to that of volcanic eruptions. This possibility has galvanized the SRM strategy that seeks to imitate volcanic eruptions by creating a stratospheric cloud that would reflect incoming sunlight, shading and cooling the planet.[13] Over time, the particles would descend into the

[10] Albedo, the fraction of solar radiation reflected by a surface or object, is often expressed as a percentage. Snow-covered surfaces have a high albedo, the surface albedo of soils ranges from high to low, and vegetation-covered surfaces and oceans have a low albedo.

[11] See "Geoengineering: Assessing the Implications of Large-Scale Climate Intervention Hearing before the House of Representatives Committee on Science and Technology," 111th Congress (2009) (Hearing Charter).

[12] Ken Caldeira and David Keith, "The Need for Climate Engineering Research," *Issues in Science and Technology*, (Fall 2010). Retrieved 25 March 2015, from http://issues.org/toc/27-1/.

[13] Paul Crutzen, "Albedo Enhancement by Stratospheric Sulfur Injections: A Contribution to Resolve a Policy Dilemma?," *Climatic Change*, 77(2006), 211–220.

lower atmosphere where they would precipitate. By some estimates, the total mass of particles needed would amount to a few percent greater than today's sulfur emissions from power plants.[14] It costs roughly US$10 billion dollars to provide cooling that counteracts the heating from a doubling of atmospheric carbon dioxide and the global climate could be altered within months of deployment.

8.1.2. *Whitening marine clouds*

Low altitude liquid clouds tend to cool the planet more than they warm it by reflecting sunlight that would otherwise reach the surface and heat up the ground. The "cloud whitening" strategy for SRM is to make low-lying clouds over the ocean more reflective to sunlight than they would otherwise be.[15] In relatively dust-free parts of the marine atmosphere, increasing the number of cloud condensation nuclei per unit volume in low-level marine clouds (which cover approximately one-quarter of the ocean surface) raises cloud albedo significantly.[16]

A variation of this SRM strategy envisions lofting a fine mist of seawater droplets over the ocean that would form a moist sea-salt aerosol.[17] The particles within the aerosol would provide sites for cloud droplets to form within the marine cloud layer. These droplets would add to the effects of natural sea salt and other small particles around which water vapor will condense and form droplets.[18]

8.1.3. *Satellites in space*

In 1997, Lowell Wood and Edward Teller proposed positioning a superfine reflective mesh of aluminum threads in space between the Earth and the

---

[14]The Royal Society (2009), *op. cit.*, p. 32.

[15]U.K. House Committee on Science and Technology, "The Regulation of Geoengineering," p. 38, H. C. 221, Fifth Report of Session 2009–2010, 18 March 2010.

[16]B. Albrecht, "Aerosols, Cloud Physics and Fractional Cloudiness," *Science*, 245(4923) (1989), 1227–1230.

[17]Stephen Salter *et al.*, "Sea-Going Hardware for the Cloud-Albedo Method of Reversing Global Warming," *Philosophical Transactions of the Royal Society A*, 366(1882) (2008), 3989–4006.

[18]The Royal Society (2009), *op. cit.*, p. 38.

sun in an effort to reduce the amount of sunlight that reaches the Earth.[19] An alternative to placing reflectors in low Earth orbit is to station them at Lagrange Point 1.[20] Lagrange points are imaginary points in space at which objects sent from Earth will stay put. Lagrange point 1 (L1) is located about four times farther from Earth than the moon and nearly one million miles from the Earth towards the sun.[21] The potential advantage of this location is that large effects can be achieved with much lighter weight scattering structures. In this stationary position, the reflectors would presumably pose less of a threat to earth-orbiting satellites than it would if deployed in low Earth orbit.

8.1.4. *Whitening earth's surface*

The ecological impact of localized radiative-forcing strategies like those described below would be profound and likely negative if the strategies were implemented on any scale large enough to be effective. The impacts could include changes to atmospheric circulation, reduced cloud cover, and reduced rainfall.[22] These impacts have not yet been fully assessed in climate models.

8.1.5. *White roofs*

Painting roofs, roads, and pavements bright reflective "white" could enhance the reflectivity of the built environment, especially during the summer in sunny regions.[23] The resulting global radiative forcing would depend on how much urban area could be brightened in this way, but estimates suggest that albedo modification on 2.3% of the planet's land

[19]Edward Teller *et al.*, "Global Warming and Ice Ages: I. Prospects for Physics-Based Modulation of Global Change," Lawrence Livermore National Laboratory, Italy, 1997.

[20]Roger Angel, "Feasibility of Cooling the Earth with a Cloud of Small Spacecraft near the Inner Lagrange Point (L1)," *Proceedings of the National Academy of Sciences of the United States of America*, 103(46) (2006), 17184–17189.

[21]Lagrange Points are five positions in an orbital configuration where a small object affected only by gravity can theoretically be stationary relative to two larger objects.

[22]See Teller *et al.* (1997), *op. cit.*, 25 [citing J. G. Charney, "Dynamics of Deserts and Droughts in the Sahel," *Quarterly Journal of the Royal Meteorological Society*, 101(428) (1975), 193–202].

[23]See Hashem Akbari, Surabi Menon, and Arthur Rosenfeld, "Global Cooling: Increasing World-Wide Urban Albedos to Offset CO_2," *Climatic Change*, 94(3–4) (2009), 275–286.

surface would yield a radiative cooling of about −0.2 watts per meter squared $(W/m)^2$.[24] The amount of solar radiation that reaches the Earth's surface is 198 W/m^2.[25] The total annual cost of implementing a "white roof strategy" on 1% of the planet's land surface would be an estimated US$300 billion, according to the Royal Society.[26]

8.1.6. *More reflective crops and grasslands*

The albedo of plant canopies varies significantly for different types of plants as a result of differences in basic leaf spectral properties, morphology, and canopy structure. Promoting the growth of specific crop and grassland species and varieties would theoretically increase the albedo of vegetated surfaces. One study considered genetically modifying leaf or canopy structure to achieve greater temperature reductions.[27]

8.2. Risks of Deploying SRM

As most of the geoengineering proposals potentially pose adverse trans-boundary environmental and human safety consequences, there would have to be universal international approval of their use before implementation. The following section briefly reviews the literature related to the risks of geoengineering, particularly SRM.

8.2.1. *Effect on rainfall*

The deployment of SRM could increase or decrease rainfall in poorly understood ways.[28] For example, a recent study found that combining

[24]There is disagreement over the amount of land that would need to be covered to produce significant levels of radiative forcing. See e.g., Akbari *et al.* (2009), *contra* Thomas Lenton *et al.*, "The Radiative Forcing Potential of Different Climate Geoengineering Options," *Atmospheric Chemistry and Physics Discussions*, 9 (2009), 2559–2608.

[25]See The Royal Society (2009), *op. cit.*, p. 25, footnote 11.

[26]*Ibid.*

[27]Andy Ridgwell *et al.*, "Tackling Regional Climate Change by Leaf Albedo Bio-Geoengineering," *Current Biology*, 19 (2) (2009). See also Joy S. Singarayer, Andy Ridgwell, and Peter Irvine, "Assessing the Benefits of Crop Albedo Bio-Geoengineering," *Environmental Research Letters*, 4 (4) (2009).

[28]Govindswamy Bala *et al.*, "Impact of Geoengineering Schemes on the Global Hydrological Cycle," *Proceeding of the National Academy of Sciences*, 105 (22) (2008), 7664–7669.

a reduction of incoming radiation with high levels of atmospheric carbon dioxide could have substantial impacts on regional precipitation — potentially leading to reductions that could create drought in some areas or increases that could cause flooding in other areas.[29] Another example involves injecting sulphur dioxide to enhance stratospheric aerosol. This SRM technique could modify the Asian and African summer monsoons, reducing precipitation and potentially affecting the food supply to billions of people.[30]

In 2007, researchers at the National Center for Atmospheric Research showed that the Mount Pinatubo eruption caused large hydrological responses, including reduced precipitation in many regions. Simulations of the climate response to volcanic eruptions have also shown large impacts on regional climate. Scientists have also shown that volcanic eruptions in the tropics produce changes in atmospheric circulation, inducing long-term winter warming trends over continents in the Northern Hemisphere, as well as eruptions at high latitudes weakening the Asian and African monsoons, causing reduced precipitation.[31] Several studies have also suggested that major regional effects could result from sulphate geoengineering, which could counteract or reinforce the impacts of climate change.[32] If sulfate aerosols were injected into the stratosphere, it could induce acid rain.

8.2.2. *Rebound warming*

There is evidence that deploying SRM over a sustained period of time would create a risk of "rebound warming" should SRM be halted abruptly. In brief, rebound warming refers to the phenomenon whereby, upon cessation of SRM, the climate rapidly returns to its pre-geoengineered state. The rapid return of temperature to that of the pre-geoengineered state would probably

---

[29] Gabriele C. Hegerl and Susan Solomon, "Risks of Climate Engineering," *Science*, 325 (5943) (2009), 955–956.

[30] Alan Robock *et al.*, "Regional Climate Responses to Geoengineering with Tropical and Arctic SO_2 Injections," *Journal of Geophysical Research*, 113 (D16) (2008).

[31] D. Lunt *et al.*, "Sunshade World: A Fully Coupled GCM Evaluation of the Climatic Impacts of Geoengineering," *Geophysical Research Letters*, 35 (2008).

[32] See supra at notes 47–50.

produce significant stresses to society and ecosystems. However, no studies have been done to explore this.[33]

8.2.3. *Unintended consequences*

The climate is a classic example of a chaotic system, or a system that never exactly repeats itself. It is therefore extremely difficult to predict the effects of changes and one must expect that minor human-induced changes in the system can lead to vastly different outcomes.[34] Concerns about the unintended consequences of attempting to "control" the climate are therefore justified, if for no other reason than our sheer lack of understanding of the complexity of the climatic system.

> When analyzing [sic] potential problems associated with geoengineering in relation to long-term climate change, the language of "risk" is often used, implying some knowledge about both potential outcomes of geoengineering technologies and their probabilities. But so embryonic are geoengineering technologies that there is commonly little knowledge yet about the nature of (potentially unwanted) outcomes and still less knowledge of probabilities. This is a situation of "indeterminacy" (or "ignorance") rather than risk. Through research and the accumulation of empirical evidence, uncertainties can sometimes be recast as risks and expressed as probabilities.[35]

In 2007, the Intergovernmental Panel on Climate Change (IPCC) concluded that geoengineering proposals are "largely speculative and unproven, and with the risk of unknown side-effects".[36] It should also

---

[33]Philip Rasch, Written Testimony before House Committee on Science and Technology, "Geoengineering II: The Scientific Basis and Engineering Challenges," 4 February 2010.

[34]G. E. James, "Chaos Theory: The Essentials for Military Applications," in *ACSC Theater Air Campaign Studies Coursebook* (Maxwell, AFB, A1a.: Air University Press, 1995), p. 8.

[35]The Royal Society (2009), *op. cit.*, p. 37 (citing Andrew Stirling, "Science, Precaution and the Politics of Technological Risk: Converging Implications in Evolutionary and Social Scientific Perspectives," *Annals of the New York Academy of Sciences*, 1128 (1) (2008), 95–110.

[36]IPCC, "Mitigation of Climate Change. Contribution of Working Group III to the Fourth Assessment Report of the Intergovernmental Panel on Climate Change," (2007c).

be noted that because some effects may be at least partially known but unintended, geoengineering research or deployment could raise questions of informed consent of those who will be affected.[37]

8.2.4. *Unilateral deployment*

The possibility that a state may unilaterally deploy geoengineering technology has been widely discussed and is said to have "cast a pall over substantive inquiry into climate intervention research and implementation".[38] Under a unilateral deployment scenario, Country X decides that its interests would be best served by deploying a SRM technology in the face of adverse climate change impacts such as rising sea levels and poor agricultural output. However, Country X is not able to gather international support for the deployment of the SRM technology and decides to go ahead on its own. As such, it is willing to act against global consensus, potentially imposing adverse effects on other states, and offending those who view geoengineering as unnatural and do not approve of its use. The scenario continues to posit that, as a result of such unilateral action, Country X faces the imposition of trade sanctions or more serious international responses including the use of force. Unilateral geoengineering is therefore viewed as a potential challenge to the international order somewhat akin to nuclear proliferation.[39] Victor *et al.* (2009) have also argued that unilateral action could create a crisis of legitimacy that would increase

[37] David Morrow, Robert Kopp, and Michael Oppenheimer, "Towards Ethical Norms and Institutions for Climate Engineering Research," *Environmental Research Letters*, 4(4) (2009).

[38] Joshua B. Horton, "Geoengineering and the Myth of Unilateralism: Pressures and Prospects for International Cooperation," *Stanford Journal of Law, Science and Policy* 4(1) (2011), 56.

[39] See, Clive Hamilton, "The Return of Dr. Strangelove: The Politics of Climate Engineering as a Response to Global Warning," (June 2010). Retrieved 30 March 2015, from http://www.clivehamilton.net.au/cms/media/documents/articles/dr_strangeloves_return.pdf. For a brief but fascinating account of the historic links between nuclear research and geoengineering research, see Clive Hamilton, "Geoengineering and the Politics of Science Bulletin of the Atomic Scientists," (2014). DOI: 10.1177/0096340214531173. Retrieved 30 March 2015, from http://mahb.stanford.edu/wp-content/uploads/2014/05/Hamilton_2014_Geoengineering-and-the-politics-of-science.pdf.

the difficulties of managing/regulating geoengineering schemes when they are implemented.[40]

8.2.5. *The moral hazard argument*

Some have argued that geoengineering may make it more difficult to address the root causes of climate change by diverting political efforts to secure binding commitments to mitigate GHG emissions.[41] While deployment of geoengineering would almost certainly have adverse consequences on international mitigation and adaptation efforts, it can be argued that research alone would trigger these consequences. David Keith, an engineering professor at the University of Calgary and an authority on geoengineering said: "The primary argument against research on SRM is fear that it will reduce the political will to lower greenhouse gas emissions."[42] This risk is especially acute with SRM. Unlike CDR that removes carbon dioxide from the atmosphere, SRM techniques do not address the underlying cause of climate change. SRM cannot replace mitigation, but only delay the need for it.

8.3. International Legal Regulation of Geoengineering Research

No multilateral environmental agreements (MEAs) and almost no international law directly addresses geoengineering.[43] However, several MEAs and some customary international law have specific application when discussing the legal environment into which any geoengineering research would be introduced.[44] This part reviews the MEAs that are likely to have the most impact on geoengineering field research.

[40]David G. Victor *et al.*, "The Geoengineering Option: A Last Resort against Global Warming?," *Foreign Affairs*, 88 (2009), 64.

[41]Albert Lin, "Does Geoengineering Present a Moral Hazard?," *Ecology Law Quarterly*, 40 (2013), 673.

[42]David Keith, "Learning to Manage Sunlight: Research Needs for Solar Radiation Management, Testimony before the U.S, House of Representatives Committee on Science and Technology," 4 February 2010, 1.

[43]Karen N. Scott, "International Law in the Anthropocene: Responding to the Geoengineering Challenge," *Michigan Journal of International Law*, 34 (2) (2013), 309, 310.

[44]Jesse Reynolds, "Climate Engineering Field Research: The Favourable Setting of International Environmental Law," *Washington and Lee Journal of Energy, Climate and the Environment*, 5(2) (2014), 417.

8.3.1. *United Nations Framework Convention on Climate Change (UNFCCC)*

The UNFCCC is the most significant MEA regarding geoengineering because of its subject matter, its global participation, and its strong institutional support. Reynolds has argued that, at a minimum, the UNFCCC supports research into CDR (including ocean fertilization) because Articles 1.7, 1.8, 4.1, and 4.2 commit treaty Parties to stabilizing GHGs through both the reduction of GHG emissions and the enhancement of sinks and reservoirs which is defined to include oceans.[45] Further, the Kyoto Protocol commits Parties to further the Protocol's objectives by researching and promoting "carbon dioxide sequestration technologies and advanced and innovative environmentally sound technologies".[46]

On SRM technologies, the UNFCCC is less clear because such technologies will not further the treaty's objective of stabilizing GHG concentrations. Reynolds argues that some passages in the UNFCCC can be interpreted as having a relatively favorable position on SRM research.[47] First, the UNFCCC calls for some urgency given the expected onset of climate change. Secondly, Article 3(3) of the UNFCCC calls for cost-effectiveness in climate policies and measures "so as to ensure global benefits at the lowest possible cost". Thirdly, Article 4(1)(f) states that Parties "shall . . . employ appropriate methods . . . with a view to minimizing adverse effects on the economy, on public health, and on the quality of the environment, of projects or measures undertaken by them to mitigate or adapt to climate change". Finally, there are a number of passages in the UNFCCC, which call for the development, and diffusion of technology and research, implying a positive stance towards research into geoengineering techniques.[48]

The UNFCCC also invokes two principles of international environmental law, the precautionary principle and the principle of common but differentiated responsibility (CDR principle), which can be argued to point favorably to geoengineering. Reynolds and Fleurke make the argument that

[45] *Ibid.*, 359.

[46] Kyoto Protocol, 11 December 1997, 2303 U.N.T.S. 148, Article 2.1(a)(iv).

[47] Reynolds (2014), *op. cit.*, p. 360.

[48] For example, Articles 4(1)(g) and (h), 5, and 9(2).

CDR research is a precautionary measure towards minimizing the causes of climate change, and SRM research a precautionary response to the serious and potentially irreversible impacts of climate change.[49]

8.3.2. *Convention on Biological Diversity (CBD)*

A recent CBD report concluded that geoengineering "could reduce the magnitude of climate change and its impacts on biodiversity. At the same time, most geoengineering techniques are likely to have unintended impacts on biodiversity." At the 2008 CBD Conference of the Parties, a decision was taken whereby parties to the CBD were urged to ensure that ocean fertilization did not take place until there was greater understanding of its risks and benefits. "Small scale scientific research studies within coastal waters" were permitted. The 2010 CBD Conference of the Parties took a further decision to expand the 2008 moratorium to cover all "climate-related geoengineering activities". The decision invites "Parties . . . to consider . . . ensur[ing] . . . in the absence of science based, global, transparent, and effective control and regulatory mechanisms for geoengineering, and in accordance with the precautionary approach and Article 14 of the Convention, that no climate-related geoengineering activities that may affect biodiversity take place, until there is an adequate scientific basis on which to justify such activities and appropriate considerations of the associated risks for the environment and biodiversity and associated social, economic, and cultural impacts, with the exception of small-scale scientific research studies that would be *conducted in a controlled setting in accordance with Article 3 of the Convention*, and only if they are justified by the need to gather specific scientific data and are *subject to a thorough prior assessment of the potential impacts on the environment*"[50] (emphasis added).

- "Conducted in a controlled setting"

The CBD does not define "controlled setting". The *Cartagena Protocol on Biosafety to the Convention on Biological Diversity* ("Biosafety Protocol"),

---

[49] Jesse L. Reynolds and Floor Fleurke, "Climate Engineering Research: A Precautionary Response to Climate Change?," *Carbon and Climate Law Review*, 2 (2013), 101–107.

[50] 10th Meeting of the Conference of Parties to the Convention on Biological Diversity, 18–29 October 2010, Decision X/33 — Biodiversity and Climate Change, at 5, U.N. Doc. UNEP/CBD/COP/DECX/33/8(w) (2010).

which was adopted in 2000 and entered into force in 2003, may provide guidance on the likely construction of "controlled setting". Article 3 of the Biosafety Protocol defines "contained use" as "any operation, undertaken within a facility, installation or other physical structure, which involves living modified organisms that are controlled by specific measures that effectively limit their contact with, and their impact on, the external environment".[51] The term "control" may therefore be construed to imply "reversibility", or the ability to cease a technology program and have its effects terminated in a short time.

- "In accordance with Article 3"

Article 3 of the CBD reaffirms the sovereignty of states over natural resources within their borders and their responsibility with respect to transboundary environmental harm.[52] Activities that have transboundary impacts are likely to be considered "large scale". Conversely, activities that have purely domestic impacts are likely to be considered "small scale".[53] Article 3 of the CBD requires parties to prevent, reduce, and control transboundary pollution and environmental harm resulting from activities within their jurisdiction and control. It also imposes the duty to cooperate in mitigating transboundary environmental risks and emergencies through notification, consultation, negotiation and, in appropriate cases, environmental impact assessment.[54] Article 3 does not necessarily prohibit activities that create a risk of transboundary harm, provided these obligations are observed. In the absence of express prohibition, parties to the CBD are required to exercise due diligence in regulating activities under their jurisdiction and control. Where an activity has transboundary implications or takes place beyond national jurisdiction (e.g., ocean fertilization in the high seas), international cooperation to regulate the activity will be necessary.

---

[51] *Handbook of the Convention on Biological Diversity*, Third Edition (Canada: Secretariat of the Convention on Biological Diversity, 2005), p. 32.

[52] *Ibid.*, p. 93.

[53] See UNEP/CBD/COP/10/L.36.

[54] Patricia Birnie *et al.*, *International Law & the Environment*, Third Edition (UK: Oxford University Press, 2009).

- "Subject to a thorough prior assessment of the potential impacts on the environment"

This condition alludes to the clause "in accordance with the precautionary approach and Article 14 of the Convention".[55] Article 14 of the CBD requires parties to "introduce appropriate procedures requiring environmental impact assessment of its proposed projects that are likely to have significant adverse effects on biological diversity with a view to avoiding or minimizing such effects".[56] A precautionary approach may apply where the impacts of geoengineering on the environment are not yet fully known but believed potentially to be serious, if not irreversible. The precautionary principle "helps to identify whether a legally significant risk exists by adding the role of scientific uncertainty . . . [but] says nothing about how to control that risk, or about what level of risk is socially acceptable".[57]

8.3.3. *The U.N. Convention on the Prohibition of Military or Any Other Hostile Use of Environmental Modification Techniques ("ENMOD")*

The ENMOD prohibits military or any other hostile use of "environmental modification techniques having widespread, long-lasting or severe effects as the means of destruction, damage, or injury to any other State Party".[58] As such, the ENMOD is the only existing treaty that directly addresses intentional intervention with the climate. The United States and nearly 70 other nations have signed and ratified this treaty but it is considered to be a dormant instrument with no supporting institutions and regular meetings.

"Environmental modification" includes "any technique for changing — through the deliberate manipulation of natural processes — the dynamics, composition or structure of the Earth, including its biota, lithosphere,

[55]Conference of the Parties to the Convention on Biological Diversity, 18–19 October 2010, Decision X/33, Retrieved 30 March 2015, from http://www.cbd.int/climate/doc/cop-10-dec-33-en.pdf.

[56]Article 14: Impact Assessment and Minimizing Adverse Impacts, para. 1(a).

[57]Birnie *et al.* (2009), *op. cit.*

[58]Convention on the Prohibition of Military or Other Hostile Use of Environmental Modification Techniques, 18 May 1977, 16 I.L.M 88, Article 1.

hydrosphere and atmosphere, or of outer space"[59] thereby including many geoengineering techniques. However, ENMOD only prohibits hostile or military use of environmental modification techniques. It does not prohibit research or the use of such techniques for peaceful purposes. Instead, the preamble of the ENNOD recognizes and encourages peaceful environmental modification. For present purposes, it can be argued that the ENMOD has the potential to play a role in facilitating and regulating geoengineering rcscarch but not in its current state of almost non-existent supervisory and enforcement mechanisms.[60]

8.3.4. *United Nations Convention on the Law of the Sea (UNCLOS)*

The UNCLOS sets out the rights and obligations of states in their marine activities, including the obligation to protect and preserve the marine environment. Some proposed geoengineering methods such as marine cloud brightening and ocean fertilization are likely to take place above or in the ocean. UNCLOS is generally supportive of marine scientific research.[61] However, the location of the proposed activity determines some of the rights and duties in relation to marine scientific research. For example, on the high seas, states and international organizations have the right to conduct research, but must have due regard for other states.[62]

- Ocean fertilization: a form of "dumping"?

UNCLOS requires its Parties to prevent, reduce, and control marine pollution by dumping, which is defined as "any deliberate disposal of wastes or other matter . . . at sea" but excludes "placement of matter for a purpose other than the mere disposal thereof".[63] The London Convention on the Prevention of Marine Pollution by Dumping of Wastes and Other Matter and its London Protocol regulate dumping at sea. The definition of dumping

---

[59] *Ibid.*, Article 2.

[60] For discussion, see Charles R. Wunsch, "The Environmental Modification Treaty," *ASILS Int'l L. J.*, 113 (1980), 128–129 (describing key shortcomings in the enforcement mechanism of ENMOD).

[61] See, for example, Articles 87(1), 88, 238, 243, and 251.

[62] Articles 87 and 257.

[63] Article (1)(1)(5).

under the London Convention and Protocol is essentially the same as the UNCLOS one.

In a 2008 decision by the parties to the London Convention and Protocol, a non-binding regulatory framework for ocean fertilization was established.[64] Under this framework, ocean fertilization is not permitted except "legitimate scientific research". The parties' 2010 decision contains an assessment framework to determine whether the scientific research is legitimate or not.[65] Researchers are required to apply to the appointed regulatory agency of their home state for approval. Controversially, there have already been some ocean fertilization experiments that did not comply with this regulatory framework.[66] In 2013, the parties to the London Protocol (but not the London Convention) approved an amendment to the Protocol that introduces a more comprehensive and binding regulatory framework.[67] The amendment requires the acceptance by two-thirds of the parties to come into force. Under this proposed framework, marine geoengineering activities listed in an annex are either prohibited or require a permit by the relevant member state. The permit process includes an assessment based on a detailed description of the proposed activity, notification of potentially affected countries, consultation with relevant parties, and the seeking of consent "from all countries with jurisdiction or interests in the region of potential impact".[68]

8.3.5. *Convention on Environmental Impact Assessment in a Transboundary Context (Espoo Convention)*

The Espoo Convention, developed through the United Nations Economic Commission for Europe (UNECE), is concerned with applying environmental impact assessment (EIA) at the international level. It sets out

[64]Resolution LC-LP.1 on the Regulation of Ocean Fertilization, 3 I.M.O. Doc. LC10/16/Annex6 (31 October 2008).

[65]Resolution LC-LP.2 on the Assessment Framework for Scientific Research Involving Ocean Fertilization, I.M.O. Doc. LC32/15/Annex6 (14 October 2010).

[66]See discussion in Reynolds (2014), *op. cit.*, p. 382.

[67]See Res. LP.4(8) on the Amendment to the London Protocol to Regulate the Placement of Matter for Ocean Fertilization and Other Marine Geoengineering Activities, 18 October 2013, Rep. of the Thirty-Fifth Consultative Meeting and the Eighth Meeting of Contracting Parties, Doc. LC 35/15, Annex 4 (21 October 2013).

[68]*Ibid.*, Annex 5, paragraph 11.

states' commitments to assess transboundary environmental impacts, to make such information available to the public and to other states, and to mitigate these transboundary environmental impacts.[69] The Convention is therefore limited to transboundary environmental harm between parties to the Convention, and does not cover effects that are domestic only or occur in a non-Party state. It can be argued that the notification and other procedural requirements imposed by the Espoo Convention will introduce transparency and allow for some degree of public participation in geoengineering research trials.

8.3.6. *The Convention on Access to Information, Public Participation in Decision-Making and Access to Justice in Environmental Matters (Aarhus Convention)*

The Aarhus Convention, like the Espoo Convention, is concerned with improving transparency and public participation in environmental governance. The duties imposed on parties to the Aarhus Convention focus on the provision of environmental information to citizens, giving citizens access to redress mechanisms when faced with environmentally harmful activity, and promoting public participation in environmental decision-making. In relation to geoengineering, should a party to the Aarhus Convention wish to carry out any research, it would be obliged to fulfill procedural duties that would provide valuable information about the research, including its potential risks and benefits, and give its citizens an opportunity to participate in the governance of the research activity (for example, at the public consultation stage in an EIA process).

8.3.7. *The Oxford Principles and the Asilomar Conference Recommendations*

Two sets of principles have been articulated as a step towards creating an international governance structure for geoengineering. The Oxford Principles, authored in 2009 by Steve Rayner, Tim Kruger, and Julian Savulescu of the Oxford Geoengineering Programme together with Catherine Redgwell

[69] See, generally, "Convention on Environmental Impact Assessment in a Transboundary Context," 25 February 1991, 1989 U.N.T.S. 309.

(then at University College London) and Nick Pidgeon (University of Cardiff), were submitted to the United Kingdom House of Commons Science and Technology Select Committee. The UK government's endorsement of the principles is the only official national policy statement on geoengineering to date.[70] The Oxford Principles formed the basis of the Asilomar Conference Recommendations, a set of principles agreed to at a gathering of geoengineering researchers in Asilomar, California, in March 2010.[71]

The Oxford Principles set out the following guidelines for geoengineering governance:

- Geoengineering to be regulated as a public good.
- Public engagement in geoengineering decision-making.
- Disclosure of geoengineering research and openpublication of results.
- Independent assessment of impacts.
- Governance before deployment.

The Asilomar geoengineering conference issued the following principles:

- Climate engineering research should be aimed at promoting the collective benefit of humankind and the environment.
- Governments must clarify responsibilities and, when necessary, create new mechanisms for the governance and oversight of large-scale climate engineering research activities.
- Climate engineering research should be conducted openly and cooperatively, preferably within a framework that has broad international support.
- Iterative, independent technical assessments of research progress will be required to inform the public and policy-makers.

---

[70] See Oxford Martin School, University of Oxford, Oxford Geoengineering Programme, Retrieved 30 March 2015, from http://www.geoengineering.ox.ac.uk/oxford-principles/history/.

[71] "The Asilomar Conference Recommendations on Principles for Research into Climate Engineering Techniques," Conference Report, November 2010. Retrieved 27 May 2015, from http://www.climate.org/PDF/AsilomarConferenceReport.pdf.

- Public engagement and consultation in research planning and oversight, assessments, and development of decision-making mechanisms and processes must be provided.

8.4. An ASEAN Response to Geoengineering

Thus far, research and discussion on geoengineering has mostly been driven by Europe and the United States while other regions that are significantly affected by climate change, such as Asia, have yet to make their perspectives heard.[72] This transatlantic dominance of the geoengineering discourse carries the risk of ignoring the needs, values, and preferences of the rest of the international community. Yet, any meaningful research trials (which most likely will have to be on a large scale to generate useful information about the impacts of geoengineering) and eventual deployment are likely to have significant impacts on a range of global concerns, including but not limited to, food security, biological diversity, and public health. It is therefore in the interests of states that currently remain passively silent on the issue to make their perspectives known and to claim their stake in the discourse of governing an emerging set of potentially risky technologies. This will also bode well for the legitimacy of any future geoengineering governance framework.[73]

Resource constraints and many other pressing policy issues may render it challenging for governments, particularly of developing countries, to divert attention and time towards geoengineering. As such, the pooling of resources to consider geoengineering at the regional level may be a way forward. One such regional organization outside Europe and North America is the Association of South East Asian Nations (ASEAN).[74] ASEAN Member States include Brunei Darussalam, Cambodia, Indonesia, Lao PDR, Malaysia, Myanmar, the Philippines, Singapore, Thailand, and Vietnam. To the best of the author's knowledge, there is no evidence in

[72]See RSIS Centre for Non-Traditional Security (NTS) Studies, 2011, Pilot Workshop on Governing Geoengineering in the 21st Century: Asian Perspectives (18–19 July 2011), Singapore, p. 4, for expression of similar sentiment.

[73]Legitimacy here referring to the acceptance and justification of shared rule by a community; S. Bernstein, "Legitimacy in Global Environmental Governance," *Journal of International Law and International Relations*, 1(1–2) (2005), 139–66.

[74]Information about ASEAN is available at http://www.asean.org.

the public domain of policy responses to geoengineering by any ASEAN Member States or ASEAN itself. Arguing that it will be in the interests of ASEAN Member States to consider the risks and benefits of geoengineering at this relatively early stage of research rather than at the later stage of deployment, this paper advocates for a coordinated effort at the ASEAN level to consider geoengineering. This coordinated effort should aim to deliver the following outcomes: (1) That the risks and benefits of geoengineering for the ASEAN region are assessed and the information is made available to the respective member states to facilitate policy discussion at the domestic level; (2) That ASEAN issues a statement on its views on geoengineering. It can be argued that none of the ASEAN Member States are likely to become active in geoengineering research and development so the statement will be from the perspective of countries which will not deploy geoengineering but will share in the risks and benefits of large-scale geoengineering projects; (3) Given the potential for geoengineering to be used for hostile purposes, the ASEAN statement should also consider the security aspects of geoengineering. This initiative also accords well with the aims and purposes of ASEAN, which include "promot[ing] active collaboration and mutual assistance on matters of common interest in the educational, professional, technical and administrative fields".[75] In this regard, lawyers and law professors in ASEAN Member States can play an instructive role by providing information and expert opinions to their home governments and/or the ASEAN Secretariat. Another pathway is via workshops as such this one conveyed at the National University of Singapore, when researchers come together as a transnational epistemic community that can assist ASEAN in reaching an informed position on geoengineering.[76]

8.5. Conclusion

This chapter has sought to provide an overview of geoengineering techniques and their associated risks. At present, the technical discourse is

[75]"The ASEAN Declaration (Bangkok Declaration)," Bangkok, 8 August 1967. Retrieved 30 March 2015, from http://cil.nus.edu.sg/rp/pdf/1967%20ASEAN%20Declaration-pdf.pdf.

[76]I am grateful to the reviewer of an earlier draft of this chapter for this suggestion.

focused on research to better understand the possibilities of eventual deployment while the policy discourse is concerned with the regulation and governance of these potentially risky technologies. Part III of this chapter considered the various international legal instruments that form the regulatory environment into which any geoengineering techniques will be introduced. These instruments include the UNFCCC and the UNCLOS while recent efforts to develop principles to govern the development of geoengineering include the Oxford Principles and the Asilomar Conference Guidelines. As stated in the introduction of this chapter, the aim of this chapter is to start a dialogue amongst academics, policy-makers, industry, and other possible stakeholders who represent views outside Europe and the United States. This chapter has also advocated for ASEAN to take the lead on this issue on behalf of its member states. Ultimately, questions such as whether geoengineering ought to be pursued and the parameters within which geoengineering research ought to take place affect all members of the international community, and not just Europe and the US. It would be unwise for ASEAN Member States not to let their voices, collective or individual, be heard on this issue.

Theme IV

Economic Interconnections

Chapter 9

Assessing Green Jobs in Taiwan: A Tri-Pillar Approach*

Fan Chien-Te and Hsu Yun-Hsiang

9.1. Introduction

Climate change creates both challenges to and opportunities for our society. On the one hand, rising sea levels, warming atmosphere and oceans, and recurring extreme weather conditions directly or indirectly caused by climate change pose severe challenges to the regions we reside in. On the other hand, opportunities emerge from these challenges. Witnessing these daunting events, there is greater call for a greener economy because these damages are unequivocally caused by a traditional economic model that highly relied on carbon energy.[1] Cities, especially densely populated ones located in coastal areas, are eager to search for a new development paradigm. As such, the notion of green growth is proposed and widely discussed by mayors and ministers from Singapore to Toronto.[2] Leaders in these cities move beyond pure regulatory approaches[3] to take a closer look

*This research is partially supported by the Ministry of Science and Technology, ROC (MOST: 103-2621-M- 007-001). We appreciate the excellent assistance from Ms. Ya Zhi Lo, graduate student of ILST, NTHU, Taiwan.

[1]IPCC's latest report notes that "Human influence on the climate system is clear. This is evident from the increasing greenhouse gas concentrations in the atmosphere, positive radiative forcing, observed warming, and understanding of the climate system". See IPCC Fifth Assesment Report (AR5), (2013). Retrieved 20 May 2015, from https://www.ipcc.ch/report/ar5/.

[2]See "3rd Annual Meeting of the OECD Urban Roundtable for Mayors and Ministers, 'Cities and Green Growth'." Retrieved 30 March 2015, from http://www.oecd.org/greengrowth/3rdannualmeetingoftheoecdurbanroundtableformayorsandministerscitiesandgreengrowth.htm.

[3]Regulatory approach has been pursued in global community since the United Nations Framework Convention on Climate Change (UNFCC) came into effect. Mitigation actions

at the environment dynamics between government, business, and industrial workers. This quest for a whole new economic model spurs a series of opportunities, in terms of "green jobs"[4] for stakeholders involved. A new set of jobs is being created to meet the demands of this greener economy to enable local businesses and communities to adapt to climate change; some are existing ones while some are newly created. These opportunities have gained attention among major cities and international organizations, ranging from the Organization for Economic Cooperation and Development (OECD), the United Nations Environment Programme (UNEP), to the International Labor Organization (ILO), that try to assess the potential as well as the scope of green jobs.[5]

The niche for promoting green jobs at municipal levels exists since they are close to the sources of job creation. Much of the production and consumption activities occur in large metropolitan areas. City managers are being placed in a better position, compared to those at national levels, to prepare the economy with greener jobs.[6] But here come the problems: what are these job opportunities and how to assess where they come from? To be clear, what is the exact content of green jobs? Current discussion can only give us a rough idea that these jobs can be something that fall in the range of limiting greenhouse gas emissions, minimizing waste and pollution, and protecting and restoring the ecosystem.[7] However, every city has its own specific labor demand. Jobs created in one place may not be the same for another. Further, some priori criteria, like "decent", that these international

in greenhouse gas emissions and adaptation to the impacts of climate change since remained the focus of the international community. After years of continuous efforts, the adoption of the Bali Roadmap among member states has started a new era for the future of climate change, though the subsequent Copenhagen Accord was far from satisfactory. Consensus toward diversified tracks and sustainable development to meet with the challenges from climate change has been reached in Durban Platform and Doha Gateway, which are two exciting breakthroughs in the global community.

[4] See UNEP, *Green Jobs: Toward Decent Work in a Sustainable, Low-Carbon World* (Kenya: UNEP, 2008).

[5] See A. Jarvis, V. Adarsh, and R. Justin, *Assessing Green Jobs Potential in Developing Countries: A Practitioner's Guide* (Geneva: ILO, 2011).

[6] See S. Hammer *et al.*, "Cities and Green Growth: A Conceptual Framework," OECD Regional Development Working Papers 2011/08, OECD Publishing. Retrieved 30 March 2015, from http://dx.doi.org/10.1787/5kg0tflmzx34-en.

[7] See ILO, *Green Jobs Becoming a Reality Progress and Outlook 2013* (Geneva: ILO, 2013).

organizations adopt to define green jobs may run against the needs at local level.[8] Some jobs, like recycling, that are key to support cities' sustainability but with low pay and poor prospects may not be considered "green" enough. To make things complicated, the stimulus package originating from the global crisis after 2009 blurs the line between "green" and "recovery". Scholars found that the focus of the "green stimulus" in some countries is more on reviving "growth" than on achieving "green".[9] The problem is that the former focus tends to seek the short-term impact on economy rather than on cutting greenhouse gas emissions.[10] All these controversies call for further clarification on green jobs.

In our view, promoting green jobs is a dynamic concept. It is the product of an ecosystem, in which government intervention, economy, and society closely interact with each other. This product is highly contingent on local conditions, especially in terms of government intervention; as the ILO has stated in its report, "outcomes for employment and incomes are largely determined by the policy instruments used and the institutions which implement them, rather than being an inherent part of the shift to a greener economy".[11] Policy choices at the local government level are determinant factors on whether jobs are genuinely "green". Besides, we can find that green jobs is a concept that lays more emphasis on its practical usage for implementation rather than for academics. In this context, constructing an operational definition would be a more suitable option. Based on this idea, we seek for indicators that can capture the system dynamics in the process toward a greener economy. We establish a tri-pillar — climate policy, economy, and society — structure (see Figure 1), embedded with indicators from these levels. The reasons we adopt this approach are three-fold: first,

[8] See ILO, *Are "Green" Jobs Decent*? (Geneva: ILO, 2012).

[9] See Y.-B. Chang, J.-K. Han, and H.-W. Kim, "Green Growth and Green New Deal Policies in the Republic of Korea: Are They Creating Decent Green Jobs?," *International Journal of Labour Research*, 4(2) (2012), 131–248.

[10] J. Houser, S. Mohan, and H.-W. Heilmayr, *A Green Global Recovery? Assessing US Economic Stimulus and the Prospects for International Coordination* (US: Peterson Institute for International Economics, 2009), PB09-3.

[11] See ILO, *Working towards Sustainable Development: Opportunities for Decent Work and Social Inclusion in a Green Economy* (Geneva: ILO, 2012). Retrieved 30 March 2015, from http://www.ilo.org/wcmsp5/groups/public/---dgreports/---dcomm/---publ/documents/publication/wcms_181836.pdf.

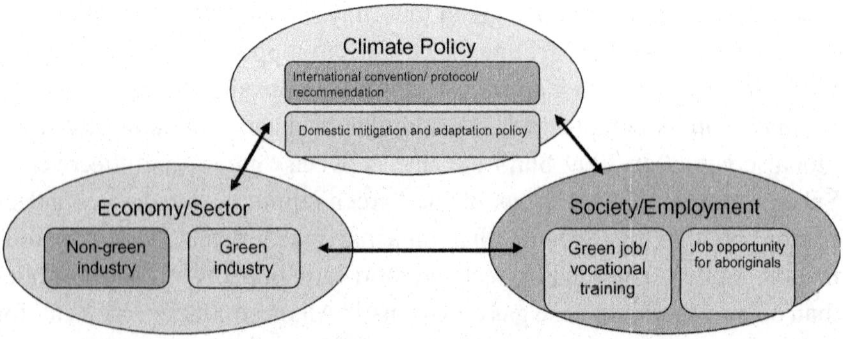

Figure 1. Linkage between Climate Policy, Economy, and Society

focusing on indicators, instead of defining what green jobs are, can help practitioners assess the source and the scope of employment in a greener economy; second, obviously, this approach can be better tailored to the local needs; third, for benchmarking the best practice, indicators can help us systematically compare with experiences in other cities or regions and provide workable solutions.

This chapter aims at using Taiwan as an example to assess green jobs at the local level. We try to adopt a tri-pillar approach, locating crucial indicators to identify green jobs for this densely populated island. The following section reviews the development from a regulatory to a policy approach in climate policy. The third section discusses the approach we intend to adopt to conduct our research. The fourth and fifth sections introduce the institutional context and research development of climate policy in Taiwan. Lastly, we lay down our research agenda, with some tentative contributions. Action to curb carbon emissions is imperative and the time we have is limited. It is our expectation that reduction targets can be firmly supported by generating green jobs at the local level.

9.2. From Regulatory to Policy Approach

Since the United Nations Framework Convention on Climate Change (UNFCC) came into effect, mitigation actions in greenhouse gas emissions and adaptation to the impacts of climate change have remained the focus of the global community. After years of continuous efforts, the adoption of the Bali Roadmap among member states began a new era for the future of climate policy, though the subsequent Copenhagen Accord was far from satisfactory. Consensus toward diversified tracks and sustainable development to meet with the challenge from climate change has been reached in the Durban Platform and Doha Gateway, which are two exciting breakthroughs in the global community.

Regional state or individual member countries are also seeking what is doable under this global consensus. Some promising policy action, instead of pursing regulatory change, has already taken place. The European Union (EU), for example, has set up a "Roadmap for moving to a low-carbon economy in 2050".[12] Based on their implementation experience over the past two decades, policy-makers in EU adopted a framework approach that tries to adjust or reframe the stakeholders' perspective towards climate change, which is shortsighted and, more often than not, characterized by bounded rationality. Two policy themes can be seen in their economic framework — "Toward a Low Carbon Society" and "Innovations, Green Growth & Jobs". The former tries to change people's style of life, work, residence, and transportation so that the structure of energy and goods consumption can be shifted, while the latter links policy with technology and capital that is necessary for affected sectors. It is clear that EU policy-makers pursue a long-term transformation strategy, stressing that this process toward a low-carbon society can promote regional economy development.[13] They realized the emergence of a low-carbon economy will simultaneously create the need for applied technologies, such as renewable energy, energy-saving building materials, electric car, smart grid, low-carbon energy supply, carbon capture and storage[14]; at the same time, capital investment, mainly from the government

[12]EU, "Roadmap to Moving to a Low-Carbon Economy in 2015," *European Commission*. Retrieved 22 September 2013, from http://ec.europa.eu/clima/policies/roadmap/index_en.htm.

[13]"The transition to a low-carbon society would boost Europe's economy thanks to increased innovation and investment in clean technologies and low- or zero-carbon energy." — See "Building a Low-Carbon Society," *Climate Action*. Retrieved 20 May 2015, from http://ec.europa.eu/clima/citizens/aworldyoulike/index_en.htm.

[14]A low-carbon economy would have a much greater need for renewable sources of energy, energy-efficient building materials, hybrid and electric cars, "smart grid" equipment, low-carbon power generation, and carbon capture and storage technologies.

sector, can stimulate growth in the manufacturing and environment service sectors,[15] which account for 1.5 million job opportunities[16] in the future. Fundamentally, EU's take on carbon reduction and mitigation signaled a strategic shift from the traditional approach, which is a vague commitment to "Bigger Carbon Reductions", toward a more substantial and long-term framework that focuses on technological change and economic incentive.

At the local level, the United States has taken a step further by initiating a regional market-based regulatory program, which is the Regional Greenhouse Gas Initiative (RGGI),[17] across nine states in 2009. RGGI embodies the notion of technological change and economic incentive in EU's framework where a man-made market is created that allows local stakeholders to sell emission allowances through auctions, which incentivizes them to invest technology in clean energy and to generate decent job opportunities. Through market mechanisms, the RGGI model does have the potential to curb emission levels and to spur energy innovation; however, the creation of decent jobs is not always guaranteed. This promise highly depends on which technology is applied and whether the workers can be trained or transferred to fill the positions created by this new technology.[18] That is, there is uncertainty in this seemingly promising program and further analysis is needed to identify the linkage between technological change and job opportunity.

---

[15]To make the transition, the EU would need to invest an additional €270 billion or 1.5% of its GDP annually, on average, over the next four decades. The extra investment would take Europe back to the investment levels seen before the economic crisis, and would spur growth within a wide range of manufacturing sectors and environmental services.

[16]Up to 1.5 million additional jobs could be created by 2020 if governments used revenues from CO_2 taxes and from auctioning of emission allowances to reduce labour costs.

[17]Further information can be accessed at http://www.rggi.org/.

[18]An interview on the National Public Radio (NPR) of coal miners at the Appalachian Mountains clearly pointed out this fact: "EPA's claim that the rule would create jobs in cleaner technologies . . . at least, not here", "They can say it's going to create new jobs, but if they don't train people, I don't see it happening because this is a fairly depressed area". Retrieved 30 March 2015, from http://www.npr.org/2014/06/03/318414819/pa-coal-area-worries-emission-rules-will-cost-economy-jobs.

9.3. A Tri-Pillar Policy Approach

These progresses and lessons point out to a vacuum in climate policy research, which is our research orientation (see Figure 1) and revolves around the question: how is institutional design in climate policy linked to energy innovation as well as job opportunities in the economy? Institutional design here refers to both international and domestic institutions. We believe that domestic policy is partly shaped by commitments at the international level, and this in turn reframes the industrial and employment structure at the local level. The question is how these three factors interact with each other, which is the key to the success of any adaptation initiative. This interaction clearly affects stakeholders' expectation and willingness for compliance in a country, but few have been able to document the dynamics between policy intervention, the industrial, and the employment structure.

The vacuum is understandable, because this field of study is multidisciplinary in nature. That is to say, to conduct research in climate policy dynamics requires knowledge from different disciplines. For example, the motivation for power plants to substitute coal with other renewable sources is to comply with emission standards, and this impacts job opportunities among coal miners, which in turn may form political pressure to revise current regulation. We can clearly identify five different disciplines — ranging from international law, public policy, energy technology to economics and labor research — that are involved, and knowledge from one of them may become the input for others.

Meanwhile, there is a need for holistic thinking for policy-making. Holistic thinking suggests that humans are integral parts of ecosystems and that a dynamic interaction exists between each part of these ecosystems. More precisely, socio-economic factors alter human society as a subsystem which directly or indirectly drive changes in ecosystems, and these changes of ecosystems thereby influence human well-being which will affect the socio-economic factors. Accordingly, socio-economic factors such as policy-making require a systematic point of view. The ontology of the system paradigm sees the world as an open, non-separable, and irreducible one. This ontological assumption is system thinking which leads to system analysis that stresses the structure and the relationships between the parts of a complex problem. Moreover, system analysis is believed to have an organizational function that is needed when different aspects of a set of problems depend on different kinds of information.[19] That is, system analysis can be a principle for the distribution of work which serves to support a cross-disciplinary project and to solve a cross-disciplinary problem.

System analysis fits the needs of climate research since it addresses the complex nature of climate change. The relationships between climate change and human society are neither simple cause-effect ones nor are they linear. These non-linear relationships and the changing patterns of unpredictability of climate change drive the urge to grasp a wider web of complexity with a new methodological approach. System analysis, with its focus on the structure and the relationships between the parts of a complex whole, may satisfy this urge. It has the great potential to bridge the different domains of knowledge in the context of general societal tendencies which call for a holistic view and cooperation between disciplines.[20] Through system analysis, the dynamics of climate policy can be fully captured if the components of the problem, the structure, and the relations between the parts of the problem are understood.

With this idea in mind, we propose a cross-disciplinary project that consists of experts from these fields, in the hope that the dynamics in climate policy can be fully captured so that the risks or opportunities arising from mitigation and adaptation actions can be minimized or maximized.

9.4. Current Trends in Taiwan: Toward Policy Actions at the City Level

In Taiwan, the policy theme or roadmap for carbon reduction at the national level is to lower emissions to zero and to fully replace traditional sources of energy with renewable ones by the year 2050. As for climate adaptation action, the National Development Council and Ministry of Science has

[19] M.O. Olsson and G. Sjöstedt, "Systems and Systems Theory," in *Systems Approaches and Their Application Examples from Sweden* (The Netherlands: Kluwer Academic Publishers, 2004).

[20] H. Liljenstrom and U. Svedin, "Bridges, Connections and Interfaces — Reflections over the Meso Theme," in *Micro Meso Macro: Addressing Complex Systems Couplings* (Singapore: World Scientific, 2005).

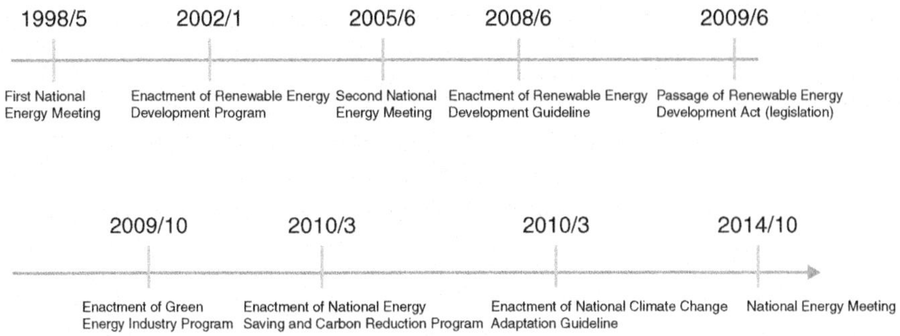

Figure 2. Timeline for Related Energy and Climate Policy in ROC (Taiwan)

enacted the National Climate Change Adaptation Guideline (see Figure 2). For all these efforts, neither carbon reduction nor adaptation action in ROC (Taiwan) has covered industrial development and employment. A well-planned climate policy, such as the one outlined by the EU's climate adaptation platform (Climate-ADAPT),[21] should anticipate the adverse effects from all levels and take appropriate action. For a country that is still seeking its effective adaptation policy,[22] this systematic perspective is undoubtedly insightful.

9.5. Articulating the Way Leading to Green Jobs

Fortunately, the deficiency in a comprehensive assessment of policy intervention to the local economy has already caught attention. Some bottom-up analytic work has started in Taiwan, which is trying to model all the relevant mechanisms and interactions. Our project is one among these bottom-up works. The main motivation for this research project is to better understand how climate policy can affect local economy, in order to derive insights from its dynamics and to draft policy response. On epistemology,

[21]"Adaptation means anticipating the adverse effects of climate change and taking appropriate action to prevent or minimise the damage they can cause, or taking advantage of opportunities that may arise. It has been shown that well planned, early adaptation action saves money and lives later." See, "Adaptation to Climate Change of European Union." Retrieved 22 September 2013, from http://ec.europa.eu/clima/policies/adaptation/index_en.htm.

[22]Currently, the National Development Council in Taiwan is in the process of drafting the Green Economy Guidelines. The final version is expected to be published in the year 2015.

we build our perspective on the concepts of green economy[23] and green jobs. These two are a reflection of the past economic development in academics and the policy response to this reflection among practitioners. Conceptually, green economy holds that a weak sustainability can be achieved if we substitute the pursuit of economic growth with a higher standard of environmental and social protection. In this substitution, an economy maintains a lower rate of growth, which is weak in terms of economic performance, but a higher sustainability, and not at the expense of intergenerational social well-being. This substitution can be realized by replacing traditional physical capital with other non-physical ones, such as human and social capital. Green economy holds that a higher quality of human resources, higher level of technology, or better job opportunities can be used to replace the consumption of natural resources in the production process. In response, green jobs outline how this green economy can function at the local level, with concrete action in creating "decent" jobs.[24] We apply these two concepts as our research theme, linking them together.

The concepts of green economy and green jobs have been in circulation for a while, and regained their prominence due to a series of green recovery or stimulus packages[25] after the economic recession in 2009. This massive injection of public funding into the economy has led research interest into the linkage between the countercyclical measures,

[23]It is well-documented that the traditional model of economic development, on the one hand, has posed severe consequences on the environment — damages ranging from exacerbating climate change, ocean acidification, to stratospheric ozone depletion, that have already exceeded environmental limits. On the other hand, in the pace of development, the asymmetric concentration of wealth to the few has led to greater inequality, undermining social mobility and economic opportunities. In view of these pitfalls in the traditional economic model, a shift in the policy paradigm of economic development has occurred, which is the green economy. See E. Barbier and A. Markandya, *A New Blueprint for a Green Economy* (UK: Routledge, 2013).

[24]Experts in the United Nations (UN) propose a "green job" framework for member countries to link economic sectors, energy, technology, job creation and training, and social protection. They aim at encouraging members to create jobs that can protect ecosystems and biodiversity, reduce energy consumption, and minimize waste and pollution. See Renner, Sweeney and Kubit, "Green Jobs: Working for People and the Environment," Worldwatch Institute, (2008), 3.

[25]See Houser *et al.* (2009) *op. cit.* and E. Barbier, "How Is the Global Green New Deal Going?," *Nature*, 464(7290), 832–833.

energy innovation,[26] and the workforce and society that can support it.[27] Methodologically, we build six modules — international law, low-carbon economy, low-carbon technology, domestic energy policy and regulation, low-carbon society and employment, and resilience in aboriginal groups and development of ethnic minorities — to embody this notion of green economy, and to capture the dynamics between policy intervention, local economy, and civic society. This research design is based on three reasons: first, the linkage between bottom (local) and upper (international) level that can be constructed;[28] second, complementarity and substitutability among these modules that can be clearly demonstrated; third, policy impact that can be identified so that the question whether the job opportunities created by interventions are "decent" enough can be answered.[29] The following are the six modules that deserve our special attention:

(1) International law dimension: identifying the green economy development in international law (including environment and trade issues in WTO and UNFCCC), drafting, and reconciling climate change strategies at the local level, in the hope that the green economy and jobs policy can be drafted;
(2) Low-carbon economy dimension: taking international trade and macroeconomic environment into account, anticipating green economy and jobs strategy among important neighboring partner countries, as the reference to promoting low-carbon economic structure in our country;
(3) Low-carbon technology dimension: assessing technology capacity of green energy or environmental service industry in our country to meet with the international competition assessment;

[26] See E. Barbier, "Linking Green Stimulus, Energy Efficiency and Technological Innovation: The Need for Complementary Policies," European Commission, 2011. Retrieved 30 March 2015, from http://transatlanticenergyefficiency.eu/sites/default/files/Green%20Stimulus_Barbier.pdf.

[27] See C. Martinez-Fernandez *et al.*, "What Role Do Local Governments and Stakeholders Have in an Inclusive Job-rich Recovery?," in *Job-Rich Growth in Asia* (France: OECD Publishing, 2011).

[28] See Chang, Han and Kim (2012), *op. cit.*

[29] See M. Wei, S. Patadia, and Kammen, "Putting Renewables and Energy Efficiency to Work: How Many Jobs Can the Clean Energy Industry Generate in the US?," *Energy Policy*, 38(2) (2010), 919–931; Hongtao Yi, "Clean Energy Policies and Green Jobs: An Evaluation of Green Jobs in U.S. Metropolitan Areas," *Energy Policy*, 6 (2013), 644–652.

(4) Domestic energy policy and regulation dimension: evaluating the content of green economy and jobs in national energy policy and regulation;
(5) Low-carbon society and employment dimension: based on development of green industry at the national level, assessing the creation of "decent" green jobs, and reviewing the training capacity in current social and employment policies;
(6) Resilience in aboriginal groups and development of ethnic minorities: assessing the impact of climate change on the development and employment opportunities among aboriginal groups.

With setting a roadmap for green development in our country in mind, we aim to plot the timelines of each module till the year 2050. This design enables us not only to explore the linkage among these modules, but to review their related policy areas in a long-term perspective. For example, domestic policy (Module 4) is affected by an international convention or agreement (Module 1), and this impacts the economic sector (Module 2) and which level of energy technology is being applied (Module 3), and in turn decides what jobs will be created (Module 5) and whether the minority has a chance to get them (Module 6). For those eager to understand deficiency, contradiction, or repetition among policy subfields, this research design can provide a systematic view, integrating knowledge on climate impacts from different disciplines in a consistent manner.

This systematic approach has important policy implications for an export-oriented country like Taiwan, especially when its trade partners, such as China, and competitors, such as South Korea, have already developed their green growth strategy based on international convention or agreement. Industrial strategy at the local level can be identified by means of this approach, and it is our plan to assess the potential for green jobs in Taiwan in this three-year project.

Six aspects of indicators are included in this tri-pillar structure, as suggested by OECD[30]: (1) International agreement or commitment on climate change; (2) Policy investment on greenhouse gas emissions

---

[30]See S. Hammer *et al.*, "Cities and Green Growth: A Conceptual Framework," OECD Regional Development Working Papers 2011/08, OECD Publishing, 2011. Retrieved 30 March 2015, from http://dx.doi.org/10.1787/5kg0tflmzx34-en.

reduction at the local level; (3) Change in green technology and innovation capacity; (4) Direct and indirect economic impacts such as changes in pollution output level, public health, and improvement in individual utility; (5) Employment outcome and skill gap; (6) Social impact on minorities and their resilience.

9.6. Next Step: Research Schedule and Contribution before the Critical Year of 2017

The year 2017 is a critical time, in that all the carbon emissions would be locked-in in existing energy infrastructure if we do not take any reduction actions, as suggested by the International Energy Agency.[31] Our research project responds to that emergency and lays down a three-year timetable (as Table 1). The first year will be a regulatory inventory for current law and

Table 1: Timetable for Research Project

First Phase (2014) Regulatory Inventory	Second Phase (2015) Green Economy	Third Phase (2016) Green Jobs
• Related international convention and UNEP recommendations on climate change. • The adjustment of international investment conventions (WTO, FCCC, GATT) toward green commerce issue. • Describing five facets of green jobs in ILO's recommendation. • Scenarios for the barriers and policy responses in 2050 under two national adaptation policies. • Institutional comparisons and recommendations.	• The development of green economy under adaptation policy. • Inventory for the industrial strategy among neighboring international trade partners; suggestion for industrial policy under green economy. • Technology content of green industry. • Industrial transformation related to green industry. • Barriers for green industry in Taiwan.	• Green jobs that support industrial transformation. • Function analysis for decent green jobs. • Know-how transfer in decent green jobs. • The formation of action plans at cabinet level.

[31] International Energy Agency, "World Energy Outlook 2012," IEA, 2012. Retrieved 30 March 2015, from http://www.worldenergyoutlook.org.

policy, both international and domestic, such as the UNEP recommendation and two adaptation policies. We aim at identifying the potential, as well as difficulties, in linking the global climate initiatives and local practices, especially in terms of barriers and policy response. In the second year, we turn to see the green economy/sector, assessing the industrial development under adaptation policy. This part of research is derived from the results of regulatory inventory in the first year. Then, we integrate previous research results and weave them into the social/employment dimension in the third year, assessing the potential in green jobs and its impact on aboriginal groups. Action plans at the cabinet level will also be proposed in this year.

This research aims at strengthening the linkage between global initiatives and local practices. One of the project results will turn into an action plan at the cabinet level, such as the Green Economy Guidelines and related governmental training or subsidy programs for green jobs. Meanwhile, the review for green industry can serve as the foundation for industry–academy cooperation.

On the other hand, domestic adaptation policy is the embodiment of international initiatives at the local level in the form of governmental practice. However, the loose connectivity between the international and the domestic level, and the lack of clarity around which policy instruments are effective in driving regional and local reform is clearly the obstacle faced by regional or member countries.[32] To strengthen the linkage between these two, we propose six policy modules to connect the international with the local from the bottom-up. It is expected that different disciplines can better interact and dialogue with each other in this research design.

In summary, the research starts with the examination on the past green economic development, and is followed by evaluating advantageous technologies in Taiwan. Through these, we are able to assess the low-carbon industries with development potential and international competitiveness and plan the job training that aims at improving the quality of employment. Further, strategies to promote a low-carbon economy and green

[32]See the research statement of JRC Peseta II project in EU. Retrieved 30 March 2015, from http://peseta.jrc.ec.europa.eu/.

employment through regulation realignment and policy adjustment are also recommended. We also need to bear in mind that any policy response will be compromised if we do not cut carbon emissions. Uncontrolled levels of greenhouse gas emissions could overwhelm any adaptive or mitigation actions we devise, in a fundamental way.

Chapter 10

Sustaining Growth, Climate Change, and Meeting Environmental Obligations: What can ASEAN Governments Do?

Euston Quah and Tan Tsiat Siong†*

10.1. Introduction

The Association of Southeast Asian Nations-5 (ASEAN-5)[1] has been undergoing strong economic growth despite today's uncertain global economic outlook. The region continues to expect a rosy outlook in the short to medium term (Table 1). Yet, in terms of Gross Domestic Product (GDP) per capita, the majority of the ASEAN-5 economies fall far behind the developed world (Figure 1). Many parts of ASEAN are still experiencing poverty (Figure 2), struggling to meet the needs of the present in areas such as sanitation and education, and ranking moderately low globally in the Human Development Index (HDI) (Table 2). As such, it is evident that strong economic growth remains paramount in improving living standards.

While economic growth raises income and generally alleviates the sufferings of the impoverished, we cannot overlook the fact that growth

*Professor Euston Quah is a Professor and Head of the Department of Economics, Nanyang Technological University. He is also the President of the Economic Society of Singapore and the Editor of the *Singapore Economic Review*.

†Tan Tsiat Siong is currently doing his postgraduate studies at the National University of Singapore.

[1]ASEAN-5 comprises Indonesia, Thailand, Malaysia, the Philippines, and Vietnam.

Table 1: Economic Growth Rate in Developing Asia and ASEAN-5

	Average Annual Real GDP Growth (%)	
Period	Developing Asia	ASEAN-5
2000–2010	8.38	5.14
2011	7.8	4.5
2012	6.64	6.10
2013	6.50	5.17
2014	6.66*	4.95*
2016	6.66*	5.62*
2018	6.51*	5.56*

*Projected
Source: IMF World Economic Outlook (April 2014).

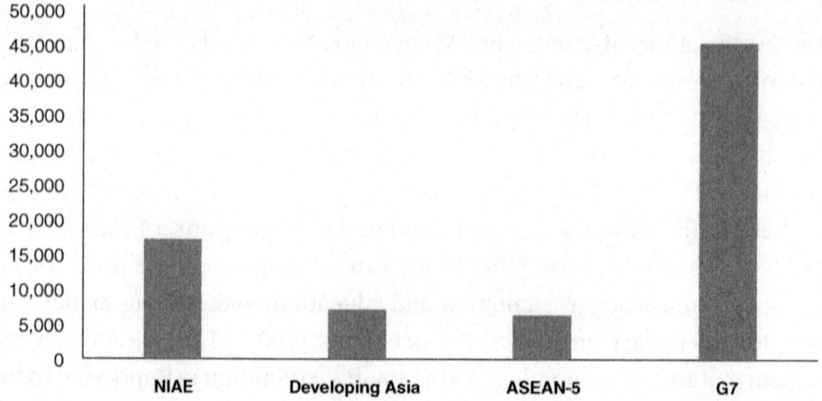

Figure 1. GDP Per Capita Based on Purchasing Power Parity
Source: IMF World Economic Outlook (April 2014).

carries its own costs and negative impacts to the natural environment and to people's health. Worsening air quality is an example of the local cost of growth, as pollution emissions are increasing steadily, particularly in ASEAN-5 due to fast economic growth and fossil fuel combustion (Karki *et al.*, 2005). On a regional scale, the damming of Mekong River by China and Laos for hydropower has endangered not only the livelihood of fishermen and downstream users, but also threatens the existence of

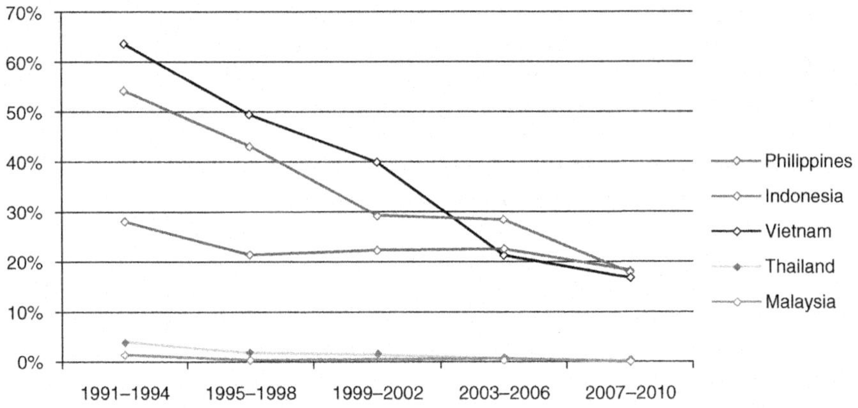

Figure 2. Poverty Rate at US$1.25 a Day (PPP) Benchmark, % of Population
Source: World Bank (2010).

Table 2: Human Development Index (HDI) of Countries

	East Asia and Pacific	South Asia	ASEAN	United States
HDI	0.683	0.558	0.669	0.937

Source: UN Human Development Report (2013).

ecological species such as the Mekong giant catfish and the Irrawaddy dolphin. In 1997, the carbon emissions from the clearing of agricultural land via slash and burn methods in Indonesia over a three-month period alone exceeded the annual carbon emissions of Europe (Quah and Chander, 2014). This large contribution to the world's carbon emissions is a global cost to every person and environment, on top of the direct costs of the resulting regional haze. Thus, the local, regional, and global costs of growth have proven to be immense, and are imposed on both current and future generations (Quah and Rae, 2010).

Growth is costly; therefore, the true success of growth hinges on whether "sustainable development" has been achieved. First coined by the 1987 World Commission on Environment and Development Report, "sustainable development" has been defined as "development that meets the needs of the present without compromising the ability of the future generations to meet their own needs" (The Brundtland Commission,

1987). Over the years, the notion of sustainable development has come to mean different things to different people, and there are various ways to operationalize the concept. Nevertheless, it is a central paradigm aimed at pursuing the twin goals of continued economic growth — meeting the needs of the present, and environmental conservation — securing the ability of the future generation in meeting their own needs.

Looking at the five largest economies of ASEAN, it has been shown that economic growth is much needed, and yet growth has been and can be costly to the environment. The same can be generalized to the other five member countries with smaller economies. Thus, working towards sustainable development is necessary in the long run. In the next section, it is first highlighted that environmental problems can be discussed from an economic perspective. Against this backdrop, characteristics of good environmental governance are broadly presented. Section 10.4 zooms into the five specific challenges encountered by ASEAN when working towards sustainable development. Further elaboration on Cost-Benefit Analysis and the Damage Schedules Approach, two suggested economic tools, are then provided in Sections 10.5 and 10.6 respectively. Section 10.7 concludes the chapter.

10.2. Economics and Environmental Problems

Environmental problems are essentially an economic issue. The fundamental economic principles of scarcity and opportunity cost dictate that that every choice made entails a trade-off; more resources allocated for a certain purpose means that the alternatives have to be sacrificed to some extent. The pursuit of economic growth to meet present needs has very often necessitated the acceptance of some degree of impact on the living environment. Conversely, preserving the living environment will always involve forgoing some measure of economic growth. Simply put — there is no free lunch in this world. The sooner society acknowledges this and becomes keenly aware of the opportunity cost of choices such as development versus environmental conservation, the easier it will be for policy-makers to make sound decisions. Understanding that environmental issues and problems are economic problems, and hence amenable to economic instruments such as cost-benefit analysis (CBA) and environment impact assessment (EIA) when choosing between projects, will go a long way towards reaching the goal of sustainable development.

10.3. Environmental Governance

The key to operationalizing the concept of sustainable development lies with good environmental governance on local, regional, and global levels. This pertains the balance of current needs and the interests of future generations.

Good local environmental management generally involves sound and pragmatic environmental policies. ASEAN governments should explore and expand the use of market solutions to solve environmental problems. For example, emissions and effluence tax, cap-and-trade systems, and deposit-refund systems are all instruments for creating incentives and disincentives for pollution control. Such mechanisms rely on the workings of the market, allowing for flexibility with regard to where and how much to reduce pollution, thereby lowering overall costs.

Secondly, it is crucial to establish baselines for environmental variables and pollutants, and to set a minimum acceptable environmental quality level as a guiding principle. When doing so, it is also important to remain cognizant to the fact that these baselines change over time in tandem with changing expectations on the quality of life.

In addition, any long-run solution to environmental problems must involve the innovation and adoption of green technology. Care must be taken to avoid possible conflicting results in this undertaking, as an attempt to fix one problem using new technologies might create another. A case in point is the introduction of methyl tertiary butyl ether (MTBE), a fuel additive used to reduce air pollution, which can inadvertently lead to the contamination of water bodies through leaking storage tanks. Also, with the use of reflective catalytic reduction to control nitrogen oxides emissions, ammonia is released.

Working with stakeholders and making them part of the solution is another characteristic of sound environmental policy. Even a well-intentioned and well-planned project can fall apart if it lacks popular support. Prior consultations with stakeholders can minimize the risk of ex-post unhappiness, and increases the likelihood of reaching an acceptable and robust solution.

The first step to improving local environmental management is for ASEAN governments to get prices right. It is imperative that prices in the market reflect the true social opportunity cost of using the resources. This means that externalities and environmental intangibles such as air quality, scenic views, and noise level have to be taken into consideration. Starting with the valuation of non-market environmental goods, governments should take stock of used and remaining environmental assets, and eventually establish green accounts to augment the Gross National Product (GNP). In evaluating projects, greater employment of appropriate and inclusive CBA, and the adoption of damage schedules (relative ranking of green goods provisions) are also warranted, especially in the rapidly developing ASEAN-5.

Lastly, environmental governance at regional and global levels should adopt a "bottom-up" approach rather than a "top-down" approach, as policies adopted within a country tend to be more effective than international or regional treaties.

10.4. Key Challenges for ASEAN Governments

In ASEAN, there lies a rather unique set of challenges in environmental governance. Five challenges are identified to have a direct or indirect bearing on achieving sustainable development. These challenges have had significant impacts on ASEAN, which will continue, and therefore require much attention and immediate action (Quah, 2010). The five key challenges for ASEAN governments currently are as follows:

- NIMBY syndrome.
- Pricing green goods.
- Solid waste disposal.
- Transboundary pollution.
- Climate change.

10.4.1. *NIMBY syndrome*

For economic growth and national development, the construction of certain kinds of facilities and infrastructure such as nuclear power plants, sewage pipelines, landfills, and airports are by general consensus, necessary. However, it is with little doubt that local residents and the immediate neighborhood surrounding the facilities in question would have to endure all negative external effects and sometimes life-threatening risks that come with them. For example, landfills are necessary for a nation's waste disposal and benefits the entire population, but only communities in close proximity to the landfills have to suffer from the pungent odor emitted, or the consequential pollution of ground water. Living near nuclear power plants also poses a threat to lives, but the entire nation benefits from the power generated. While the nation enjoys the goods and services produced by these facilities, the host communities bear much of the associated burden or social costs, and as such they would prefer to see these facilities located elsewhere. Aptly named as Not In My Backyard (NIMBY) facilities, this asymmetric distribution of costs and benefits among stakeholders is a major characteristic.

A cost-benefit analysis of NIMBY facilities using a national accounting yardstick rather than a local or regional one, would in many cases, lead to the acceptance of the facility, thereby satisfying the Kaldor-Hicks efficiency rule.[2] For cases that fail the cost-benefit test, the proper decision is to not build the relevant facility (Quah and Tan, 2002).

Examples of ASEAN's experience with the NIMBY syndrome in 2011 include: protests against a proposed rare earth refinery in Kuala Lumpur, Malaysia, and protests against a nuclear power plant at Balong, Indonesia. Imposing NIMBY facilities on unwilling recipients is one of the most difficult challenges faced by governments today. It requires the reconciliation of perceived risks, benefits, and costs, in order for community opposition to be overcome.

NIMBY facilities have been the subject of much debate in Europe and North America, but there has only been limited research and analysis of the ASEAN experience. There is, however, no doubt that the NIMBY syndrome will continue to increase in importance in ASEAN, as recent trends show increasing dissonance amongst potential NIMBY host communities. Community recipients are no longer silent or even trusting, and government as well as private sector enterprises now find that the public demand a voice

[2]Kaldor–Hicks efficiency is an outcome in which is possible for those that are made better off to compensate those that are made worse off such that ultimately, everyone is made better off in a Pareto optimal outcome.

in the decision-making processes which they used to entrust to others. This could well be attributed to greater educational opportunities and higher literacy rates. Globalization too has ushered in the Internet generation, wider information dissemination, more extensive and investigative media coverage, and the like. This has forced governments to become more transparent and accountable for their decisions, and firms and businesses to become more responsible for their actions. With rapid economic growth and greater energy demand, the expansion of infrastructure and facilities creates more NIMBY issues. Lastly, rising affluence in ASEAN would possibly mean that people expect a higher quality of life, of which a better environment plays a part.

Ideally, NIMBY facilities should be sited where people are least likely to mind it, i.e., where there is the lowest social cost — an economically efficient outcome. Even so, after the site is identified, it is still difficult to win over the local populace, thereby resulting in the failure of planned projects stemming from insufficient support from those immediately affected. Public involvement, together with mitigation and compensation auctions during a siting decision process, is a potentially useful and successful strategy. Public policy research should thus aim to reduce intrasocial conflicts by innovating new conflict resolution instruments.

One way to site NIMBY facilities is via the compulsory acquisition of land by the government, accompanied with market compensation. Often, the biggest difficulty in designing a compensation package to address the asymmetric cost-benefit distribution is that of eliciting truthful preferences — at what level of compensation are people truly willing to accept the NIMBY facility? A viable alternative would be to make use of the sealed-bid auction mechanism to select host communities and compensate them (Quah and Tan, 2002; Quah and Yong, 2008). In a sealed-bid auction, potential hosting communities or local governments submit sealed bids indicating the minimum compensation sum they would be willing to accept from a higher level government to locate a NIMBY facility within their vicinity. The bids are then compared and the facility will be built in the jurisdiction that submitted the lowest bid. The winning community (hosting the NIMBY facility) will then receive compensation equal to their bids. If all the bids are perceived to be too high, plans for the facility are either shelved or reduced in scale. Alternatively, a re-bid is conducted or new sites are proposed. The losing communities (not hosting the NIMBY facility) then make a penalty payment proportionate to their submitted bids. This prevents strategic bidding, whereby communities exaggerate their bids in an attempt to avoid hosting the facility. The sealed-bid auction therefore provides some means of assessing the actual and perceived externalities from NIMBY facilities, and ensures that the optimal location is selected in terms of minimizing social cost. Besides, the sum of payments from non-host communities will exceed the compensation to the winner, and the excess money can be put into a special fund. This special fund can be used to further compensate the host community, or for contingencies and matters pertaining to externalities imposed by the NIMBY facilities.

For NIMBY facilities that involve threats to health and human life, mitigation attempts and increased engagement with the local public might be more productive. The bribing effect suggests that monetary compensation may be seen as a form of attempted bribe on the part of the federal government. Identifying which items are compensable, and which are not, are further issues that require careful investigation.

Compensation schemes also raise the issue of equity. Attempting to elicit a community's acceptance price for hosting a NIMBY facility may result in a biased concentration of such facilities in relatively poor neighborhoods. Poor communities are willing to accept lower monetary compensation to have a NIMBY facility sited in proximity. In view of such complexities, ASEAN governments need to learn from the North American and European experiences of facility siting.

10.4.2. *Pricing green goods*

Central to the idea of sustainable development is the need to place monetary value on green goods and services provided by the environment. Monetization of environmental assets provides a means to measure the intensity and degree of concern for the environment. To ignore such valuations, or to ascribe descriptive features and list them, as seen in many cost-benefit studies and environmental impact analyzes, is to essentially reduce them to a monetary value of zero, i.e., neither a cost nor a benefit. This non-pricing in the use of environmental assets can be argued to be tantamount to taking the environment for granted. The carrying capacity of the environment may become ignored to the extent that it satisfies excess demand. For example, there is no incentive to protect the ozone layer since there is no price charged for its depletion.

Valuing and measuring environmental "intangibles", however, is not an easy task. The lack of ownership rights to environmental assets implies zero market transactions, and hence no market price to indicate economic values. As such, environmental economists attempt to seek how much people are willing to pay to conserve or to improve the environment. Alternatively, for environmental degradation, people might be asked to reveal the compensation sum required for their welfare loss. The prices of green goods are computed using various valuation methods that ASEAN governments should pick up. Valuation methods include the contingent valuation method, hedonic pricing, travel cost pricing, benefit transfer, pair-wise comparisons, and other methods.

Indeed, many of these so-called "intangible" environmental goods are now capable of being measured and valued, but valuation methods have their own limitations. Revealed preference approaches, in which people are not explicitly asked via surveys, require complete and efficient markets. Such a condition is unlikely to be fulfilled in developing ASEAN economies. Conversely, stated preference approaches, in which people place a value on non-market goods via surveys, are subjected to behavioral biases. ASEAN governments need to evaluate which of these approaches are more appropriate for them to achieve accurate data, given their individual constraints and circumstances.

Without monetary values, many policy decision-makers tend to appeal to a subjective element for valuation, usually their own value judgments. If the political system of "one man one vote" were to be applied in decision-making involving environmental improvement or degradation, then a measure of the intensity of preference for alternative proposals would be lost. If a decision on projects to improve the environment is made on the basis of a "numbers game", then the value of environmental improvement will consequently be incorrectly measured, and a wrong decision will be made.

10.4.3. *Solid waste disposal*

Another important environmental concern faced by many countries in the world today, including ASEAN, is the management of solid waste.

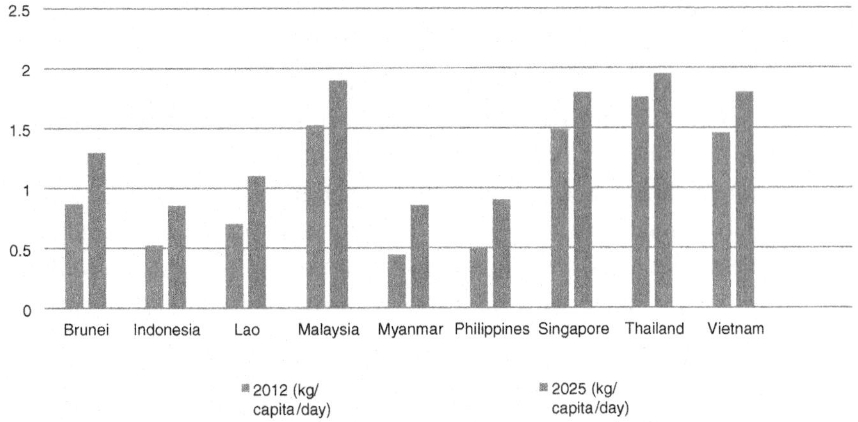

Figure 3. Current and Projected Municipal Solid Waste of ASEAN Countries (Kg/Capita/Day)

Source: World Bank: Urban Development Series; Knowledge Papers no. 15 (2012).

Rising affluence leads to increased consumerism, leading to increased waste generation (Figure 3). The problem is compounded by "throw-away" societies, where there is excessive production of disposable goods. The issue at hand is thus to correct the imbalance between the amount of waste that is produced, and the means and space necessary to dispose of them.

In the case of solid waste disposal by landfills, the solution is a costly one, especially for small countries. This is due to space constraints, as well as amenity and external problems arising from the build-up of methane gas released from decaying organic matter, and pollution from run-offs into water streams. In addition, there are problems of odour, soil contamination, and visually disturbing landscape aesthetics. Incineration, an alternative method, brings about problems of air pollution and incomplete disposal. In view of landfills and incinerators, the facility siting issue also becomes a major NIMBY problem.

Considering the various options, ASEAN governments should pursue waste reduction. This requires deploying an optimal mix of technology advancement, education, a willing and complying public, control regulations, and the proper application of economic instruments such as taxation policy and optimal pricing strategies. The effectiveness, enforcement problems, and cost justification associated with implementation remain as issues to be dealt with.

There is also the commonly neglected fact that waste reduction as an active policy in the long run, makes it increasingly difficult for households and firms to produce enough recycled materials from waste. In other words, there is a strong substitutive relationship between waste reduction, recycling, and incineration. Recycling and incineration activities may compete for financial and political support and more importantly, compete in the same domain with waste disposal. For many large cities, incinerators require heavy financial investment and this may make funds unavailable for recycling facilities, with the latter requiring subsidies. Also, private incinerators often require a contractually guaranteed volume of waste. If recycling and waste reduction are encouraged policies, there may be a shortfall of operators. These are significant issues in need of careful assessment and, perhaps for each jurisdiction, a cost-benefit analysis.

Recycling also brings about its own set of problems. Take the case of newsprint recycling. It is an established fact that producing 100 tons of de-inked fibre from old papers would produce 40 tons of sludge, which must be disposed somehow. While the total volume of material is reduced, the remaining concentrated form is more costly to dispose of properly. It has also been argued in a report from Resources for the Future, that the most likely effect of newsprint recycling appears to be a smaller, rather than larger, forestry inventory (Jenkins *et al.*, 2000). This is because most newsprint comes from trees that have been planted expressly for this purpose. If recycling increases, less of these trees will be planted.

10.4.4. *Transboundary pollution*

As economies in ASEAN continue to grow, increasing occurrences of perennial transboundary pollution have had severe economic ramifications on both victim and perpetrator countries. Typically, benefits from polluting industries accrue to one city (or country) but costs are imposed on others. That is to say, pollution crosses borders. The challenges of transboundary pollution have to be handled properly in order to avoid regional conflicts that may compromise growth and stability in ASEAN.

An example of transboundary pollution in ASEAN is the regional haze caused by forest fires in Indonesia, be it by local or foreign companies. Slash-and-burn practices perpetrated by palm oil plantations, transmigration settlers, and other agriculture industries to clear land have troubled Indonesia and its regional neighbors for years. The resulting haze from the forest fires has become an almost annual occurrence, imposing heavy costs on Singapore, Malaysia, Thailand, Brunei, and of course Indonesia herself. In 1997, the cost of the transboundary haze was estimated to be US$4.5 billion for Indonesia, Singapore, and Malaysia (Quah, 2012). This figure includes losses from tourism and recreation, and damages to environment and health.

Another example of transboundary pollution is the case of the Mekong River pollution by-products from paper mills in Yunnan; untreated domestic wastewater as well as industrial and agricultural run-offs from northeast Thailand; and sewage discharge from Cambodia and Laos, have resulted in serious pollution of the river. Agro-pollutants have been found in fish in Cambodia, and parts of the river have reported toxic levels of pesticides and high levels of mercury. Farmers' incomes and the health of citizens are subjected to significant risks.

The issue of transboundary pollution is difficult to manage because it involves multiple sovereign nations. Conventional tools such as imposing taxes, which are normally used to deal with externalities within national boundaries, cannot be enforced in this case. Furthermore, subsidies and side payments are difficult to implement. Coasian solutions involving the assignment of tradable pollution rights can lead to optimal control of pollution, but without a supranational authority to assign and enforce pollution rights, it is not a feasible option.

There are in fact incentives for victim countries to make side payments to polluters (to stop polluting) given that there are net *ex-post* benefits for both polluter and victim countries (Table 3). Such redistribution of gains can minimize pollution and reduce future losses.

To illustrate, Table 3 shows two options of agricultural methods:

(1) Using slash-and-burn, which is financially cheaper (X_0) but more polluting.
(2) Using other methods, which are financially more expensive (X_1) but less polluting.

For the polluters to switch from Option 1 to Option 2, polluters have to be paid more than the difference in financial costs of these two options ($Z > X_1 - X_0$). On the other hand, victim countries are only willing to

Table 3: Conditions of Side Payment

Example: Haze Pollution from Slash-and-Burn	
Conditions	
Option 1: Slash-and-burn.	Cost $= X_0$
Option 2: Non-slash-and-burn.	Cost $= X_1$
Gains from side payments/damages avoided by victim country/countries	$= K$
Amount of side payments from victim country/countries	$= Z$
Rule	
• Subsidize when $K > Z$	
• If $Z > (X_1 - X_0)$, polluters would have the incentive to take up Option 2, as it would result in lower cost than Option 1, and still gain.	
Amount of side payments governed by:	
• Difference between slash-and-burn cost and non-slash-and-burn cost $(X_1 - X_0)$.	
• Side payments required to avoid damages, whichever is smaller.	

pay polluters this amount Z if the gains from having less pollution is larger $(K > Z)$.

However, the issue of moral hazard may arise, whereby polluters threaten to pollute in order to receive subsidies. It also removes the incentives for countries to improve on their technology and lower costs.

There is no single economic model that resolves or fits all cases of transboundary pollution. Generally, an international or regional agreement is needed to resolve these issues, and game theoretic research offers reasons for optimism. Game theory tells us that a regional agreement is possible through gainsharing and the creation of binding agreements, which eventually leads to a cooperative outcome (Missfeldt *et al.*, 2002). The Chander–Tulkens rule states that the cost of controlling pollution should be shared by affected countries in proportion to their damages from pollution (Chander and Tulkens, 1992). For example, an earlier study estimated the relative impacts of the Indonesian haze on Indonesia, Malaysia, and Singapore to be 93.8%, 5.1%, and 1.1% respectively (Quah and Siddiqui, 2004). Hence, the cost of controlling and preventing fires in Indonesia should be shared in this proportion. The cost of controlling pollution includes the costs of issuing subsidies to improve technologies that reduce pollution; monitoring and enforcement measures; responses to pollution sources such as putting out forest fires, and so on. Three characteristics have to be satisfied for a successful regional agreement (Quah and Chander, 2014):

(1) The agreement must be individually rational, i.e., no country should be worse off if the agreement is implemented.
(2) No group of countries can be better off by leaving the agreement and having their own separate agreement.
(3) It should not be possible to draw up alternative agreements such that no country is worse off, but some country/countries is/are better off, i.e., the agreement should exhaust all gains from controlling pollution.

However, a regional agreement is still susceptible to the problem of free riding, where some countries rely on others in the negotiation process and avert potential transaction costs. In the event of high transaction costs, or the absence of a bilateral relationship between countries, cooperation is theoretically impossible. The strategic intervention model can then come into play (Figure 4). For instance, where there is unidirectional pollution and the victim country has no direct bargaining power over the polluter, the victim country can leverage on a third party to gain indirect bargaining power over the polluter, thereby mitigating the problem. This scenario is

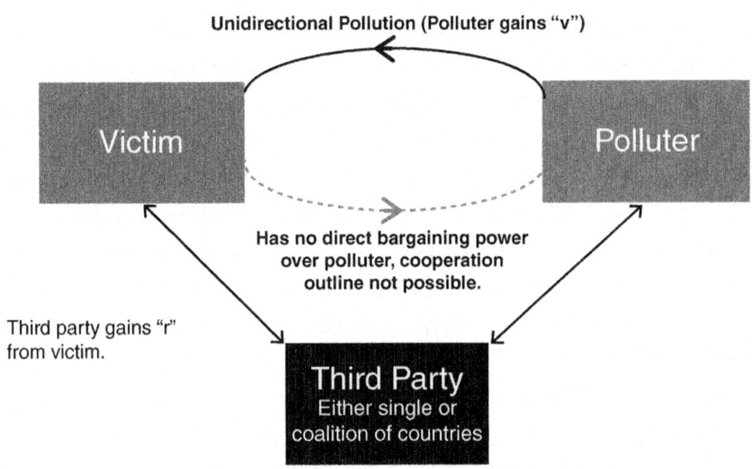

Figure 4. Strategic Intervention Model

possible given that the victim country has bargaining power over the third party, who in turn has bargaining power over the polluter.

The presence of a world/regional governing body to enforce trade embargos or trade sanctions would reduce sovereignty issues. However, weaknesses of such institutions raise the transaction costs of cooperation. Furthermore, weak enforcement and the presence of uncertainty can also affect the success of the outcome. In fact, to tackle the Indonesian haze problem, ASEAN came up with an agreement on transboundary haze pollution, but Indonesia has yet to ratify the agreement. An ASEAN joint haze monitoring system was also developed, but countries are reluctant to release sensitive information (Feng, 2014). There is a need for ASEAN to create a stronger and more coordinated regulatory agency, so as to strictly enforce the prosecution of pollution. It is also important to examine the level of integration and coordination among stakeholders and institutions in both the polluting country and its victim countries, and to regularly estimate and monitor damages suffered by victim countries.

10.4.5. *Climate change*

Trying to tackle any environmental challenge is incomplete without considering the challenges posed by climate change. The climate change issue presents a classic demonstration of the famous "tragedy of the commons" problem, where countries act based on their self-interest in deciding on the level of carbon emitted into the common atmosphere. While there has been much discussion about the science of climate change and its consequences, there is still inadequate policy discussion on the economics of climate change, or concrete suggestions to mitigate or adapt to its impending negative impacts.

ASEAN's economic growth and industrialization, coupled with its expanding energy requirements largely satisfied by fossil fuels, has inevitably led to a great spike in carbon emissions. About 90% of ASEAN's primary commercial energy requirement is fulfilled by fossil fuels. The rapid increase in the use of information and communication technologies also translates to a substantial rise in electricity input (Lean and Smyth, 2010). According to the ASEAN Centre for Energy, energy consumption in ASEAN is expected to increase from 200 million tons of oil equivalent (MTOE) in 2000 to approximately 580 MTOE in 2020. Rapid economic growth, if left unchecked, can be detrimental to the climate and eco-security of our planet.

While ASEAN is certainly not the largest contributor of carbon dioxide in the world, carbon emissions in the region are increasing. Moreover, ASEAN is highly vulnerable to climate change. An increase in the frequency of tropical storms, cyclones, and flooding can have significant impacts on ASEAN, which is highly reliant on agriculture. Large coastal areas, vulnerable to flooding, also mean that a rise in sea level can be disastrous. It has been forcasted that two thousand Indonesian and Filippino archipelagic states will be submerged by 2030 (Lean and Smyth, 2010). Furthermore, an increase in temperature in the already warm equatorial region can lead to higher incidences of cardiovascular and respiratory diseases, dengue fever, and malaria. These risks are even greater given the high population density in many parts of ASEAN.

Climate change is a complex issue due to the heterogeneous preferences and costs among developed and developing countries. The different stages of development among ASEAN countries imply asymmetric priorities and attitudes towards present and future trade-offs that are difficult to reconcile, even if there is a possible convergence of incentives to reduce emissions (Quah, 2010b). Generally, developing countries prefer non-environmental goods such as transportation and education, and place high importance on dealing with more immediate issues such as poverty and sanitation. On the other hand, developed countries prefer environmental goods, as environmental amenities lead to greater life satisfaction. The Copenhagen Accord recognized this situation of "a common goal but differentiated priorities" among developed and developing countries, and this point was reiterated at the United Nations Climate Change Conference 2013 in Warsaw.

To reduce carbon emissions, developed countries incur high costs of replacing and retrofitting current infrastructure. In contrast, developing countries face a lower opportunity cost of building more energy efficient infrastructure. Whether developing countries are willing to invest in a better environment of the future due to lower costs, or indulge in non-environmental goods and meet present needs, ultimately depends on how they weigh the long- and short-run trade-offs. The differences in marginal abatement costs further aggravate this predicament.

There is also the issue of equity. Developing countries would like to see developed countries shoulder the liability for past emissions, while developed countries urge developing countries to bear future responsibility of limiting carbon growth. The historical burden of developed countries versus the future responsibility of developing countries has been a subject of frequent discord. Climate change is a result of past and present emissions, but the consequences are borne by future generations. Across time and across nations, the discrepancy between countries becomes obvious.

Due to heterogeneity between countries, it is difficult to establish uniform cuts or benchmarks across the board. Global treaties that set a single benchmark for environmental quality for all countries are bound to fail.

The quandary faced by ASEAN governments is compounded by uncertainties that surround the impact of climate change. For example, there could well be possible misattribution of weather phenomena. Some have claimed that climate change is part and parcel of Earth's natural cycles. Uncertainties complicate the formulation of sound economic policies that rely on climate estimates to determine both the welfare-maximizing route to sustainable development, and the dollar value of damages from climate change. Creating good policies requires accurate and convincing estimations, and there is a need for more studies and estimates related to the timeline and magnitude of damage brought about by climate change.

Moreover, media reports are customarily biased, with their focus usually skewed towards the damage from climate change, with the benefits made less obvious. A warmer world might lead to increased yield for certain types of crops, more cultivable land, reduced heating expenses, reduced mortality in colder seasons, and the melting of ocean ice sheets which would open new sea lanes and create shorter transport routes in the Arctic. Greater accessibility to the Arctic could possibly yield new discoveries of raw materials with lower extraction costs. Being oblivious to the benefits of climate change would result in overestimation of the net costs of global warming. If the net costs of climate change are not as large as many have presumed to be, perhaps the solution is to find the best way to adapt to, rather than mitigate, the changing climate.

Long-term mitigation strategies include moving towards a low-carbon economy, taxing carbon emissions, cap-and-trade systems, and switching to green technology. These should be complemented and balanced with other adaptive strategies in the middle term.

Comparing mitigation strategies, carbon taxes allow policy-makers to control carbon price, is less costly than cap-and-trade systems, and has fewer opportunities for rent-seeking, financial finagling, and corruption. The potential revenue gains can also be used to cut other distortive taxes. Alternatively, cap-and-trade systems allow control over emission qualities, but results in price volatility. Such systems require the linking of regional/global markets and the setting up of an exchange where trade occurs. However, due to the complexity of collaboration, high transaction costs, risk of corruption, and income disparities across ASEAN nations, the cap-and-trade system may be a less suitable option for ASEAN.

Switching to green technology poses its own problems. ASEAN's adoption rate of green technology will be very different across countries, due to the lack of funds and lack of intellectual property protection in the less developed countries. As such, governments can improve intellectual property protection, but the key lies in enforcement rather than stricter laws. Richer members of ASEAN can also help poorer members, as it might be in the interest of some rich and highly vulnerable countries to transfer funds to poorer nations to reduce carbon emissions. Governments can also create private–public partnerships by granting green energy firms preferential tax status or monopoly rights.

Adapting to a warmer world, an alternative to climate change mitigation, includes measures such as constructing anti-flooding infrastructure amidst rising sea levels, developing flood/drought-resistant agriculture, managing water resources, improving climate forecast systems, building better drainage systems, or even dealing with poverty to increase the adaptive capacity of people.

For example, looking at the case study of Singapore, the government has begun to pursue various different adaptation strategies. They include raising the minimum reclamation levels for newly reclaimed land by one meter since 2011, piloting a flood forecasting system, expanding capacities of drainage systems through 20 projects within five years, expanding water desalination and reverse osmosis, and others. The key, however, still lies in striking a balance between adaptation strategies and longer-term mitigation.

All in all, ASEAN's success in battling climate change depends on the current population's preference for environmental goods — to what extent are current generations willing to sacrifice their material affluence for future generations. It is also important not only to adopt mitigation policies against climate change, which takes a longer time and can be difficult to achieve, but also to complement mitigation strategies with adaptive measures.

10.5. Appropriate and Inclusive Cost-Benefit Analysis

There is a pragmatic principle to be noted when it comes to environmental policy — money spent on addressing environmental problems is money diverted away from other equally pressing needs such as healthcare, education, foreign aid, etc. Governments have limited funds for both developmental and environmental objectives. In order not to be carried away with only environmental concerns, more cost-benefit analyzes for significant proposals should be conducted so we can more clearly assess which projects should be undertaken. Actively adopting CBA promotes a proper balance of different government objectives and allows for the assessment of projects that yield the largest net benefits. This improves public decision-making based on relative contributions to the "social profit" of projects. We can say the proposed project is economically and socially efficient only when social benefits exceed social costs. This then provides good justification to society for using its resources for the proposed project.

For CBA to be inclusive, it must be complete and capture all effects. For example, economic development requires the construction of infrastructure, which creates externalities and affects non-market goods (diminishing use and non-use values, detrimental effects on cultural heritage, aesthetic pleasure, etc.). The benefits of pursuing economic growth in the present could be coupled with environmental degradation and external costs not only to third parties but also to future generations. All such costs and benefits have to be included in the CBA (Quah and Mishan, 2007).

People often forget that there are two costs involved in pollution or environmental degradation to society. The cost encompasses both the costs of pollution (climate change, rising sea levels, health problems, social tension) as well as expenditure towards reducing the damage from pollution and controlling pollution (investment in infrastructure and technology,

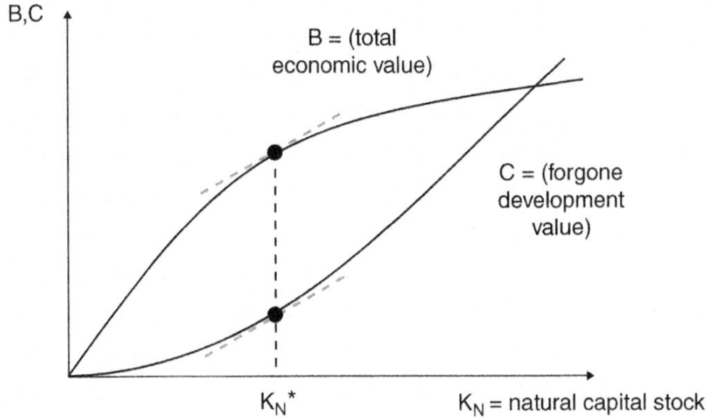

Figure 5. Optimal Stock of Natural Capital

B = benefit from increasing natural capital stock, benefits that accrue as use and non-use value.

C = cost of increasing the natural capital stock, and this cost is the forgone benefit from using the natural assets for some other purpose.

$K_N{}^*$ = optimal stock.

research and development, enforcement and regulation expenditure). It is the policy of an appropriate CBA to minimize total costs by comparing the incremental cost for both types of costs and balancing these costs. For example, the goal of governments is not to eliminate pollution, but to achieve the optimal level of natural capital stock (Figure 5). Similarly, dealing with climate change is not merely a matter of environmental protection, it also a question of finding the optimal balance between the costs and benefits of doing so.

Although appropriate and inclusive CBAs are encouraged, conducting CBAs in developing countries in ASEAN can be problematic. Developing countries have less complete goods, financial markets, and factors of production (Quah, 2013). Distorted markets create more errors in revealed preferences, and it is also harder to determine discount rates. Higher levels of disguised employment in the agricultural sector translate to higher levels of household production, and wages will not be completely reflective of the true opportunity cost of labor (*ibid.*). The problems of inequity and wealth distribution also complicate matters. CBA is a minimum criterion, but inequity issues cannot be neglected.

10.6. Damage Schedules Approach

Cost-benefit analysis and environmental impact analysis may be too time-consuming and costly for developing ASEAN nations with limited budgets, especially since they require the valuation of non-market goods. Governments may also shun CBAs due to constraints in market efficiency, education literacy, and time preference choices. Alternatively, ASEAN governments can adopt the damage schedule method, an approach that relies on society's preferences to rank examples of environmental loss and economic activity (Figure 6). This approach is cost-effective, practical, and reasonably reliable (Chuenpagdee, 1996). The approach can be summarized in six fundamental steps as follows (Quah, 2006):

(1) Develop hypothetical, but realistic scenarios of different levels of damages to the resources or different levels of activities that can cause such damages.

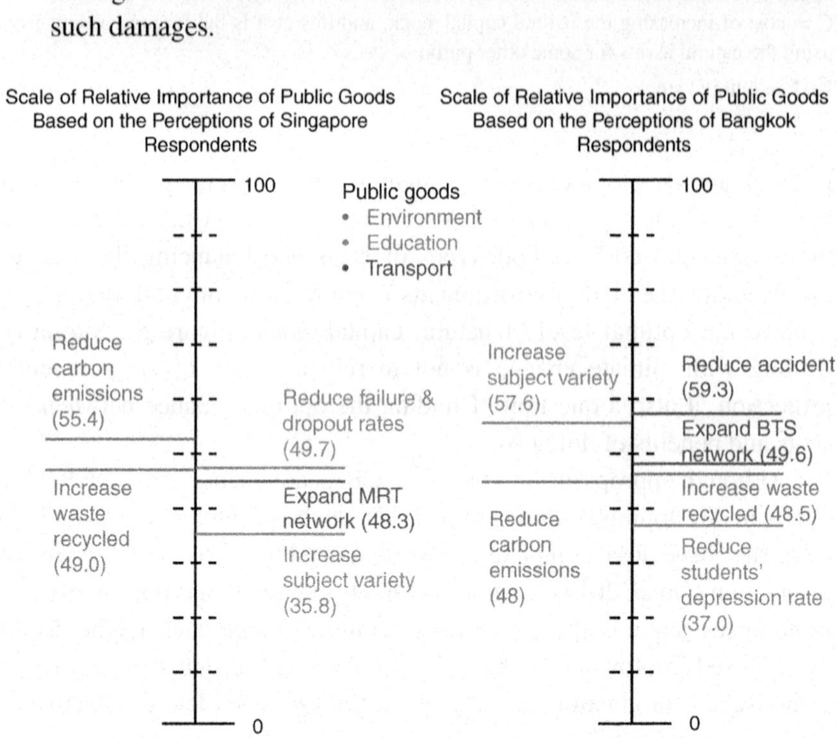

Figure 6. Damage Schedule Approach in Comparing Public Goods

Source: Oug and Quah (2014).

(2) Use the paired comparison method to present these scenarios (generally as a questionnaire booklet).
(3) Conduct the survey, asking respondents to complete the survey on his/her own.
(4) Analyze the data using the variance stable rank sum method.
(5) Test for significant differences between the rankings of relative resource importance obtained from various interest groups, and aggregate responses appropriately.
(6) Suggest policy responses in accordance with the relative importance of the resources.

10.7. Conclusion

To conclude, the goal of achieving sustainable development is one that has been proven to be highly elusive. Current discussions on this vague concept are mostly too theoretical, without much headway in pragmatism which is required for policy-making. Much can be improved upon by ASEAN governments through the use of economics instruments, particularly in tackling the five key challenges elaborated in this chapter.

Environmental governance requires the commitment to the three dimensions of sustainable development: economic growth, social improvement, and environmental protection. As countries continue to assess the progress and implementation gaps in meeting these commitments, there is also a constant need to prepare for and address new and emerging challenges. For example, it is estimated that the world's population will grow to 8.5 billion by the year 2020. According to the United Nations, there will be more than 30 so-called "megacities" with at least 10 million residents, and most of these cities will be in the less developed countries of the world (UN, 2011). This emergence of megacities, possible concentrated in Asia, including ASEAN, requires governments to come up with urban solutions.

To reiterate a few important points that have been covered: Locally, individual governments should adopt good environmental management policies. Regionally and globally, a "bottom-up" approach is preferred over a "top-down" approach. ASEAN countries can and must do more to cooperate with advanced countries to get technology transfers and encouragement for adopting good environmental policies, particularly to tackle the five key challenges: NIMBY syndrome, pricing green goods, solid waste disposal, transboundary pollution, and climate change.

References

Chander, P. and H. Tulkens. "Theoretical Foundations of Negotiations and Cost Sharing in Transfrontier Pollution Problems." *European Economic Review*, 36 (2/3) (1992), 388–399.

Chuenpagdee, R. "Assessing Environmental Values: The Damage Schedule Approach." *Economy and Environment Program for Southeast Asia*, (October 1996).

Dasgupta, S., B. Laplante, H. Wang, and D. Wheeler. "Confronting the Environmental Kuznets Curve," *Journal of Economic Perspectives*, 16(1) (2002), 147–168.

Feng, Z. K. "S'pore Frustrated at Slow Pace in Tackling Haze." *The Straits Times*, 3 April 2014.

International Monetary Fund. "World Economic and Financial Surveys: World Economic Outlook." IMF, April 2014.

Jenkins, R.R., S.A. Martinez, K. Palmer, and M.J. Podolsky. "The Determinants of Household Recycling: A Material Specific Analysis of Recycling Program Features and Unit Pricing," Discussion paper 99-41-REV, April 2000. Resources for the Future. Retrieved 4 August 2014, from http://www.rff.org/Documents/RFF-DP-98-41-REV.pdf.

Karki, S.K., M.D. Mann and H. Salehfar. "Energy and Environment in the ASEAN: Challenges and Opportunities." *Energy Policy*, 33 (2005), 499–509.

Lean, H. and R. Smyth. "CO_2 Emissions, Electricity Consumption and Output in ASEAN." *Applied Energy*, 87(6) (2010), 1858–1864.

Missfeldt, Fanny. "Game Theoretic Modelling of Transboundary Pollution." *Journal of Economic Surveys*, 13(3) (2002), 287–321.

Ong, Q. and E. Quah. "Welfare Perceptions of Public Expenditure on Environmental and Non-Environmental Goods." *Theoretical Economic Letters*, 4 (2014), 457–464.

Quah, E. Comment on Gary Hufbauer and Jisun Kim, "Reaching a Global Agreement on Climate Change: What are the Obstacles?," *Asian Economic Policy Review*, 5(1) (June 2010a), 39–58.

Quah, E. "Economic Diversification and Environment Security: Key Challenges for Asian Governments." *CSPS Strategy and Policy Journal*, 1 (2010b), 81–94.

Quah, E. Comment on Gooch Liz, "Malaysia Haze Points to a Regional Problem." *The New York Times*, 12 June 2012.

Quah, E. "Cost-Benefit Analysis in Developing Countries: What's Different?." In Michael A. Livermore and Richard L. Revesz (eds.), *The Globalization of Cost-Benefit Analysis in Environmental Policy* (UK: Oxford University Press, 2013).

Quah, E. and A.I. Siddiqui. "Modelling Transboundary Air Pollution in Southeast Asia: Policy Regime and the Role of Stakeholders." *Environment and Planning A*, 01/2004, 36(8) (2004), 1411–1425.

Quah, E. and C. Chander. "Tackling Haze with Cost Sharing." *The Straits Times*, 13 June 2014.

Quah, E. and E.J. Mishan. *Cost-Benefit Analysis*, Fifth Edition (UK: Routledge Publishers, 2007).

Quah, E. and J.S. Yong. "An Assessment of Four Popular Auction Mechanisms in the Siting of NIMBY Facilities: Some Experimental Evidence." *Applied Economics*, 40(7) (2008), 841–852.

Quah, E. and K.C. Rae. *Pursuing Green Growth in Asia and the Pacific* (US, Cengage Learning, 2010).

Quah, E. and K.C. Tan. *Siting Environmentally Unwanted Facilities* (UK: Edward Elgar Publishing, 2002).

Quah, E., K.C. Tan and E. Choa. "Use of Damage Schedules in Environmental Valuation." *Applied Economics*, 38 (2006), 1501–1512.

The Brundtland Commission. "Our Common Future." Report of the World Commission on Environment and Development, 1987. Retrieved 13 July 2014, from http://www.un-documents.net/wced-ocf.htm.

United Nations. "Population Distribution, Urbanization, Internal Migration and Development: An International Perspective." UN, 2011. Retrieved 28 July 2014, from http://www.un.org/esa/population/publications/PopDistribUrbanization/PopulationDistributionUrbanization.pdf.

United Nations. "Human Development Report 2013." UN, 2013. Retrieved 9 July 2014, from http://hdr.undp.org/en/2013-report.

World Bank. "Poverty and Equity Data." World Bank, 2010. Retrieved 10 June 2014, from http://data.worldbank.org/topic/poverty.

World Bank. "Urban Development Series; Knowledge Papers no. 15." World Bank, 2012.

Theme V

Some Case Studies

Chapter 11

Options for Adaption to Climate Change

Richard L. Ottinger, Wang Pianpian†, and Kristen M. Motel‡*

11.1. Overview: Adaptation Measures in Developing Countries

In order to tackle climate change, the Intergovernmental Panel on Climate Change (IPCC) provided a portfolio of measures: mitigation, adaptation, and constant research.[1] Although Article 10 of the Kyoto Protocol underlined the importance of adaptation,[2] adaptation to climate change had obtained limited attention in the early negotiations of climate talks.[3]

*Dean Emeritus, Pace Law School, Chair of Energy Law and Climate Change Specialty Group of the IUCN World Commission on Environmental Law.

†Pianpian Wang is a graduate of Environmental Law Program (LLM, 2013) at Pace University School of Law. Prior to that she studied environmental law in Sun Yat-sen University, China.

‡Kristen Motel is a law student at Pace University School of Law and is expected to obtain a JD with a Certificate in Environmental Law in 2015.

[1]Intergovernmental Panel on Climate Change (IPCC), "Contribution of Working Group to the Forth Assessment Report," 225 (2007), Chapter 3. Retrieved 28 March 2013, from http://w.ww. ipcc.ch/pdf/assessment-report/ar4/wg3/ar4-wg3-chapter3.pdf. The whole report can be seen at http://www.ipcc.ch/publications_and_data/ar4/wg3/en/contents.html.

[2]"Article 10 (b) Formulate, implement, publish and regularly update national and, where appropriate, regional programmes containing measures to mitigate climate change and measures to facilitate adequate adaptation to climate change . . .". See "Kyoto Protocol to the United Nations Framework Convention on Climate Change." Retrieved 6 April 2015, from http://unfccc.int/resource/docs/convkp/kpeng.pdf.

[3]The reasons are as following: "[f]irst, the focus of the international climate change negotiations has largely been on mitigation (perhaps because attention to adaptation could been on viewed as 'giving up on mitigation') . . . Second, adaptation is largely undertaken at the local scale . . . Third, it is difficult to generalize the ways that individuals or communities are likely to adapt to specific impacts." See IPCC at 225.

In 2010, the Cancun Session of the Conference of Parties (COP) of the United Nations Framework Convention on Climate Change (UNFCCC) highlighted the equal importance of adaptation just as mitigation.[4] Since then, increasing attention has been drawn to adaptation practice by the international society.

Typically, adaptation can be broken down into three steps:[5] (1) realizing the effects of climate change; (2) designing the activity for remediation of climate change consequences; and (3) implementing the adaptation measures. The most important "adaptive step" is decision-making and the way in which the activity is designed.[6] Developing countries have already realized how climate change affects them. Change of temperature, precipitation, and extreme events such as superstorms and sea level rise already have threatened them, in particular their water, agriculture, and coastal zones.[7] Their adaption programs "must be an integral part of a development policy process that ensures mainstreaming of climate adaptation in all relevant sectors of society while not forgetting the other multiple drivers (social, economic, and environmental problems)".[8] In other words, the ultimate goal of the adaption program design should create a link between the adaptation measures and overall sustainable development.

---

[4] See United Nations Framework Convention on Climate Change (UNFCCC), "The Cancun Agreement," 2010. Retrieved 17 March 2013, from http://cancun.unfccc.int/adaptation/. Before the Cancun Agreement, "in 2001, at COP 7 in Marrakesh, Parties established the LDC work programme to develop national climate change mechanisms and build capacity, including NAPAs: Through National Adaptation Programmes of Action LDCs identify and report their adaptation needs". Also, "in 2006, at COP 12 in Nairobi, the Subsidiary Body for Scientific and Technological Advice (SBSTA) was mandated to undertake a 5 year project to address impacts, vulnerability and adaptation in relation to climate change — the Nairobi work programme". See UNFCCC, "Adaptation." Retrieved 28 March 2013, from http://unfccc.int/focus/adaptation/items/6999.php.

[5] Pieter Terpstra, "The Difficulty of Defining Adaptation Finance." Retrieved 30 November 2013, from http://www.wri.org/blog/difficulty-defining-adaptation-finance.

[6] *Ibid.*

[7] See UNFCCC, "Climate Change: Impacts, Vulnerabilities and Adaptation in Developing Countries," 18–26, 2007. Retrieved 6 April 2015, from http://unfccc.int/resource/docs/publications/impacts.pdf.

[8] Ole Mertz, Kristen Halsnaes, Jørgen E. Olesen, and Kjeld Rasmussen, "Adaptation to Climate Change in Developing Countries," 747 (2008). Retrieved 4 April 2015, from http://www.researchgate.net/profile/Jorgen_Olesen2/publication/23966336_Adaptation_to_climate_change_in_developing_countries/file/9fcfd50a940e0129ee.pdf.

The measures will need to be different for each country since "adaptation needs to be context specific and dependent on the resources available to the impacted domain".[9]

Therefore, developing countries need to conduct analyzes of climate change impacts and vulnerability assessments in top-down and bottom-up adaptation planning.[10] This is because

> [i]t is now increasingly recognized that, for poor communities, adaptation approaches that are rooted in local knowledge and coping strategies, and in which communities are empowered to take their own decisions, are likely to be far more successful than top-down initiatives.[11]

However, adaptation cost and adaptive capacity are two barriers in developing countries. On a regional basis, the East Asia and Pacific Region bears the highest adaptation cost according to some economic analyzes.[12] With regard to adaptive capacity, developed countries have more strengths on adaptation development than developing countries because of their greater resources and often better governance systems, technical support, and quality of national institutions.

Various partnerships that aim to improve adaptive capacity and resilience have been carried out in developing countries. However, it is a long-term process and takes substantial time to see the results. Under the pressure of climate change, instantly effective adaptation measures would be helpful to developing countries. This chapter proposes a series of adaptation paths that can be promoted and replicated under different circumstances regardless of different governance structures.

---

[9]Aondover Tarhule, "Climate Change Adaptation in Developing Countries: Beyond Rhetoric," 173–174, 2012. Retrieved 4 April 2015, from http://cdn.intechopen.com/pdfs-wm/25931.pdf.

[10]Valerie Nelson, Richard Lamboll, and Adele Arendse, "Climate Change Adaptation, Adaptive Capacity and Development Discussion Paper," 3 (2008). Retrieved 4 April 2015, from http://climateanddevelopment.nri.org / background_papers / nelson_lamboll_and_arendse_climate_change_adaptation.pdf.

[11]IIED, "Participatory Learning and Action: Community-based Adaptation to Climate Change," 3. Retrieved 4 April 2015, from http://pubs.iied.org/pdfs/14573IIED.pdf.

[12]World Bank, "The Cost to Developing Countries of Adapting to Climate Change: New Methods and Estimates," 4 (2010). Retrieved 4 April 2015, from http://siteresources.worldbank.org/INTCC/Resources/Executivesummary.pdf.

Table 1: Summary of Adaptation Options

Programs / Compared Items	Adapting Sector	Implementing Organization/ Project Developer	Staff Training Needed?	Funding Size	For Urban or Rural	Program Cycle
Living Shorelines	Water	Government and citizens	Yes	Initial plan development training cost funded by the government. Implementation funded by the government and citizens	Rural or urban areas with no "hard" shoreline protection structures	A short installation period with results that can be seen within a short–mid period of time
Rebuilding Oyster Reefs	Water	Government, NGOs, or research/ educational institution	Yes	Initial creation cost and subsequent monitoring funded by the government, NGO, or institution	Both	A moderate installation period with results that can be seen within a short–mid period of time
Flood Gates and Flood Protection Systems	Water	Government or private corporation	Yes	Initial equipment and training cost funded by the government or corporation	Urban	A short planning and training period with results seen during the initial use

(Continued)

Table 1: (*Continued*)

Programs / Compared Items	Adapting Sector	Implementing Organization/ Project Developer	Staff Training Needed?	Funding Size	For Urban or Rural	Program Cycle
Citizen Monitoring	Water	Government, NGOs, or research/ educational institution and citizens	Yes	Initial equipment and training cost funded by the government or institution	Both	A moderate set-up time and ongoing program. Requires consistent coordination between citizens and funding organization
Displacement Management	Water	Government	Yes	Continuous funding required to evacuate, transport, and resettle populations	Both	Long-term and ongoing program, becoming more frequent with increased storm events
Green Infrastructure	Water	Government and NGOs	Yes	Grant sources include federal, state, and private funding	Both	Started in 2001 and requires consistent input and maintenance
Micro-Hydropower	Energy	Government and NGOs	Yes	Funded by the local government or non-profit group.	Both. Rural would be simplest	Initial training and start-up is about one year and requires consistent maintenance and training

Programs / Compared Items	Adapting Sector	Implementing Organization/ Project Developer	Staff Training Needed?	Funding Size	For Urban or Rural	Program Cycle
Rooftop Solar and Community Distributed Generation	Energy	Religious Group (NGOs)	No	Funded by the religious group	Both	A short installation period with long pre-planning and technical preparation
Strengthening Power Transmission	Energy	Utility Companies	Yes	Funded by the utility company	Urban	A mid–long term upgrade
Grow Crops in a New Way	Food	Government	Yes	Funded by local governments	Both	Results can be seen within a short–mid-long period
Coordination of the Water–Energy–Food Nexus	Water, Energy, and Food	Government	No	Funded by local governments	Urban	Consistent coordination between departments

11.2. Adaptation Practices by Sectors

One striking characteristic of developing countries is that agriculture accounts for a large part of gross domestic product and employment.[13] Thus, there are particular concerns within the agricultural sector, and it is believed that adaptation centered on agriculture in developing countries would have chain effects on their water and energy as well. In this section, 10 case studies that focus on water, energy, and food present effective adaptation measures for these important sectors.

Since developing countries generally have a lower capacity to adapt to climate changes, simple, economical, and innovative principles are prerequisites in the selection of these case studies.

WATER CASE STUDIES

Case 1: Living Shorelines to Protect Coastal Communities from Flooding and Erosion

Coastal habitats act as the front line for communities against tropical weather events.

Shoreline restoration is important to protect marshes that serve as a protective barrier for coastal communities against storm surge and waves during severe weather events.[14] Shoreline restoration also helps prevent coastal erosion, which can dramatically alter the landscape of a community and cause damage to cities and towns. Elements of shoreline restoration projects include mangrove restoration and other headland restoration measures like the restoration of dunes and beaches, marsh creation, implementation of oyster barrier reefs,[15] and built shoreline protection measures like sea walls and breakwaters.[16]

---

[13] Alok Kumar, "Lecture 2: Major Characteristics of Developing Countries," 4, University of Victoria. Retrieved 6 April 2015, from http://web.uvic.ca/~kumara/econ420/characteristics-dev.pdf.

[14] La. Coastal Prot. and Restoration Auth., "Featured Projects." Retrieved 6 April 2015, from http://coastal.la.gov/our-work/projects/featured-projects/ (last).

[15] See Case Study 2: Oyster Barrier Reefs.

[16] La. Coastal Prot. and Restoration Auth., *op. cit.*

A. How do living shorelines work?

Mangrove destruction has left many coastlines vulnerable to increased storm surge and erosion. An effective shoreline protection measure is to rebuild mangrove belts in grids to prevent coastal erosion. Erosion can be slowed by building grids of mangroves, by first starting with trellises of small permeable bundles made of branches and bamboos.[17] This allows for maximum sediment capture while minimizing the wave force, helping to restore marshlands around each bundle.[18] These grids create low-energy conditions that restore land so that mangroves can be planted and eventually recolonized onto the belts.[19] By building mangroves in grids, space is freed up for other coastal activities like aquiculture, recreation, and fishing. Furthermore, evidence shows that restoring mangroves in grids also helps restore crab populations.[20]

Currently, this practice is being implemented in the Pacific Island Nations of Sri Lanka, Indonesia, and in Thailand, Malaysia, and India, and also in western African countries like Guinea Bissau and Sierra Leone.[21] In some locations the practice is being implemented using a "Bio Rights Microcredit Scheme" that combines these environmental improvements

[17]Gerard van Raalte, "BwN for Coastal Resilience. A Pilot on Rehabilitation of a Mangrove-Mud Coast in Timbul Sloko (Java, Indonesia)," Building with Nature (2 February 2014). Retrieved 6 April 2015, from https://publicwiki.deltares.nl/display/BWN/BwN+for+coastal+resilience (explaining a pilot study on rehabilitation of a mangrove-mud coast).

[18–19]*Ibid.*

[20]See M.E. Walton, L. Le Vay, J.H. Lebata, J. Binas, and J.H. Primavera, "Assessment of the Effectiveness of Mangrove Rehabilitation Using Exploited and Non-Exploited Indicator Species," *Biological Conservation*, 138 (2007), 180–188.

[21]These mangrove practices are being implemented by non-profit agencies like Deltares and Wetlands International. See Deltares — Retrieved 30 June 2014, from http://www.deltares.nl/en; Wetlands International — Retrieved 30 June 2014, from http://www.wetlands.org/Whatwedo/Mangrovesforcoastalresilience/tabid/174/Default.aspx. Wetlands International has partnered with the Indonesian government service BPDAS-PB of the Ministry of Forestry to coordinate the engagement of government agencies in data gathering, fieldwork, and policy dialogues regarding mangrove restoration. The Indonesian Government Centre for Marine Resource Survey is undertaking a detailed mapping of mangrove status and distribution in Indonesia. See Pieter Van Eijk, "Mangrove Capital," Wetlands International. Retrieved 30 June 2014, from http://www.wetlands.org/LinkClick.aspx?fileticket=gu1BlW9FeFo%3D&tabid=56.

with the goal of poverty reduction.[22] Funding is given to communities with the promise that they will undertake mangrove restoration.[23] The environmentally-friendly economic development resulting from these projects is further funded, and a condition for receiving the financial support is an agreement that the countries will refrain from returning to practices that damage the mangroves.[24] This scheme enables local communities to manage mangroves sustainably.

In addition to restoring mangroves, shorelines restoration projects, sometimes known as "living shorelines", are being implemented in many different countries and locations.[25] This process involves altering and restoring shorelines to protect them from erosion using "non-hard" techniques like plantings, beach nourishment, breakwaters, sills, and low profile oyster reefs. The "hard" techniques used in the past, like seawalls, bulkheads, and rock revetments, are expensive and do not adjust to changing weather needs because they do not allow wetlands to retreat

---

[22]Wetlands International, "Bio-Rights: Micro-Credit for Nature Conservation." Retrieved 30 June 2014, from http://www.wetlands.org/Aboutus/Howwework/Biorights/tabid/2732/Default.aspx.

[23–24]*Ibid.*

[25]In the United States, states with active shoreline restoration projects include Delaware, Florida, Louisiana, Maryland, New York, North Carolina, Texas, and Virginia. Louisiana has managed to benefit over 110,000 acres of coastal habitat through restoration. See La. Coastal Wetlands Planning & Restoration Act Program, "CWPRA Legislation." Retrieved 30 June 2014, from http://lacoast.gov/new/About/. New York City is conducting wetland restoration projects at beach sites to improve natural flooding patterns. See Jonathan D. Cybulski, "Transportation Infrastructure Resiliency," 10 (Harvard University, Volpe Center Report No. DOT-VNTSC-14-02). Retrieved 6 April 2015, from http://www.bv.transports.gouv.qc.ca/mono/1149029.pdf. In Belgium, 40% of the coastline is protected using soft solutions such as planting brushwood, sand fences, and planting of Marram Grass. See Toon Verwaest, "Nourishment Practices in the North Sea Countries," 2000. Retrieved 6 April 2015, from http://www.masterpiece.dk/UploadetFiles/11320/25/Nourishment_practices_in_the_North_Sea_Countries_may_2000.pdf. (Along the Black Sea, wetlands on the Danube River floodplain are being restored by constructing salt marshes along sea dikes to mitigate the power of incoming waves.) Canada's shoreline restoration programs in the Caribou Islands, Nova Scotia, and in Point Ellice Park in Victoria, British Columbia, include, "soft armoring" and "green shores" which build brush walls and add biomass. See The Arlington Group Planning and Architecture Inc., "Sea Level Rise Adaption Primer," 2013. Retrieved 6 April 2015, from at http://issuu.com/scbc/docs/slr-primer.

towards the land as sea level rises.[26] Additionally, the machinery used to build "hard" techniques can be disruptive and cause damage to coastal wetlands.

Living shorelines can be created through non-structural stabilization or hybrid stabilization, depending on wave energy. Usually for the portion of the shoreline closest to land, riparian vegetation is restored, banks are graded, and fiber logs are put into place.[27] Closer towards the water, a tidal marsh is created and vegetation is further enhanced.[28] Some projects will also use beach nourishment, dune restoration, and marsh toe revetment.[29] Closest to the water, marsh stills and marshes with groins are created.[30] Depending on the wave energy of the area, there can be a living breakwater or oyster reef located offshore to further protect the coast.[31]

B. Implications for further replication

Implementing a shoreline restoration program has many benefits; most importantly, it is a method to ensure less property loss during storms by preventing bank erosion. Living shorelines have a low technology cost because they provide a natural solution. These programs are also aesthetically pleasing and conducive to recreation activities because they protect the beach.

Aside from minimizing coastal erosion and providing protection during storms, restored shorelines help improve coastal water quality by filtering impurities and stormwater run-off during rain events. Furthermore, the living shorelines restore marine habitat and spawning grounds for marine life, improving the local fishing industry. When storms create removal of shading and vegetation from shoreline they can cause shallow

---

[26]NOAA, "Alternate Shoreline Stabilization Methods." Retrieved 30 June 2014, from http://coastalmanagement.noaa.gov/initiatives/shoreline_stabilization.html; N.C. Coastal Fed'n, "Terminal Groins 101." Retrieved 6 April 2015, from http://www.nccoast.org/Content.aspx?key=79e38ba1-e221-4033-a2b4-86f17b374a7f&title=Terminal+Groins. North Carolina no longer allows hard structures to be built because the state recognizes they are no longer a sustainable solution. See NOAA Coastal Servs. Ctr., "North Carolina Sea Wall Ban Stands up to Legal Challenge," March/April 2000. Retrieved 6 April 2015, from http://www.csc.noaa.gov/magazine/2000/02/nc.html.

[27]Va. Ctr. for Coastal Res. Mgmt., "Living Shorelines Program." Retrieved 30 June 2014, from http://ccrm.vims.edu/livingshorelines/.

[28–31]*Ibid.*

water temperatures to rise, depleting oxygen and adversely affecting fish. Consequently, shoreline restoration improves wildlife habitat by providing cover and keeping temperature and oxygen at sustainable levels.

It can be difficult to implement a shoreline restoration program in areas where the shoreline is privately owned. When land is privately owned, it is up to individual property owners to make informed decisions about managing their shorelines. If there are buildings directly adjacent to the beach, it is more difficult to locate an area suitable for restoring wetlands. Shorelines restoration will also not be useful in areas where most of the shoreline is already hardened; the seawalls would first have to be demolished.

Each site must be evaluated for its distinct features, which requires data collection from fundamental monitoring. For example, restored shorelines are not the best protection method for areas with high wave energies. When deciding to implement shoreline restoration, it is helpful to have private contractors specializing in marine areas trained to correctly plant the vegetation and install the techniques. Also, when considering whether to build mangrove belts, the coastal conditions must be conducive to mangrove growth. "[M]angroves can only be successfully restored if the abiotic conditions are optimal (bathymetry, fresh water, and sediment input)."[32]

Living shorelines programs are typically implemented by a Coastal Planning Commission. The Commission develops the program after a Shoreline Management Plan is created to identify areas of importance and agencies with authority and responsibility for implementing the program.[33] The Plan requires data and monitoring of the prioritized sites to identify the areas of high priority. The Shoreline Management Plan is then adopted, usually by town boards and carried out by municipal officials or local citizen's organizations.[34] These municipal officials or local citizen's organizations, where available, encourage landowners to implement living shorelines by communicating the protection and benefits previously mentioned. This communication can be in the form of pamphlet

[32]van Raalte (2014), *op. cit.*

[33]Va. Ctr. for Coastal Res. Mgmt., *op. cit.*

[34]*Ibid.*

distribution, town hall meetings, or through information posted on websites. Successful shoreline restoration programs have incorporated free training for the local citizens' organizations, landowners, and shoreline contractors, a list of local native plants that will thrive in tidal areas, a planting and cost guide for these plants, public demonstration areas where citizens can see how the shorelines are being restored, and a decision tree for the local government agency responsible for making the decisions.[35]

Shorelines restoration programs require multigovernment agency collaboration. These programs typically involve many divisions of local government including Building and Zoning, Fish and Game, Tourism, Economic, Recreation, and Environmental Protection departments.[36] These departments ideally should be involved in the creation of the Shoreline Management Plan, evaluate its effectiveness, and raise any concerns they have about the program implementation.[37] Specific responsibilities of each agency should be outlined in the Plan.

Case 2: Oyster Reefs to Protect Coastlines

Oyster reefs protect the coastline from erosion by slowing wave velocity and storm surge. Up to 90% of the energy from wind-generated waves can be absorbed by these reefs.[38] Reefs not only preserve the coastline, they protect coastal communities from property damage and shield them from the economically destructive effects of beach erosion. Creating oyster reefs not only protects the shoreline, but it also restores estuary ecosystems and provides more stock for commercial fishermen. Adult oyster reefs can significantly increase coastal water quality by filtering it.

[35–36] *Ibid.*

[37] In the United States, Virginia has a "networked program", meaning the program relies on a network of state agencies and local governments to administer enforceable laws and regulations that protect wetlands, dunes, fisheries, subaqueous lands, and air and water within Virginia's coastal zone. See Va. Ctr. for Coastal Res. Mgmt., "Living Shorelines Program Frequently Asked Questions." Retrieved 30 June 2014, from http://ccrm.vims.edu/livingshorelines/faq.html.

[38] NOAA, "Coral Reef Conservation Program, Coastal Protection." Retrieved 13 May 2011, from http://coralreef.noaa.gov/aboutcorals/values/coastalprotection/; http://www.nature.org/ourinitiatives/regions/northamerica/unitedstates/louisiana/oyster-reef-restoration-in-louisiana.xml.

A. How does oyster reef creation and restoration work?

Oyster reefs are built to act as a barrier against coastal erosion to break incoming waves on the reef. Juvenile oysters are raised in a hatchery and then "attach, set, and grow on shell substrate".[39] New reefs constructed from marl, a sedimentary rock made out of limestone, can be constructed using bags of recycled oyster shells.[40] Oyster larvae attach to the structure and grow while additional oyster larvae attach to older shells, creating a substantial reef.

As part of the New York and New Jersey Baykeeper's Oyster Restoration Program, microscopic oyster larvae are released into 450 gallon saltwater tanks containing bags of clam shells to start the reef.[41] The larvae attach and grow on the shells at the hatchery, and once the oysters have grown for two months, they are released onto the new oyster reefs.[42] The new reefs are usually deployed off of a pier.[43] The structures are placed sub-tidally, because two to three feet of water must remain above the oysters during the lowest tide to ensure oyster survival.[44] In New York City's Soundview Park, 125 cubic yards of shells were placed by a barge and deposited in the intertidal region around existing rock outcroppings.[45] The shell reef provides substrate on which new oysters can attach.

Louisiana has three oyster reef projects taking place in Vermillion Bay, Grand Isle, and St. Bernard Parish that are being built to protect the coastline from erosion and absorb wave energy during storms.[46] Besides restoring the coastline after Hurricane Katrina, Louisiana also needed to stimulate its commercial fishing economy. Seventy percent of the world's oysters come from the Gulf of Mexico, and oyster reefs are a viable way to protect and increase the oyster population and create commercial fishing jobs.[47]

---

[39]New York New Jersey Bay Keeper, "Oyster Restoration Program." Retrieved 30 June 2014, from http://nynjbaykeeper.org/resources-programs/oyster-restoration-program/.

[40–45]*Ibid.*

[46]Louisiana Coastal Protection and Restoration Authority, "Featured Projects." Retrieved 30 June 2014, from http://coastal.la.gov/our-work/projects/featured-projects/.

[47]The Nature Conservancy, "Restoring Oyster Reefs for People and Nature." Retrieved 30 June 2014, from http://www.nature.org/ourinitiatives/regions/northamerica/unitedstates/louisiana/oyster-reef-restoration-in-louisiana.xml.

Louisiana has put Louisiana State University in charge of monitoring the conditions and health of theses oyster reefs.[48]

B. Implications for further replication

The success rate for oyster reef creation or restoration programs has been high.[49] For example, the oyster project at Naval Weapons Station Earle in New Jersey had 90% of the oysters on the reefs survive the winter.[50] There is an initial start-up cost to this program; however, once the reefs are placed into the water, the only continuing cost is for monitoring reef success. Part of that start-up cost includes testing oyster support structures to determine which structures will have the highest oyster survival and growth rates. Finding the initial funding to implement the program can be a challenge.

The water quality in which the oyster reefs will be placed must be taken into consideration when choosing a location for the reefs. Creating oyster reefs in waters that are contaminated will create not only a public health issue, but will also prevent fishermen and citizens from harvesting and profiting from contaminated oysters.[51] However, as with any water where shellfish harvesting is restricted or prohibited, posted signs and educating the local population can be an effective way to prevent the ingestion of

[48] *Ibid.*

[49] In North Carolina's Alligator River an oyster reef was created in Swanquarter National Wildlife Refuge to break incoming waves, prevent storm damage, and control erosion. The project also provides more oysters to help North Carolina's shellfish economy and improves offshore wildlife habitat. See The Nature Conservancy, "Building Oyster Reefs off the North Carolina Shore: Good for Plates and People." Retrieved 30 June 2014, from http://www.nature.org/ourinitiatives/regions/northamerica/unitedstates/northcarolina/explore/oyster-reef-restoration-in-north-carolina-1.xml. Texas's oyster reef creation at Mad Island Marsh Preserve has also been successful. See The Nature Conservancy, "Southern Solutions for a Global Problem." Retrieved 30 June 2014, from http://www.nature.org/cs/groups/webcontent/@web/@southcarolina/documents/media/prd_014253.swf. New York is investigating installing offshore breakwater systems using oyster reefs at Totten and next to Great Kills Harbor. See New York City Department of City Planning, "Urban Waterfront Adaptive Strategies," 2013. Retrieved 6 April 2015, from http://www.nyc.gov/html/dcp/pdf/sustainable_communities/urban_waterfront.pdf.

[50] New York New Jersey Bay Keeper, "Oyster Restoration Program," *op. cit.*

[51] The Nature Conservancy, "Restoring Oyster Reefs for People and Nature," *op. cit.*

contaminated food. Patrolling the waters in boats is another way to ensure contaminated oysters will not be harvested.[52]

Case 3: Flood Control Structures

Flood control structures are an effective way to protect buildings and communities from during precipitation events. When planned and used correctly, water control structures can reduce the level of clean-up after storms, helping cities recover from severe storms more quickly. When using flood control structures, protection priority should be given to essential emergency resources, specifically hospitals, communication and data centers, schools, and fire departments. In the United States, flood control structures are being installed in coastal cities to protect from rising floodwaters.[53] Globally, many cities are incorporating the use

---

[52]New York New Jersey Bay Keeper, "Oyster Restoration Program," *op. cit.*

[53]New York City is installing 25 tidal floodgates along Beach Chanel Drive and other waterfront areas to reduce the risk of flooding and make them more resilient. See New York City, "Coastal Protection Report: Mayor Bloomberg Details Progress on Resiliency Projects Outlined in the City's Long-Term Plan to Protect City against the Effects of Climate Change on Hurricane Sandy Anniversary," 29 October, 2013. Retrieved 30 June 2015, from http://www1.nyc.gov/office-of-the-mayor/news/348-13/mayor-bloomberg-details-progress-resiliency-projects-outlined-the-city-s-long-term-plan-to/#/0. This investment prevents water from flooding the streets and overpowering the storm drains. In low-income areas like Coney Island Commons, the housing is being redesigned to include deployable flood panels that will minimize damage in the future. See New York City, "Coastal Protection Report," *op. cit.* An integrated flood protection system involving flood gates and raised entrances to tunnels that are vulnerable to flooding is being implemented in Hunts Point, East Harlem, lower east side Manhattan, Hospital Row, Red Hook, Staten Island, and Newton Creek areas of New York City. See New York City, "Coastal Protection Report," *op. cit.* (Floodgates are also being used to protect the subway system and public transit hubs.) In New Orleans, the city is building a perimeter defense system consisting of storm surge barriers in front of the city. Robert Reid, "Defending New Orleans," Civil Engineering, Novermber 2013. Retrieved 6 April 2015, from http://www.asce.org/uploadedFiles/Awards/OPAL_Awards/ce_Inner%20Harbor_Navigation_Canal_Surge_Barrier.pdf. In North Carolina, water control structures that have flashboard risers and tidal gates are being used on the Alligator River to restore hydrologic regime and prevent saltwater intrusion. The Nature Conservancy, "Climate Change Adaptation: Alligator River National Wildlife Refuge." Retrieved 30 June 2014, from http://www.nature.org/images/climate-change-adaptaion-in-the-alligator-river-nwr.pdf.

of water control structures into their emergency and future planning efforts.[54]

A. How do flood control structures work?

There are many types of flood control structures that can be used. Flashboard risers are structures installed in a drainage outlet that allow water in the drainage outlet to be raised or lowered as needed; when the risers are lowered, the drainage occurs less quickly.[55] When more flashboards are added to the riser, the drainage rate is decreased and the water level in the surrounding area will rise.[56] Flashboard risers lessen the amount of water running through the storm drain, so the intensity and frequency of

[54] Indonesia has started a project to build a five-mile long sea wall along Jakarta's coast to protect homes and businesses after last year's rain season caused $580 million in damages. See Adriana Nina Kusuma and Gayatri Suroyo, "Indonesia to Build $263 Million Sea Wall to Protect Capital," *Reuters*, 9 October 2014. Retrieved 6 April 2015, from http://www.reuters.com/article/2014/10/09/us-indonesia-infrastructure-idUSKCN0HY16720141009. In Venice, Italy, the MOSE Project is currently underway. This project involves installing locks to protect Venice from flooding. The locks will be activated when the water is 3.6 feet above normal levels. Sylvia Poggioli, "MOSE Project Aims to Part Venice Floods," National Public Radio, 7 January, 2008. Retrieved 6 April 2015, from http://www.npr.org/templates/story/story.php?storyId=17855145. Only four of the anticipated 78 gates have been installed so far due to economic issues in the country. Poggioli (2008), *op. cit.* When completed, the floodgates will extend more than a mile, blocking three inlets to the lagoon. See Italy, "Coastal Flood Risk Italy." Retrieved 30 June 2014, from http://www.climateadaptation.eu/italy/coastal-floods/. In Tokyo, Super Levees are used to protect the city during severe rain events. See Cameron Allan McKean, "Tokyo Building Enormous 'Super-Levees' to Hold Back Its River," Next City, 30, 4 December 2013. Retrieved 6 April 2015, from http://nextcity.org/daily/entry/tokyo-is-building-enormous-super-levees-to-hold-back-its-river. The levees are designed for extreme events in dense urban areas. The levees consist of a broad river embankment that can withstand overflow to prevent destruction. The super levee is 10 meters high and 300 meters wide, and resistant to seepage and earthquakes. See McKean (2013), *op. cit.* Le Thu Trang, "Flood Adaptive Cities, Preliminary Project Report," May 2013. Retrieved 6 April 2015, from at http://www.deltainterventions.com/wp-content/uploads/2012/09/p4report_Trang-Le.pdf.

[55] See U.S. Fed. Emergency Mgmt. Agency, *Floodproofing Non-Residential Buildings*, Chapter 4, "Other Flood Protection Measures," 31 July 2013. Retrieved 6 April 2015, from http://www.fema.gov/media-library-data/2c435971150193efc6a6ba08f2403863/P-936_sec4_508.pdf.

[56] *Ibid.*

combined sewer overflows is reduced, and this minimizes health risks due to flooding.

Conversely, flood or tidal gates can be adjusted to control or stop water entry from a storm surge and prevent the intrusion of floodwaters while also deflecting flood borne debris.[57] Permanent floodwalls are the simplest and most cost-effective to design and construct. These are usually made of reinforced concrete or masonry, sometimes with steel reinforcing bars embedded in the core.[58] Floodwalls are usually placed away from buildings so that structural modifications are not required.[59] Permanent floodwalls are the best suited for geographical areas that experience high storm surge and rapid flowing water.

Temporary floodwalls can be easily assembled and are moved into place and anchored where necessary.[60] The barriers are then filled with water or sand to hold them in place. In the United States on the Mississippi River, invisible floodwalls are utilized.[61] The "invisible floodwalls" are steel walls that can quickly be put up at the first warning of a flood, and when not needed, they can be taken down and stored.[62] These steel floodwalls are also being deployed in New York City, particularly in areas that experienced catastrophic flooding during Hurricane Sandy.[63] The Verizon Headquarters Building has invested in a nine-foot high deployable wall, and the NYU Langone Medical Center and the New York and New Jersey Port Authority Buildings have also invested in temporary walls.[64]

[57–60] *Ibid.*

[61] *See* Brian Johnson, "Flood Wall Project Pays off Early for St. Paul's Downtown," Finance & Commerce, 29 September 2010. Retrieved 6 April 2015, from http://finance-commerce.com/2010/09/flood-wall-project-pays-off-for-st-paul-downtown-airport/; Flood Control America, "Removable Flood Wall Project — St. Paul Downtown Airport." Retrieved 6 April 2015, from http://floodcontrolam.com/projects/flood-wall-st-paul/.

[62] Flood Control America, "How the Floodwall Works." Retrieved 30 June 2014, from http://floodcontrolam.com/works/. Flood Control America is the manufacturer of the temporary floodwalls used on the Mississippi River.

[63] David W. Dunlap, "A Modern Flood Barrier Aims to Protect Verizon's Landmark Building," *The New York Times*, 30 October, 2013. Retrieved 6 April 2015, from http://www.nytimes.com/2013/10/31/nyregion/a-modern-flood-barrier-aims-to-protect-verizons-landmark-building.html?ref=nyregion&_r=0.

[64] Dunlap (2013), *op. cit.* Heather Holland, "NYU Langone Medical Center Builds Flood Wall-Shielded Emergency Room," *DNAinfo N.Y.*, 24 October, 2013. Retrieved 6 April 2015,

B. Implications for further replication

Flood duration is a critical consideration in choosing the right flood control structure for a specific community. The longer the wall will be in contact with the floodwaters, the more potential there is for seepage and a need for backfill.[65] Gravity floodwalls are most appropriate for areas needing only low-level protection.[66]

Agency and community planning is an essential part of adaptation measures. For example, floodwalls may only protect the low side of a building because they must tie into high ground. "When a building is protected by a floodwall, underground utilities that serve the building and run under the floodwall must be considered so the utilities do not become conduits that allow floodwaters to pass through or under the floodwall and into the building."[67] During emergency planning, local governments can consider making it mandatory for any redevelopment in urban areas to have flood control structures in place.

In planning to use temporary barriers, the barrier must be the appropriate size for the site. It is also important to train emergency personnel and conduct drills so that they are knowledgeable about how to deploy the barriers. An effective monitoring system is essential when using temporary floodgates because authorities need enough time to deploy the temporary barrier. Proper storage of the barriers and cleaning them after use will ensure their effectiveness for future events.[68] These temporary barriers are most effective for river floodwaters and not for coastal floods; there has not been enough data about whether saltwater hinders their performance.[69] Additionally, temporary walls that fill with water may float instead of protect in denser saltwater.[70]

All flood control measures require consultations with engineers to determine the correct location and method of installation, and also to assess that the structure will provide adequate protection. This is essential because

from http://www.dnainfo.com/new-york/20131024/kips-bay/nyu-langone-medical-center-builds-flood-wall-shielded-emergency-room; Press Release, "Port Authority of N.Y. & N.J: Port Authority Continues Aggressive Efforts to Rebuild Facilities Following Superstorm Sandy and to Prepare for Upcoming Hurricane Season," 29 May 2013. Retrieved 6 April 2015, from http://www.panynj.gov/press-room/press-item.cfm?headLine_id=1794.

[65] See U.S. Fed. Emergency Mgmt. Agency, *Floodproofing Non-Residential Buildings*, *op. cit.*

[66–70] *Ibid.*

once the barriers are breached they provide no protection. As an additional benefit, tidal gates can also restore the hydrogeology of interior wetlands while preventing saltwater intrusion.[71]

A drawback to using permanent floodwalls is the high cost and large quantity of materials required to construct them, especially if the height needed is significant.[72] When placed on a river, permanent structures can affect the "hydrological regime both up and down stream, and often exacerbate flooding in other places".[73] There is a suspicion that permanent flood control measures, such as channelization and levee construction, may greatly magnify flood stages.[74] Flooding conditions "may also be transferred downstream as water is accelerated through" the flood control system.[75] Additionally, human engineering with permanent flood control measures has the potential to deepen and narrow rivers.[76] Furthermore, an aesthetic drawback to water control structures is that they may completely surround a structure.

Case 4: Using Citizen Monitoring as an Early Warning System to Minimize Damage from Storms

Monitoring programs are an easy and effective way to minimize damage from future storms by monitoring conditions and using early warning systems. Citizen monitoring programs can be used to monitor water levels, water quality, sediment dynamics, storm patterns, wave strength, and erosion. Citizens who live on the shoreline are best situated to conduct monitoring during minor storms because they usually will be present to observe a variety of weather conditions and patterns.

A. How does citizen monitoring work?

Citizen monitoring programs allow citizens to monitor sea level rise and weather conditions when they are provided with the necessary equipment and training. The programs set up a system for data collection and results are

---

[71–72] *Ibid.*

[73] Graham A. Tobin, "The Levee Love Affair: A Stormy Relationship?," *Water Resources Bulletin*, 31 (1995), 359.

[74] Robert E. Criss and Everett L. Shock, "Flood Enhancement through Flood Control," *Geology*, 29 (2001), 875, 878.

[75] Tobin (1995), *op. cit.*, 359, 361.

[76] Criss and Shock (2001), *op. cit.*, 875, 876.

submitted to a research program or government agency. Allowing citizens to monitor shoreline changes and weather conditions saves considerable cost because government agencies or academic institutions do not have to repeatedly travel to the shoreline. Additionally, the data collected provides a more comprehensive picture about weather patterns and shoreline alteration. The collected data can be shared with neighboring communities and with the state for use in local emergency, economic, and tourism planning. Sharing data prevents the duplication of efforts, saving time and valuable municipal resources.

In the United States, government agencies like the Environmental Protection Agency ("EPA"), and non-profit resource-protection agencies,[77] have recognized the value of citizen monitoring programs. The EPA has Citizen Science Grants that provide citizens with smart phones and basic surveillance equipment to monitor water pollution and air quality.[78] The EPA provides these individuals with training and educational brochures to distribute to other citizens in their community. The trained citizens periodically hold local town hall meetings to discuss their findings with the public and with the EPA. These forums act as question-and-answer sessions for the public, while also informing the municipality and state officials about the current status and challenges. In the United States, coastal states like Maine have volunteer citizens monitoring for coastline erosion.[79]

---

[77] New York-New Jersey Harbor and Estuary Program, "Citizen Science." Retrieved 30 June 2014, from http://www.harborestuary.org/citizenscience.htm; Riverkeeper, "Citizen Science Water Quality Monitoring Workshops." Retrieved 30 June 2014, from http://www.riverkeeper.org/news-events/events/other-events/citizen-science-workshop-1309/. University of Rhode Island, "Watershed Watch." Retrieved 30 June 2014, from http://www.uri.edu/ce/wq/ww/.

[78] EPA, "Citizen Science Grants." Retrieved 30 June 2014, from http://epa.gov/citizen science/ (discussing citizen air monitoring programs in Newark, New Jersey and in Tonawanda and the Bronx, New York).

[79] NOAA Office of Ocean and Coastal Resource Management, "Support Research, Monitoring, and Decision Support Tools." Retrieved 30 June 2014, from http://coastalmanagement.noaa.gov/initiatives/shoreline_ppr_research.html#1. Maine is using citizens to conduct monthly beach profiles to better understand seasonal erosion rates. Teams measure beach elevation, collect sediment samples, take pictures of the beach to document erosion, and report their data to geologists at the University of Maine and the Maine Geological Survey. See State of Maine, "Beach Profiling Project." Retrieved 30 June 2014, from http://www.geology.um.maine.edu/beach/.

Maryland incorporates citizens monitoring with training for coastal property owners to teach them how to implement shoreline erosion control practices.[80]

In Nicaragua, the CAFENICA Meteorological Station at Miguel Angel Ortez uses data provided by coffee farmers to monitor weather patterns.[81] This data is used for strategic public planning and agricultural planning. The farmers are also trained in adaptation measures to help deal with anticipated droughts and dry seasons.[82] In Kenya, a Michimikuru case study was used to train students to monitor conditions on the rivers.[83] Five riverbank scouts were trained and certified on water and soil conservation.[84] In Jakarta, citizen monitoring is being used to develop flood maps and create an early warning flood detection system.[85]

B. Implications for further replication

Citizen monitoring programs are ideally replicable in shoreline areas inhabited by a local population to minimize the need to travel to the coast to conduct monitoring. Someone continuously residing on the shoreline is

[80]NOAA, "Education and Outreach Campaigns: Maryland's Shore Erosion Control Technical Assistance Program." Retrieved 30 June 2014, from http://coastalmanagement.noaa.gov/initiatives/shoreline_ppr_edu.html. Maryland Department of Natural Resources, Chesapeake and Coastal Service, "DNR Mobil Restoration Map." Retrieved 30 June 2014, from http://www.dnr.state.md.us/ccs/restoration.asp; Maryland Department of Natural Resources, "Mobil Restoration Project Input Form and Mobil App, About." Retrieved 30 June 2014, from http://www.mddnr.org.

[81]Café Direct, "Adaptation for Smallholders to Climate Change — How Nicaraguan Coffee Farmers Adapt to Climate Change — the PRODECOOP Case Study." Retrieved 30 June 2014, from http://www.adapcc.org/en/nicaragua.htm (case study, operating plan, and presentations available online).

[82]Farmers are trained in adaptation measures like how to manage stormwater by installing irrigation ditches to prevent landslides, using rainwater reservoirs, and storing water to be used in an installed pump for the community fountain. Café Direct (2013), *op. cit.*

[83]Café Direct, "Pilot Groups — Kenya." Retrieved 30 June 2015, from http://www.adapcc.org/en/kenya.htm; Café Direct, "Michimikuru Tea Growers." Retrieved 30 June 2015, from http://producersfoundation.org/projects/michimikuru-tea-growers/.

[84]Café Direct, "Michimikuru Tea Growers," *op. cit.*

[85]Asian Disaster Preparedness Center, "Flood Preparedness Initiatives of High-Risk Communities of Jakarta," 2010, 4–7. Retrieved 7 April 2015, from http://www.adpc.net/igo/category/ID226/doc/2013-c28Jbn-ADPC-Safer_Cities_27.pdf.

best situated to collect the data. Additionally, citizens living in vulnerable coastal communities have a high interest in preparing their communities and protecting their property.

When implementing a citizen's monitoring program, it is necessary to have distributable technology, like tidal gauges, thermistors, dynamometers, and turbidity tubes, which can be provided by a local government, a non-profit agency, or a university.[86] These devices help citizens measure sea level rise, water quality, wave force, and water temperature. The issuers of the equipment must initially train citizens on their use and how to accurately collect the data. Those instruments will also require occasional maintenance and calibration, which can be done by an appointed lead citizen, the municipality, or a local university. Initially, getting members of the public to volunteer to collect data can be difficult, so the preliminary organization of the program can be time-consuming. It is also necessary to have a location available to host monthly or quarterly town hall meetings so that data can be presented.

All monitoring should be coordinated by a municipal or state agency, or by an academic institution. This coordination will assure a uniform method of data submission, organization, and publication. The data should be made publically available. Its use should help to aid developers and governments in conducting coastal vulnerability and risk assessments on future development projects.[87] Ideally, the results of citizen monitoring should be incorporated into a database or online tool that neighboring municipalities and the state can access. Online entry forms, like the one for

---

[86]For an example of items to take under consideration when implementing a citizen monitoring program, see Ellin *et al.*, "North Carolina Sentinel Site Cooperative: Report on the Sea-Level Rise Research and Monitoring Coordination Workshop," 36–38, 42–44 (NOAA Technical Memorandum NMFS-SEFSC650, 2013). Retrieved 7 April 2015, from http://oceanservice.noaa.gov/sentinelsites/pdf/NC_SSC_SLR_Research_Coordination_Wkshop.pdf.

[87]Caribbean Community Climate Change Center, "Caribbean Adaptation to Climate Change Project (CPACC)." Retrieved 30 June 2014, from http://www.caribbeanclimate.bz/closed-projects/1997-2001-caribbean-planning-for-adaptation-to-climate-change-project-cpacc.html (CPACC is a database designed to foster sustainable development in the Caribbean through making data available to developers, the government, and the community on which adaptation measures will best protect the development and how to implement these measures).

the Maine Beach Profiling Project,[88] allow volunteers to input their data online to cut down on paperwork and faster transmission.[89] It is important to note that data collected may be site-specific and not transferrable to neighboring communities.

Case 5: Green Infrastructure to Manage Gray Water

Gray Water is stormwater, regular rainwater, and melted snow that runs off streets, lawns, and other urban sites. It can be used to supply all non-drinking water needs; in some countries it is widely used for agriculture. In Israel, all buildings are required to have rainwater collectors on their roofs for use as drinking water. However, since it can pick up and transport chemicals, nutrients, sediment or other pollutants, and debris during heavy rain and snowmelt events, it can adversely affect water quality in the receiving water bodies. With the trend of urbanization and high frequency of storm weather in the context of climate change, Gray Water management has captured people's attention.

Traditional Gray Water management design has been focused on collecting Gray Water in piped networks and transporting it off site as quickly as possible, either directly to a stream or river, to a large Gray Water management facility (basin), or to a combined sewer system flowing to a wastewater treatment plant.[90] Besides these concrete facilities (also called "Gray infrastructures"), a new type of method named green infrastructures that can address the concern of Gray Water run-off has been promoted in many cities in the United States. The project of Greenseams implemented by the City of Milwaukee, Wisconsin is a good example.

A. How does the Greenseams project work?

The Greenseams project was initiated in 2001. A "Greenseam", used by the City of Milwaukee, defines Greenseams lands as linkages throughout the

[88]State of Maine, "Beach Profiling Project." Retrieved 30 June 2014, from http://www.geology.um.maine.edu/beach/.

[89]NOAA Office of Ocean and Coastal Resource Management, "Support Research, Monitoring, and Decision Support Tools." Retrieved 30 June 2014, from http://coastalmanagement.noaa.gov/initiatives/shoreline_ppr_research.html#1; State of Maine, "Beach Profiling Project." Retrieved 30 June 2014, from http://www.geology.um.maine.edu/beach/.

[90]U.S. EPA, "Stormwater Management." Retrieved 30 June 2014, from http://www.epa.gov/greeningepa/stormwater/.

utility service area watersheds that collectively work in sync to manage large amounts of water. After the Metropolitan Milwaukee Sewerage District (MMSD) acquired all such lands "through voluntary sales and protected [them] with a permanent conservation easement...". Program staff "identified properties containing water-absorbing (hydric) soils and pinpointed them as priorities due to the amount of water they could hold".[91] These properties have an estimated 1.3 billion gallon storage capacity.[92]

Utilizing plants is another important element of the program. Trees can break up raindrops and slow rainfall down before hitting the ground. Also, the deep and long roots of trees can hold water as well as decelerate erosion along stream banks. Nearly 100,000 trees have been planted under this project.[93] These trees are all planted in a way that can maximize their storage function.

Besides the technical portion, the Greenseams program favors local communities' participation. The first requirement for this community involvement is that, since local land owners possess most of target properties of the project, it is crucial to reach out and establish a partnership with them. Next, contribution from local citizens helps to educate local communities and teach the value of the Greenseams project.

In the beginning, the project was not an immediate success. Local municipalities and landowners had a hard time understanding why a sewerage district would be buying land for preservation purposes, and furthermore, what the district would do with the property acquired.[94] In order to penetrate local communities, it took several years to educate the

---

[91] Jessica Hrobar, "Innovative Program Connects Wastewater Utilities and Nature," Water World. Retrieved 14 June 2014, from http://www.waterworld.com/articles/uwm/articles/print/volume-4/issue-4/features/innovative-program-connects-wastewater-utilities-and-nature.html.

[92] *Ibid.*

[93] MMSD, "Greenseams." Retrieved 14 June 2014, from http://www.mmsd.com/en/flood management/Greenseams.aspx.

[94] Maragaret Buranen, "Milwaukee Goes Green and Gray for Stormwater Management." Retrieved 14 June 2014, from http://www.stormh2o.com/SW/Articles/Milwaukee_Goes_Green_ and_Gray_for_Stormwater_Manag_16762.aspx.

public about why sewer district was implementing the Greenseams Program and what the long-term vision was.[95]

Currently, the Greenseams project covers more than 2,600 acres of land.[96] After 13 years of being in place, the program serves as a model in many cities in the United States on Gray Water management.

B. Implications for further replication

There are some common barriers to introducing green infrastructure into Gray Water management, such as limited land and technical support, barriers in codes and rules, and funding.[97] Greenseams has been able to minimize these impacts, which could imply further replication in developing countries.

How to choose a location?

Greenseams is concentrating on projects in the most urban areas within four local major watersheds, where most suburban growth is expected.[98] Also, these locations contain hydric soil and water-absorbing vegetation.[99] The core idea is to maximize the value of green infrastructure. Suburban areas in developed or developing countries are sometimes those that have insufficient public facilities to deal with Gray Water. Instead of clearing the land to build traditional Gray Water treatment facilities, green infrastructure

---

[95] *Ibid.*

[96] MMSD, "Greenseams," *op. cit.*

[97] See Sarah A. Hammitt, "Toward Sustainable Stormwater Management: Overcoming Barriers to Green Infrastructure," 2010, 19–22. Retrieved 7 April 2015, from http://stuff.mit.edu/afs/athena/dept/cron/project/urban-sustainability/Stormwater_Sarah%20Madden/Shammitt_thesis_final.pdf. See also Oregon Environmental Council, "Barriers to Overcome," Chapter 4. Retrieved 14 June 2014, from http://www.oeconline.org/our-work/water/stormwater/stormwater%20report/barriers.

[98] Maragaret Buranen, "Milwaukee Goes Green and Gray for Stormwater Management." Retrieved 14 June 2014, from http://www.stormh2o.com/SW/Articles/Milwaukee_Goes_Green_and_Gray_for_Stormwater_Manag_16762.aspx.

[99] Jessica Hrobar, "Innovative Program Connects Wastewater Utilities and Nature," Water World. Retrieved 14 June 2014, from http://www.waterworld.com/articles/uwm/articles/print/volume-4/issue-4/features/innovative-program-connects-wastewater-utilities-and-nature.html.

made full use of the existing natural resources.[100] For developing countries, it is relatively easy to pick up candidate locations that fit the above standards.

How to obtain land and funding?

In Milwaukee, all land was acquired through voluntary sales and protected with a permanent conservation easement. The project staff consistently broadened property sources by holding open house meetings, mailing out packets of information to addresses from a GIS database, and personally meeting with prospective sellers who were curious about the program.[101] With regard to funding, about half the grants have come from a state program that facilitates cooperation between local governments and private non-profit organizations. The other half of the funding has come from the U.S. Fish and Wildlife Service, other state programs, and local sources.[102]

For developing countries, it is not a difficult challenge to acquire land compared to developed countries, while limited funding is a much greater barrier. Similar to the Greenseams project, cities in developing countries can utilize available grant funding to support, promote, and maintain green infrastructure. Under some circumstances, cities or higher levels government should provide special funding for climate-related programs.

Case 6: Managed Retreat and Relocation

During severe weather events, the permanent or devastating loss of coastal areas often prevents people from being able to return to their homes and communities. Planning the retreat and relocation of those affected by these events is essential to ensuring the well-being of a country's population.

A. How does managed retreat and relocation work?

This adaptation mechanism could provide a sustainable way to adapt to climate change by removing people from harm's way. Managed retreat

[100] Regarding the cost-benefit analysis between gray and green infrastructure, see U.S. EPA, "Water: Green Infrastructure Cost-Benefit Resources." Retrieved 14 June 2014, from http://water.epa.gov/infrastructure/greeninfrastructure/gi_costbenefits.cfm.
[101] Hrobar, "Innovative Program Connects Wastewater Utilities and Nature," *op. cit.*
[102] *Ibid.*

involves the retreat of a coastal population from areas identified as vulnerable or being threatened by adverse weather events. This departure can be to permanent resettlement locations or can be temporary during the adverse weather event and rebuilding period thereafter. However, depending on the level of destruction, even temporary displacement can have a long duration. Long-term resettlement may require relocation across national borders, which can be done by creating an agreement with neighboring countries that directly addresses this issue.[103] The retreat can also be through a nationally run voluntary relocation program or a buyout program.[104] For example, the Maldives promotes resettlement from outer islands to principal islands by establishing a "sovereign wealth fund to buy new land elsewhere as an insurance policy in the event of wholesale displacement, and to consolidate the population dispersed over 200 islands onto 10–15 islands."[105]

---

[103]Dr Shakeel Kazmi, "Climate Change: Human Rights in the Times of Climate Displacement," 1 April 2012, 125 (SJD dissertation, Pace University School of Law). Retrieved 7 April 2015, from http://digitalcommons.pace.edu/cgi/viewcontent.cgi?article=1013&context=lawdissertations.

[104]In the United States, Harris County, Texas has formed a "voluntary home buy-out" program to relocate households living in flood plains. See Harris County Flood Control District, "Voluntary Home Buyout." Retrieved 30 June 2014, from http://www.hcfcd.org/buyout.asp?flash=yes. Vietnam has implemented a "Living with Floods" program to resettle 20,000 landless and poor households living in regularly flooded areas to nearby but less endangered areas. See Benoit Mayer, "The International Legal Challenges of Climate-Induced Migration: Proposal for an International Legal Framework," *Colo. J. Int'l Envtl. L. & Pol'y*, 22 (2011). Retrieved 7 April 2015, from http://www.colorado.edu/law/sites/default/files/Mayer%20%28Corrected%29-S.pdf. Mozambique is promoting voluntary resettlement from floodplains to higher ground. See also Alex de Sherbinin, "Government-Led Displacement and Resettlement Due to Climate Change," Columbia Univ. Ctr. for Int'l Earth Science Info. Network (Planet under Pressure Conference, 28 March 2012). Retrieved 7 April 2015, from http://www.planetunderpressure2012.net/abstracts/D3041307.pdf.

[105]This fund is financed by diverting a portion of the country's tourist revenue to buying land in India, Sri Lanka, and Australia. Uma Kothari, "Political Discourses of Climate Change and Migration: Resettlement Policies in the Maldives," *Geographical Journal*, 180(2) (2014), 130–140. Retrieved 7 April 2015, from http://onlinelibrary.wiley.com/store/10.1111/geoj.12032/asset/geoj12032.pdf?v=1&t=hxfgpk1t&s=6307bc8e0c6220d1b11a4b5e54c5e25fd69bac88 (the Maldives also tries to entice migration to principal islands by increasing connectivity, transportation, and building accommodations on the main islands). Alex de Sherbinin, "Assessing Displacement and Resettlement Due to Climate

Climate-related migrants are people who are forced from their homes due to the effects of climate change, including natural disasters and coastal flooding or sea level rise.[106] Under the United Nations, this displaced population is not entitled to refugee status.[107] There is no international fund that exists to support climate displacement refuges.[108] This situation can have significant impacts on the rights of those relocating and can affect funding available to the countries receiving displaced communities. Right now, the term refugee is being heavily debated and a possible universal right to housing is being used as support for awarding climate-related migrants protection.[109]

B. Implications for further replication

Retreat can be done in four stages, starting with the planning process. It is important to plan the resettlement process and choose a dedicated government agency that will be vested with authority and resources to run the program. The program should have clearly stated goals and objectives for all stages of the process and a detailed plan to achieve them. These objectives should address issues like living standards for those resettled and establishing a fund for large-scale displacement in the event that a major natural disaster displaces a large number of people.

The second stage is relocation planning, including evacuation plans to orderly move communities out of harm's way. To evacuate entire communities, evacuees are told to meet at a dedicated location, usually

---

Change Mitigation and Adaptation Projects," 15, Columbia Univ. Ctr. for Int'l Earth Science Info. Network (IAIA Symposium on Climate, 15–16 November 2010). Retrieved 7 April 2015, from http://www.ciesin.columbia.edu/confluence/download/attachments/92799474/deSherbinin_IAIAsymposium_nov2010.pdf.

[106] See generally Michael B. Gerrard and Gregory E. Wannier, "Threatened Island Nations: Legal Implications of Rising Seas and a Changing Climate," 2013; National Geographic, "Climate Refugee." Retrieved 30 June 2014, from http://education.nationalgeographic.com/education/encyclopedia/climate-refugee/?ar_a=1

[107] Kazmi (2012), *op. cit.* See also Jane McAdam, UNHCR, "Climate Change Displacement and International Law: Complementary Protection Standards," 2011. Retrieved 30 June 2014, from http://www.unhcr.org/4dff16e99.html.

[108] Kazmi (2012), *op. cit.*

[109] *Ibid.* Silke Marie Christiansen, "Environmental Refugees: A Legal Perspective, the Terminology Debate," 9, 2010, citing Richard Black, "Environmental Refugees: Myth or Reality?," 2001, 2.

a school or a church, where buses are used for transportation.[110] As a model for relocating communities, planners can look to the methods used by countries resettling communities to accomplish major dam-building projects.[111] Evacuation planning also includes identifying the location of where people can resettle. This location must be planned in advance, in coordination with other countries if extraterritorial resettlement is contemplated, or through purchasing or preparing land in a different location for resettlement locally.

The third step involves initiating economic development in the areas identified for permanent relocation to which refugees are to be resettled. When planning for relocation of people inside the same country, there are important considerations that must be taken to prevent deterioration of the resettlement areas and impacts on the rest of the country. The initial drop in living standards of those relocated following relocation should also be addressed in this plan and must focus on preventing the creation of slums and overcrowding.[112] This third step also includes planning with international relief agencies for temporary food, medical care, and shelter. In the longer term, plans need to be made for employment opportunities and adequate access to grocery stores, markets, healthcare, and education facilities. Additionally, this step should focus on community formation and activities to improve living standards for those resettled.

---

[110]U.S. Fed. Emergency Mgmt. Agency, "Emergency Declarations and Sheltering Evacuees." Retrieved 30 June 2014, from http://www.fema.gov/emergency-declarations-sheltering-evacuees.

[111]Dam projects with major resettlement components include Mali's Manantali Dam, see Peter Bosshard, "A Case Study on the Manantali Dam Project," International Rivers (1 March 1999). Retrieved 30 June 2014, from http://www.internationalrivers.org/resources/a- case-study-on-the-manantali-dam-project-mali-mauritania-senegal-2011; Egypt's Aswan Dam, see Penn. State Univ., "Problems with the Aswan Dam," https://courseware.e-education.psu.edu/courses/earth105new/content/lesson06/04.html; Costa Rica's Arenal Dam, see United Nations Environment Programme: Dams and Development Project, "Costa Rican Expropriation Law and Inter-American Development Bank Requirements for Development Projects: The Arenal Dam," Development Submission. Retrieved 30 June 2014, from http://www.unep.org/dams/documents/ell.asp?story_id=95; and Sri Lanka's Pimburetewa Dam, see Thayer Scudder, "A Comparative Survey of Dam-Induced Resettlement in 50 Cases," Cal. Tech., 2005. Retrieved 7 April 2015, from http://www.hss.caltech.edu/~tzs/50%20Dam%20Survey.pdf.

[112]Kazmi (2012), *op. cit.*

Community formation includes things like places of worship with religious compatibility, community centers, and the creation of social groups.

Finally, after each weather event requiring retreat and temporary or permanent displacement, an after-action review must be conducted to assess the procedures in place and pinpoint where improvements can be made. Those improvements should be implemented to create a sustainable redevelopment process.

There are drawbacks to managed retreat. First, and most apparently, it can be politically difficult to implement. People do not want to leave their homes and communities, and this can cause a lack of participation in voluntary resettlement. And a location receiving the refugees will also need to plan for their provisioning and treatment. Next, retreat and relocation require a reliable monitoring system to remove people safely before inclement weather. This means having early warning and detection methods available, such as installing flow regulators along rivers.[113] After the permanent relocation, it is best where possible for the receiving country to provide in advance for land, facilities, and support to accommodate the refugees.[114]

These measures create strain on the existing infrastructure in the receiving location that must be assessed and addressed. Managed retreat to domestic areas commonly involves moving populations to current internal

[113]The Bangladesh government has installed flow meters along rivers. See Shardul Agrawala *et al.*, "Development and Climate Change in Bangladesh: The Sundarbans, Adaptation Options Available for Management of Coastal Flooding," Organization for Economic Cooperation and Development, 2003, 35. Retrieved 7 April 2015, from http://www.oecd.org/dataoecd/46/55/21055658.pdf.

[114]In 2008, Maldives' President Nasheed announced plans to create a fund from revenue generated by tourism, to be used to acquire land to relocate its population in countries such as India and Sri Lanka. Kazmi (2012), *op. cit.* The right to housing in the Maldives in threatened by a scarcity of land and is still looking to permanently relocate victims of the 2004 Tsunami. United Nations News Centre, "Maldives: Climate Change Threatens Right to Housing, Says UN Expert," 26 February 2009. Retrieved 7 April 2015, from http://www.un.org/apps/news/story.asp?NewsID=30026&Cr=housing&Cr1=climate#.U7X-814_594. With the help of federal funding, the State of New Jersey has acquired homes in flood-prone areas to prevent resettlement. Press Release, New Jersey Dept. Envt'l Prot., "Christie Administration Announces $26.3 Million in Federal Aid for Post-Sandy Buyouts of 89 Residential Properties in Woodbridge," 6 February 2014. Retrieved 7 April 2015, from http://www. nj.gov/dep/newsrel/2014/14_0009.htm.

cities. This exerts pressure on the city's infrastructure and economy, and also creates the potential for overcrowding. This may lead a country to plan for permanent relocation outside of the country. This kind of settlement plan requires agreements with the receiving countries and provision for multinational cooperation and funding. When identifying countries with which to collaborate, it is important to try to choose countries that have the same or similar language, education systems for children, religion or places of worship that are available, and job opportunities.[115] Furthermore, some countries have their own internal policies for climate-related migrants and an application process that citizens may need help in fulfilling.[116]

ENERGY CASE STUDIES

Case 7: Renewable Energy to Provide and Secure Power to Communities during Severe Weather Events

Renewable energy can provide a way to secure a power supply for communities in areas vulnerable to power outages or that are dependent on power supplied from a far distance. Many renewable resources with a microgrid can be used to supply power during major storms. This case study focuses specifically on using micro hydropower for providing energy to communities immediately after severe events, obviating the need for grid electricity.[117]

---

[115]Kazmi (2012), *op. cit.*

[116]New Zealand accepts climate-related migrants but has requirements regarding age and career abilities for applicants applying for refuge. Immigration New Zealand, "Knowledge Base." Retrieved 30 June 2014, from http://www.dol.govt.nz/immigration/knowledgebase/item/1129. Kazmi (2012), *op. cit.*

[117]This method would have been ideally utilized after Superstorm Sandy in New York where parts of the Rockaway community, totaling 100,000 people, were without power for over a month after the storm. New York City, "Hurricane Sandy after Action Report," May 2013. Retrieved 7 April 2015, from http://www.nyc.gov/html/recovery/downloads/pdf/sandy_aar_5.2.13.pdf. After the storm, many people did not have a way to call for help. A power source to charge cell phones would have connected the devastated community with emergency responders, making rescue and medical aid operations easier. The New York City "Hurricane Sandy after Action Report" specifically lists better coordination of relief to affected areas and a better capacity to deal with loss of power as recommendations on how the city can improve emergency preparedness. Similarly, these units could have been utilized during Hurricane Irene in Vermont where roads were washed out and emergency

How does micro hydropower work?

Micro hydropower consists of small turbines and a generator to produce electricity. The turbines only require a small flow of water that can be provided by a natural source. The amount of flow required, approximately two gallons per minute, or water with a small elevation drop[118] in its course of flow, will generate sufficient electricity to run a micro hydropower unit.[119] This running water source could be a Gray Water run-off ditch, which could handle the requisite flow after a severe precipitation event. Additionally, multiple Gray Water run-off ditches could be constructed to slow the flow of just one ditch, and additional units can be installed, maximizing the amount of power that can be captured.

This electricity can be delivered from the power unit to locations up to a mile away. These units can also be connected to power networks or grids, but more frequently they are used as stand-alone power for remote areas. The harnessed power is stored in portable rechargeable batteries and supplied by transporting these batteries to communities, which minimizes expensive connection costs.[120] The output from the turbine shaft can also directly be used for mechanical power.[121]

Operation and maintenance of these systems can be carried out by trained local craftspeople. Since local citizens are trained and the units are small, micro hydropower is relatively inexpensive to operate. Areas of low income can also benefit, not only from power security and job creation, but

crews could not access the town of Wilmington, calling it "cut-off." *CNN*, "Irene Leaves Damaging and Deadly Floods, Rushing Waters," 28 August 2011. Retrieved 30 June 2014, from http://www.cnn.com/2011/ US/08/28/tropical.weather/.

[118]A small drop in elevation that is approximately two feet is sufficient to harness energy. See "Practical Action." Retrieved 30 June 2014, from http://practicalaction.org/micro-hydro-power.

[119]Appalachian State University, "Fact Sheet: Microhydro," 2007. Retrieved 30 June 2014, from http://appropriatetec.appstate.edu/sites/appropriatetec.appstate.edu/files/microhydro_factsheet6.pdf; Additionally, this "run of river system" has little impact on the environment because water passing through generator goes back into the stream and it is utilizing a renewable resource. See "Practical Action," *op. cit.*

[120]Appalachian State University, "Fact Sheet: Microhydro," 2007. Retrieved 7 April 2015, from http://appropriatetec.appstate.edu/sites/appropriatetec.appstate.edu/files/microhydro_factsheet6.pdf; See "Practical Action," *op. cit.*

[121]See "Practical Action," *op. cit.*

also by improved lighting, communications, refrigeration to store medicines and food, and economic stimulation.[122]

In mountainous regions of Peru and Kenya, power is generated from micro hydropower units with turbines in mountain rivers, generating power from water flowing down a cliff after precipitation events.[123] Sri Lanka has 200 mini development projects to take advantage of rivers that form during the flood season.[124] Other countries are looking into installing micro hydropower units to secure power to communities.[125]

B. Implications for further replication

Provision must be made to charge batteries for the system, usually through solar energy or wind turbines. A training program must be established for the workers who will be operating and maintaining the system.[126] Community members must be willing to be trained and take on full responsibility for operating and maintaining the power system. A fee is generally charged for using the electricity. The local community can employ craftspeople to maintain the systems. The available water sources and sources that appear

---

[122–123] *Ibid.*

[124] Hydro Review, "Sri Lanka Micro Hydro Rehabilitation Projects Receive Boost from New ADB Loan," 18 January 2013. Retrieved 7 April 2015, from http://www.hydroworld.com/articles/2013/january/sri-lanka-micro-hydro-rehabilitation-projects-receive-boost-from.html.

[125] Russia has a growing interest in using micro hydropower in the northern Caucasus. RUSHYDRO Investment Program is funding these projects, including the Barsuchkovky and Stravopol developments. See "Water, Power & Dam Construction, RusHydro's Small Hydro Projects Selected at RES Auction," 19 June 2014. Retrieved 7 April 2015, from http://www.waterpowermagazine.com/news/newsrushydros-small-hydro-projects-selected-at-res-auction-4297728. In Assam, India, 149 small hydropower projects are being planned. HydroWorld.com, "India's Assam State Planning Massive Small Hydro Expansion," 1 July 2012. Retrieved 7 April 2015, from http://www.hydroworld.com/articles/ print/volume-20/issue-4/departments/small-hydro.html. The Philippines is investigating installing mini hydropower sites in rural areas. H. Liu, D. Masera, and L. Esser (eds.), "World Small Hydropower Development Report 2013 Executive Summary," 14–15, United Nations Industrial Development Organization; International Center on Small Hydro Power. Retrieved 7 April 2015, from http://www.smallhydroworld.org/fileadmin/user_upload/pdf/WSHPDR_2013_Executive_Summary.pdf.

[126] An added benefit of preparing a community with micro hydropower is the creation of jobs.

during precipitation events must be assessed to determine the best location for the micro hydropower unit.

The units are typically initially funded by a non-profit or emergency management government agency. Some communities decide to utilize the power more frequently than just during storm events, and end up charging users for this during regular non-severe conditions. This generates income and revenue for the community to fund the operation and maintenance of the equipment.

A region with hills or varying topography is ideally suited for this adaptation method. Additionally, the technology is still developing for further improvements in equipment designs. Countries can communicate with one of the organizations actively involved in helping to set up micro hydropower units to receive more information.[127]

Case 8: Rooftop Solar and Distribution Generation

Growth in distributed generation indicates the beginning of a revolution in the way that electricity is generated and transmitted in the world. The current dominant production and distribution model consists of large, centralized power plants delivering electricity across a large area grid to consumers. In contrast, rooftop solar photovoltaic (PV) systems generate power to be utilized on-site or nearby — power travels a much shorter distance. Although it is not necessarily enabling a home or business to be totally energy self-sufficient, having solar panels still increases the degree of customer independence and energy security substantially.

With the rapidly decreasing cost of solar power technology, rooftop solar provides widely affordable clean energy.[128] In order to seize this

---

[127]United Nations Industrial Development Program (UNDP), "Small Hydropower." Retrieved 30 June 2014, from http://www.unido.org/en/what-we-do/environment/energy-access-for-productive-uses/renewable-energy/focus-areas/small-hydro-power.html. Practical Action is a non-profit organization that partners with UNDP to implement micro hydropower solutions in Peru, Sri Lanka, Kenya, and Zimbabwe. See "Practical Action," *op. cit.*

[128]According to a UNEP-backed report, the average price of crystalline silicon solar modules fell by 30% or more in 2012 in the solar PV industry. See REN 21, "Renewable Status 2013: Global Status Report," 2013, 47. Retrieved 7 April 2015, from http://www.unep.org/pdf/GSR2013.pdf. Another report estimates that the solar PV manufactures are expected to fall another 6% during 2014, continuing the downward trend set in places since 2008,

favorable condition, some large solar projects are expanding in developing nations.[129] The renewable solution can be developed on ordinary rooftops on a community basis rather than with large solar fields. The example of the extensive rooftop solar programs in Austin, Texas demonstrates their benefits.

A. The St. David's Episcopal Church Project

St. David's Episcopal Church (the Church), located in the heart of downtown Austin, worked with a local utility company and a solar-focused engineering company and created a rooftop solar project on an "unprecedented scale".[130] The Church project's uniqueness provides an example of how to integrate community rooftop solar into an existing utility grid network.

Accelerating adoption of customer-installed photovoltaic solar panels is an eroding utility sales and revenues.[131] Therefore, some American electric utilities see this solar growth as a threat to their businesses, prompting them to cut back on incentives for solar power. However, Austin Energy, the partner utility company in this project, embraces the rooftop solar trend by introducing a new way to calculate the utility bill once its customers go solar. Customers that choose solar in Austin Energy's service area continue to pay a monthly energy bill based on how many kilowatt-hours of electricity they consume. Yet those customers are given a credit called "value of solar tariff" for each kilowatt-hour their solar systems generate. That credit is subtracted off the customers' total monthly electricity bill. This credit

bringing the overall cost to a record low of US$0.20 per watt. See Joshua S. Hill, "Solar PV Production Costs to Drop In 2014," Clean Technica. Retrieved 14 June 2014, from http://cleantechnica.com/2013/11/22/solar-pv-production-costs-drop-2014/.

[129] See Zachary Shahan, "Giant Solar Power Projects Now Hitting Developing World," *Abb-Conversations*, 7 April 2014. Retrieved 14 June 2014, from http://www.abb-conversations.com/2014/04/giant-solar-power-projects-now-hitting-developing-world/.

[130] See Marita Mirzatuny, "Austin Unveils Texas-Sized Rooftop Solar Array to Power Downtwon Church," Environmental Defense Fund, 4 March 2014. Retrieved 14 June 2014, from http://blogs.edf.org/texascleanairmatters/2014/03/04/austin-unveils-texas-sized-rooftop- solar-array-to-power-downtown-church/.

[131] Giles Parkinson, "Rooftop Solar: Does It Really Need the Grid?," Renew Economy, 23 April 2014. Retrieved 14 June 2014, from http://reneweconomy.com.au/2014/rooftop-solar-does-it-really-need-the-grid-10987.

price is calculated based on "expected fuel savings, offset power plant construction, transmission and distribution savings, and environmental benefits".[132] With this new price mechanism, the conflict between rooftop solar installation and utility companies is largely relieved.

Another challenge for the project was how to connect the solar array to the grid. For years, the design of Austin Energy's downtown distribution system limited the size of rooftop solar to a maximum of 20 kilowatts — less than one-seventh the size of St. David's array.[133] This technical barrier was solved by working closely with Meridian Solar, an engineer company as well as the Church project developer.

With the implementation of the project, the Church successfully reduced its electricity bill and had enough saved money to donate to charity work.

B. Implications for further replication

Nearly 1.3 billion people do not have access to electricity and more than 2.6 billion people still rely on the traditional use of biomass for cooking. Across developing countries, the average electrification rate is 77%, while it increased to around 91% in urban areas with the trend of urbanization.[134] The situation of energy access in developing countries has two characteristics: (1) a large amount of population live without modern energy access and (2) demand for electricity in urban areas is expected to continue to increase with rapid urban population growth, which will place an ever-increasing strain on existing electrical production, generation, transmission, and distribution systems.

With many developing countries struggling to make utilities economically sustainable and meet increased energy demand, the time is right for governments to re-evaluate their energy strategies and begin to take advantage of the benefits that renewable distributed generation can provide.[135]

---

[132]See Mirzatuny (2014), *op. cit.*

[133]*Ibid.*

[134]See International Energy Agency, "Energy Access Database." Retrieved 14 June 2014, from http://www.worldenergyoutlook.org/resources/energydevelopment/energyaccessdatabase/.

[135]The main benefits of renewable distributed generation can be viewed from two perspectives. For energy consumers, it can increase power supply quality with greater

Nevertheless, for both developed and developing countries, this new paradigm energy system is facing potential conflict with traditional commercial electricity companies. It is crucial to understand how to ally with utility companies rather than exclude them from the blueprint at an early stage.

There were three key parties to the Austin Church project partnership: the utility company (Austin Energy), the private engineering company (Meridian Solar), and the Church. The Church, as a non-commercial entity, served as a coordinator and liaison to all parties to ensure project success. Similar entities like schools, hospitals, and governments at the local level can replicate the Church project with similar programs. These community entities have a similar strong social influence and can afford renewable system installation more easily compared to one single household.

With regard to the technical challenges, developing countries should start investing human capital and education in renewable distributed generation. In the Church project, the utility was very progressive and the technical developer was an engineering company with more than a decade of experience on renewable energy, which means understanding relevant technologies fully enabled it to solve the connection difficulty smoothly.

Case 9: Strengthening Power Line Transmission

High winds and seasonal storms have increased in both frequency and intensity, which causes more frequent damage to electrical systems, especially overhead utility lines.[136] One answer is to start replacing central

flexibility and lower costs. Particularly the advantage of flexibility is more obvious during storms and other extreme weather events. For the grid-side, renewable distributed generation can reduce cost and energy loss during transmission and distribution; also, its increased reliability has siting, planning, and construction benefits. See U.S. Agency For International Development, "Increasing Energy Access in Developing Countries: The Role of Distributed Generation," 2004, 13–18. Retrieved 7 April 2015, from http://www.bcse.org/files/Increasing_Energy_Access.pdf.

[136]A new report shows a ten-fold increase in major power outages (those affecting more than 50,000 customer homes or businesses), between the mid-1980s and 2012 in the US. Some of the increase was driven by improved reporting. Yet even since 2003, after stricter reporting requirements were widely implemented, the average annual number of weather-related power outages doubled. Non-weather-related outages also increased during that time, but weather caused 80% of all outages between 2003–2012. See Climate Central, "Blackout:

station power production requiring lengthy and vulnerable transmission with distributed power production with mini-grid distribution to customers as described above.

Where bolstering power transmission is essential, replacing or updating the aging or failing electricity infrastructure usually can strengthen a country's power transmission system.

Common strengthening strategies include:[137] (1) Technology-based upgrades. Updating power lines through measures such as rebuilding transmission lines to increase capacity, underground cable replacement, and reconditioned circuitry. Some power companies have replaced aluminum transmission lines with composite materials to strengthen transmission lines and reduce sagging of the lines; and (2) Management-based upgrades that provide better monitoring of transmission line conditions and anticipation and addressing of transmission vulnerabilities. Also, many utility companies have focused on existing crew training programs, which can provide an opportunity for staff to respond to emergencies more quickly and with higher efficiency. Working with arborists to implement a vegetation management plan including properly trimmed tree limbs can be important to preventing outages from falling limbs.

A. How do the projects work?

Florida Power & Light Company (FPL), as the largest electric utility in Florida and one of the largest rate-regulated utilities in the United States, expanded its investment in major upgrades to nearly 100 main power lines and other infrastructure to further strengthen its electric grid against severe weather.[138] The shining point of this plan is to help communities within

---

Extreme Weather, Climate Change and Power Outage," 2014, 3. Retrieved 7 April 2015, from http://assets.climatecentral.org/pdfs/PowerOutages.pdf. Also, the situation is similar at a global scale. Major grid failures have been doubled since 2000. See James McLinn, "Major Power Outage in the US, and around the World," 2009, 1. Retrieved 7 April 2015, from http://paris.utdallas.edu/IEEE-RS-ATR/document/2009/2009-06.pdf.

[137] See Richard J. Campbell, "Weather-Related Power Outages and Electric System Resiliency," 2012, 8–13. Retrieved 7 April 2015, from http://www.fas.org/sgp/crs/misc/R42696.pdf.

[138] See Florida Power & Light Company, "FPL Plans to Strengthen Record Number of Power Lines in 2014," FPL Newsroom. Retrieved 14 June 2014, from http://newsroom.fpl.com/2014-02-24-FPL-plans-to-strengthen-record-number-of-power-lines-in-2014.

FPL's service territory recover more quickly from major storms and improve everyday reliability for its customers.

To achieve this goal, the plan is focused on upgrades to power lines and equipment serving essential community facilities — including hospitals, police and fire stations, 911 centers, water treatment plants, grocery stores, pharmacies, and gas stations — that are crucial to people and businesses as they recover from damaging storms.

A program proposed for implementation in Indonesia, called West Kalimantan Power Grid Strengthening Project, aims to allow for power trading between two countries to import more adequate and secure power to the electricity grid.[139]

Large numbers of customers in West Kalimantan, Sumatera, and the other outer islands of Indonesia do not have access to grid-connected power supply. Their existing power generation, which heavily relies on traditional fuels, is facing extreme increases in electricity demand. Therefore, PLN, the state-owned power utility in Indonesia, plans to import hydropower-generated electricity from neighboring Sarawak, Malaysia by building an 83 km 275 kV transmission line from its Bengkayang substation to the border with Sarawak, Malaysia. On the Malaysian side, SESCO, the state owned power utility in Sarawak, will build about a 42 km 275 kV transmission line from Mambong substation to the border with West Kalimantan.

B. Implications for further replication

Although these two projects require substantial grant funding, they highlight some important messages regarding power line-strengthening.

First, a power line-strengthening plan should concentrate on essential community/public facilities. Since developing countries have limited funding resources, the priority plan should encompass infrastructure that has strong social influence as much as possible, which also resonates with the idea of development of renewable distributed generation as stated in Case 8.

[139]See Asian Development Bank, "West Kalimantan Power Grid Strengthening Project." Retrieved 14 June 2014, from http://www.adb.org/projects/41074-013/details.

Second, the plan must highlight regional cooperation. In the case of the West Kalimantan Project, the hydro-resource is the main factor that stimulated the cooperation between two countries.

FOOD CASE STUDIES

Case 10: Grow Crops in a New Way

Disproportionate rains and rising temperatures are affecting the abundance and distribution of harvested agricultural species, according to the Fifth IPCC report. Although some benefits have been brought by climate change in some high latitude regions, negative impacts of climate trends have been much more common than positive ones.[140] Severe negative impacts on the production of crops like wheat and maize are estimated for low latitude countries.[141]

These changes in food system drivers give rise to changes in food security outcomes. Based on the Fifth Assessment Report of IPCC, the largest numbers of food insecure countries will be found in South Asia.[142] Therefore, adaptation has significant meaning to low latitude developing countries in this area which heavily rely on agriculture in their domestic production.

Currently, food production is the most common entry point for adaptation. Two ways can increase agricultural resilience: (1) New crops grown in the same spot; and (2) New crops grown in new growing areas.

New crops grown in the same spot

Due to water scarcity and other climatic changes, original crops that grow in one location may reduce their yield and even become hard to grow. Therefore, the first solution aims to take advantage of new agricultural technologies to cope with the new weather patterns and to maintain a stable production yield by introducing different crops that can adapt to the climate.

---

[140] See IPCC, *The Fifth Assessment Report* — "Chapter 7. Food Security and Food Production Systems," 2013, 2. Retrieved 7 April 2015, from http://ipcc-wg2.gov/AR5/images/uploads/ WGIIAR5-Chap7_FGDall.pdf.
[141–142] *Ibid.*

According to a climate assessment report developed by the State of Connecticut,[143] dairy farming is one of the top five industries in the state that will be most affected by climate change. This result is due to high daytime temperatures and absence of nighttime cooling, causing stress on cows and decreasing milk production.[144] Yet this kind of weather is perfect for growing grapes. In this context, the government decided to promote the replacement of dairy farming with vineyards in affected areas.

A. How does the project work?

Under the legal requirements at both the state and federal level, for a bottle to be labeled as a Connecticut wine, producers must obtain a certain amount of grapes from state growers.[145] Good ingredients determine the quality of wine. As a result, it was necessary to refine the quality of locally grown grapes.

The Connecticut government partnered with the Extension School of University of Connecticut to carry on research to provide training programs to new grape growers, in particular, those dairy farmers facing declining dairy farming and were looking for new income resources. A Scaling Up Program for beginning farmers, funded by U.S. Department of Agriculture (USDA), offered a three-year outreach and training program for new and beginner vineyard farmers in Connecticut.[146] Under this program, a group of experts will work closely with a number of beginning

---

[143] The top five most imperiled agricultural planning areas or features in Connecticut were maple syrup, dairy, warm weather produce, shellfish, and apple and pear production. See Adaptation Subcommittee of Connecticut, "The Impacts of Climate Change on Connecticut Agriculture, Infrastructure, Natural Resources and Public Health," 2010, 4. Retrieved 7 April 2015, from http://www.ct.gov/deep/lib/deep/climatechange/impactsof climatechange.pdf.

[144] See Adaptation Subcommittee of Connecticut (2010), *op. cit.*

[145] The state law requires vineyards to use a minimum of 25% state grown grapes for their total wine production, and federal labeling laws are more strict having a requirement of 75% of grapes grown in the state for any bottle to be labeled as a state wine, with an even more stringent requirement of 95% of grapes grown on the particular vineyard site in order to include it on the label. See Jeremy L. Jelliffe, "An Economic Analysis of Wine Grape Production in the State of Connecticut," 2012, 2. Retrieved 7 April 2015, from http://digitalcommons.uconn.edu/cgi/viewcontent.cgi?article=1393&context=gs_theses.

[146] See UConn Extension, "Scaling up Program," 2013. Retrieved 7 April 2015, from http://ipm.uconn.edu/documents/raw2/364/Scaling%20Up%20Program.pdf.

farmers over three years, helping them solve initial barriers as well as challenges. It is worth mentioning that the training program is not only about growing operations, but also includes farm planning management and conservation.[147] But the Connecticut government does not stop there.

In order to enable its agricultural market consumption with grapes, Connecticut established the Connecticut Farm Wine Development Council, a volunteer council under administrative authority of the Connecticut Department of Agriculture, to promote local wineries. All licensed farm wineries are eligible to participate in the council's activities.[148] Two main responsibilities that the Wine Council have are marketing Connecticut-produced wine and advocating the passage of supporting legislation.[149]

What the Wine Council does is crucial. On one hand, it paves the road to sell the grapes and adds more confidence to farmers (including beginning farmers) to shift their focus to new crops. Its policy and legislation campaigns also can provide more certainty to the growing of new crops.

With these efforts, the wine industry in Connecticut projects a steady growth rate of 3.9%, which is higher than the national average.[150]

B. Implications for further replication

First, academic institutions bridge knowledge gaps by selecting suitable plants for location and training. In fact, a vast body of resources is available from all sorts of organizations for small farmers. Unfortunately, many of the small farmers and ranchers are unaware of how to utilize the information and opportunities. From the case above, it can be seen that the Scaling Up Program hosted by the University of Connecticut has played a key role in providing farmers with knowledge support.

Another similar Connecticut program, called Small Farmers' Outreach, created by the Lincoln University Cooperative Extension, aims to educate

[147] *Ibid.*

[148] Connecticut Department of Agriculture, "Connecticut Farm Wine Development Council: Moving the Industry Forward." Retrieved 23 June 2014, from http://www.ct.gov/doag/cwp/view.asp?Q=532870&A=1401.

[149] *Ibid.*

[150] See Jelliffe (2012), *op. cit.*

farmers by visiting their sites on an appointment basis and giving customized advice for free.[151] With the professional diagnosis, the owner usually can raise income and yield production.

Similar examples are happening in some developing countries already. It is significant to summarize the merits of this type of practice.

The importance of the Connecticut examples is highlighted in developing countries. In Zimbabwe, due to increasingly unpredictable weather driving up the price of maize meal, people in many regions started planting sweet potatoes during the rainy season, which could assist small-scale subsistence farmers to feed their family as well as sell any excess. In 2000, the Horticultural Research Institute (HRI), Biotechnology Trust of Zimbabwe, and academic research institutes including the University of Zimbabwe's Department of Crop Science, spearheaded efforts to create sweet potato varieties using tissue culture techniques for commercial production.[152]

Second, it was important to develop a complete chain of distribution for the new crop, which is a critical point that usually is neglected in developing countries. In the case study, the Wine Council developed a set of marketing strategies, from broadening local consumption to attracting tourists, to improve the market share.[153] Agricultural adaptation in developing countries generally is concentrated on crop growing and seed resilience. If the distribution chain cannot be enhanced, crop replacement cannot be truly successful.

Apart from the distribution and trading, in a case like Zimbabwe where a major crop will be replaced, people's eating habits may require some change as well. An educational campaign that aims to teach people how to utilize sweet potatoes would be beneficial as well.

---

[151] See Lincoln University, "Cooperative Extension." Retrieved 23 June 2014, from http://www.lincolnu.edu/web/cooperative-extension/cooperative-extension.

[152] Edward Mutandwa, "Performance of Tissue-Cultured Sweet Potatoes among Smallholder Farmers in Zimbabwe." Retrieved 23 June 2014, from http://www.agbioforum.org/v11n1/v11n1a05-mutandwa.htm.

[153] See CT Farm Wine Development Council, "Strategic Marketing Plan." Retrieved 7 April 2015, from http://www.ct.gov/doag/lib/doag/boards_commissions_councils/ct_farm_wine_delvp_council/strategic_marketing_plan.pdf.

New crops grown in non-traditional areas

A range of potential adaptation options exist across all food system activities, not just in food production; benefits also exist from potential innovations in food processing, packaging, transport, storage, and trade. Climate change effects seriously threaten traditional food systems that are centered on industrial agricultural production.

On one hand, industrial agriculture requires a vast scale of land to grow crops and raise livestock, which means the owners suffer more losses than small-scale agriculture production under the climatic stress. On the other hand, resources and infrastructure required to operate an industrial agricultural system in some developing countries are impractical for small-holder farmers. Thus, new growing areas need to be identified and developed, for instance by encouraging agriculture in every corner of cities.

Currently, urban agriculture is spreading all over the world. Based on the core idea utilizing every available fragment of earth in the city to grow crops, urban agriculture has reconstructed a new type of food system, from food production, to food processing and distribution, which can help the agriculture sector adapt to climate change as well as relieve food insecurity more effectively.

A. How do the projects work?

Growing food in non-traditional locations is a challenge for the government. Without proper comprehensive planning, urban agriculture could develop detrimentally, leading to various forms of environmental pollution.

Informed by European practices, the Seattle Green Factor program is a scoring system that displays landscape requirements developed by the Seattle Department of Planning and Development. The Department guides developers and designers in improving and increasing planted areas in new developments.[154] A scoring system encourages green features, while allowing the flexibility to experiment with various combinations of elements to meet a specific minimum Green Factor score.[155]

[154] See Seattle's Department of Planning and Development, "Seattle Green Factor." Retrieved 23 June 2014, from http://www.seattle.gov/dpd/cityplanning/completeprojects list/greenfactor/whatwhy/.
[155] *Ibid.*

B. Implications for further replication

The meaning of the Seattle scoring system is not merely limited to creating space for promoting urban agriculture. Its significance lies in integrating urban agriculture into comprehensive urban planning and future development. Meanwhile, its benefits can be maximized with combining other planning elements.

As mentioned above, generally, adaptation measures are adopted on a local and program basis, which indicates that comprehensive planning from an overall perspective may be absent. Since urban areas in developing countries are expanding rapidly, an orderly urban planning regime is required to coordinate different strategies and programs. The Seattle scoring system sets up minimum scores for new development, while it lays out alternatives as flexible choices. It is relatively easy to replicate for developing countries.

11.3. Interaction between Sectors: The Water–Energy–Food Nexus

In Section 11.2, we displayed a series of adaptation practices based on a sector basis analysis (water, energy, and food). These sectors do not always function independently. On the contrary, water, energy, and food interact with each other actively, providing a significant additional path to design adaptation measures.

A. Origin of the nexus

In 2011, a nexus approach was first proposed by the German Federal Government at the international level[156] to "integrate management and governance", to "support the transition to a Green Economy", to improve

[156]In 2011, the first nexus international conference was held in Bonn, Germany. It was organized by the German Federal Government as "a specific German contribution to the UN Conference on Sustainable Development 'Rio2012'". A conference paper called "The Water, Energy and Food Security Nexus: Solution for the Green Economy" lays the primary foundation on the background knowledge of a nexus approach. See The Water, Energy and Food Security Resource Platform, "The Bonn Nexus Conference." Retrieved 22 January 2014, from http://www.water-energy-food.org/en/whats_the_nexus/bonn_nexus_conference.html.

"resource use efficiency", and to support "greater policy coherence" in the context of the Rio+20 summit.[157] The World Economic Forum presented a conceptual framework on the water–energy–food nexus in the same year as well.[158] In 2012, the International Centre for Integrated Mountain Development developed a framework that focuses on the nexus approach to better protect the environment and eco-services of the Himalayas and South Asia.[159] Each of these three nexus concepts has its own emphases, but the core value of the nexus approach, common to all three, is about how to utilize scarce the resources of water, energy, and food more effectively,[160] and to improve policy coordination and harmonization, to design a more effective regime compared to individualized methods.

Three indispensable elements, water, energy, and food, establish a link that has vibrant interactions. It is clear to see that water is crucial for food production and energy generation, and that energy is required for water distribution and modern agricultural measures,[161] and both are required for food production. Sometimes there are conflicts, as between food production and energy development. For example, biomass and solar energy require large tracts of land that may deprive the areas available for

[157]The Water, Energy and Food Security Resource Platform, "The Bonn Nexus Conference," *op. cit.*

[158]See The World Economic Forum Water Initiative, "Water Security: The Water–Food–Energy–Climate Nexus," 2011. Retrieved 7 April 2015, from http://www3.weforum.org/docs/WEF_WI_WaterSecurity_WaterFoodEnergyClimateNexus_2011.pdf

[159]See International Centre for Integrated Mountain Development, "Contribution of Himalayan Ecosystems to Water, Energy, and Food Security in South Asia: A Nexus Approach," 2012. Retrieved 7 April 2015, from http://www.circleofblue.org/waternews/wp-content/uploads/2012/07/icimod-contribution_of_himalayan_ecosystems_to_water_energy_and_food_security_in_south_asia-_a_nexus_appr-1.pdf.

[160]The conference paper demonstrated that most conflicts happened between institutions associated with allocation and distribution of limited resources; thus, it stressed that a new nexus approach should improve "sectoral" resource productivity and lead to "overall" benefit sharing. See the "Water, Energy and Food Security Nexus: Solution for the Green Economy," *op. cit.*, 11–12.

[161]As for the interactions between water, energy, and food, the conference paper and some reports all give a detailed analysis and demonstration. See "The Water, Energy and Food Security Nexus: Solution for the Green Economy," 18–27. See also Peter Hanlon, Robin Madel, Kai Olson-sawyer, Kyle Rabin, and James Rose, "Food, Water and Energy: Know the Nexus," 8–9. Retrieved 7 April 2015, from http://www.gracelinks.org/media/pdf/knowthenexus_final_051513.pdf.

food production.[162] However, the complementary effect of energy and food exists as well. Growing food at some proper locations can reduce energy requirements by alleviating heat island effects. However, each interaction indicates a possibility to maximize the resource use efficiency as well as adaptation performance.

B. Adaptation and the nexus

Since "climate policies can impact on water, energy, and food security, and adaptation can in fact be maladaptive if not well-aligned in a nexus approach and implemented by appropriately interlinked institutions",[163] the water–energy–food nexus is important to climate change adaptation.

Although a number of global and regional conferences, workshops, and meetings on this nexus were held in 2011–2012, especially during the presentation phase for Rio+20 in June 2012,[164] there is still no comprehensive cross-sectoral adaptive approach recognized by UNFCCC or its Conference of Parties. One of the most related documents to adaptation is "Adaptation Assessment, Planning and Practice: An Overview from the Nairobi Work Programme on Impacts, Vulnerability and Adaptation to Climate Change." Yet, it is noted that the nexus approach is only mentioned there and proposed by the International Water Management Institute (IWMI) "for assessing the role of water and adaptive capacity of agro-ecosystem".[165]

---

[162]A report conducted by a Spain-based non-profit organization showed that increased demand for biofuel production has driven farmers and forest communities from Brazil to Indonesia off their land, threatening those people's well-being and food security. See GRAIN, "Land Grabbing for Biofuels Must Stop," 2013. Retrieved 7 April 2015, from http://www.grain.org/article/entries/4653-land-grabbing-for-biofuels-must-stop.pdf. This phenomenon has not just appeared in Brazil; another policy brief indicated that it is a global issue. See IIED, "Agricultural Land Acquisitions: A Lens on Southeast Asia," April 2012. Retrieved 7 April 2015, from http://pubs.iied.org/pdfs/17123IIED.pdf?.

[163]See "The Water, Energy and Food Security Nexus: Solution for the Green Economy," *op. cit.*, 8.

[164]Livia Bizikova *et al.*, "The Water-Energy-Food Security Nexus: Towards a Practical Planning and Decision-Support Framework For Landscape Investment and Risk Management," 2013, 5. Retrieved 7 April 2015, from http://www.iisd.org/pdf/2013/wef_nexus_2013.pdf.

[165]See United Nations Framework Convention on Climate Change, "Adaptation Assessment, Planning and Practice: An Overview from the Nairobi Work Programme on Impacts, Vulnerability and Adaptation to Climate Change," 2010, 41. Retrieved 7 April 2015, from http://unfccc.int/resource/docs/publications/10_nwp_adap_assess_en.pdf.

The regional level is active with regard to integrating the nexus approach into adaptation thinking. The Middle East and North Africa (the MENA region) use water as a key driver to develop a strong capacity for anticipating and adapting impacts of climate change, such as to prevent food shortages. This regional cooperation has obtained some important results, including an action plan adopted by the Arab Ministerial Water Council (AMWC), implementation of climate change innovation measures through the Arab Center for the Studies of Arid Zones and Dry Lands (ACSAD) guidelines, and the creation of manuals for the region's national water ministries and institutions to adapt to climate change in their strategies and action plans.[166]

The nexus approach was adopted to help the Mekong River Basin adapt to climate change as well. In the early 2000s, the adaptive approach to manage this region was still sector-based, yet the trans-sector and transboundary approach inspired by the nexus started in 2010.[167] Due to similar impacts of climate change that countries located along the Mekong River Basin are facing, this burgeoning nexus approach can advance information sharing among Cambodia, Lao PDR, Thailand, and Vietnam.

As for the national level, it is common to see that countries integrate the nexus into adaptation measures in three ways: (1) Some countries, such as Ukraine, attempt to "increasing interdependencies between food, energy, and water security" by launching a joint research project created by national research institutes and international organizations;[168] (2) Countries like

[166]See The Water, Energy and Food Security Resource Platform, "Adaptation to Climate Change in the Water Sector in the MENA Region (ACCWaM)." Retrieved 22 January 2014, from http://www.water-energy-food.org/en/practice/view__1108/adaptation-to-climate-change-in-the-water-se. Stockholm Environment Institute also has research in the MENA region in the context of the nexus approach. See Magnus Benzie, Marion Davis, and Holger Hoff, "Climate Change, Water and Energy in the MENA Region: Why a 'Nexus' Approach is Crucial for Mitigation and Adaptation," 2012. Retrieved 7 April 2015, from http://www.sei-international.org/mediamanager/documents/Publications/Climate/SEI-DB-2012-MENA-climate-nexus.pdf.

[167]See Nguyen Huong Thuy Phan, "A Nexus of Adaptation: Climate Change Implications on the Water–Food–Energy Nexus of the Lower Mekong Basin." Retrieved 7 April 2015, from http://www.mrcmekong.org/assets/Events/Mekong2Rio/4.2b-A-nexus-of-adaptation-Implications-of-climate-change-on-water-food-energy-nexus-Phan.pdf.

[168]There are five non-profit institutes of the National Academy of Sciences of Ukraine (NASU) and International Institute for Applied Systems Analysis (IIASA). See IIASA,

Japan provide funding for nexus pilot projects as research programs;[169] and (3) Ghana's National Climate Change Strategy, known as the "Akropong Approach", concentrates on the connection between climate adaptation and landscapes, as well as addresses the nexus of food security, water, forest, and energy within a landscape context.[170] It is worth mentioning that the three ways illustrated above can be used at the same time as well.

In fact, the local/city level is the most robust level by introducing the nexus process in its adaptation regime design and governance. The main reason is because adaptation measures are place-based, which makes it easier for local level to find nexus solutions. Also, it is necessary to understand that the local government stands on the front line with regard to adaptation measures. The government at this level enjoys high flexibility to adjust and accept new governance models.

C. Types of the nexus in practice

Unlike conventional policy-developing ways that operate mostly in isolation from one another (which easily leads to conflicts), the nexus approach can largely reduce policy contradictions and resolve conflicts between governmental agencies. Additionally, practices under different contexts can have complementary effects. In other words, even though adaptation measures may only touch one or two elements from the appearance, all natural resources of the nexus would be embraced by appropriate policy designs.

"Integrated Modeling of Robust Solutions for Food, Energy, and Water Security Management." Retrieved 22 January 2014, from http://www.iiasa.ac.at/web/home/research/researchProjects/Integrated-Modeling-of-Robust-Solutions-for-Food–En.en.html.

[169]Currently, there is a five-year project called the "Futurability Initiatives" run by the Research Institute for Humanity and Nature that tries to analyze the nexus from the perspective of human-environment security, particularly how to "optimize governance and management with these inter-connected needs". The project picked the coastal region of the Ring of Fire as the pilot subject. This project is more comprehensive and almost reaches every aspect. See Research Institute for Humanity and Nature (RIHN) Core Research Project, "Human-Environmental Security in the Asia-Pacific Ring of Fire: Water-Energy-Food Nexus." Retrieved 22 January 2014, from http://www.chikyu.ac.jp/wefn/english/index.html.

[170]See José Urrea y Lucy Holt, "Combined Solutions to Strengthen Adaptation," CIGAR Research Program on Climate Change, Agriculture and Food Security, 21 November 2013. Retrieved 22 January 2014, from http://ccafs.cgiar.org/blog/combined-solutions-strengthen-adaptation#.UuBFznCS2IU.

Table 2: Sample Practices of Leading Elements

Leading Elements of the Nexus	Sample Practices
Water	Basin Resource Management (Reservoir and Dam Operation)
Energy	Power Plants, Biofuel
Food	Urban Agriculture, Green Infrastructure, Food Waste

It is recognized that water is the priority element to emerge from the water–energy–food nexus approach and it is the prerequisite of the nexus (see Table 2). In some cases, one or two elements can be highlighted and severed as a leading element to link adaptation governance. For example, the nexus sometimes can be broken into the water–energy or water–food sectors. The energy element stands out with regard to energy generation from power plants and dam operation. While in the process of food production, particularly in urban agriculture, the food element takes the position to better maximize the benefits of the nexus. However, it is impossible for the water element to be absent from the nexus.

Here is an example from San Francisco. Similar to many cities in the US, the lead agency of urban agriculture in San Francisco is the Recreation and Park Department (RPD). Yet other agencies' active participation should be highlighted as well.

First, the Public Utilities Commission (PUC), as San Francisco's energy regulator, gets involved in the urban agriculture very actively. The PUC does not only provide funding to support two farms on land it owns, but also serves as a key agency to work with RPD to develop urban agriculture. Meantime, the PUC is the responsible department for water protection as well.[171] In order to build urban agriculture, the PUC regulates water conservation requirements for urban agriculture, including interpreting San Francisco's Water Efficient Irrigation Ordinance.[172] This design is significant to bond the interests of two agencies, which encourages the food–water–energy nexus linkage.

[171] See Services of the San Francisco Public Utilities Commission, "About Us." Retrieved 2 February 2014, from http://www.sfwater.org/index.aspx?page=161.

[172] See Services of the San Francisco Public Utilities Commission, "Water Efficient Landscape." Retrieved 2 February 2014, from http://sfwater.org/index.aspx?page=689.

Second, RPD houses a program for the city that would coordinate the work of a wide range of agencies to "generally enhance and increase urban agriculture in San Francisco".[173] On one hand, this program can embrace relevant agencies besides the water and energy department to participate in developing a sustainable system. On the other hand, it can solve the capital problem. The new ordinance requires city agencies to advocate for state and federal funding, collect data related to urban agriculture, support gleaning programs, identify opportunities to use urban agriculture for job training and employment, and ensure that existing farm and garden spaces are used fully. In order to reach these goals, RPD hires full-time staff to coordinate the work. The underlying reason is that many city agencies supported urban agriculture, but when the public had difficulty getting assistance, there was no person or agency to take a clear responsibility.

Besides positioning the staff, the agency is tasked with providing accountability. The program is required to send annual reports to the Board of Supervisors that include evaluation of a variety of metrics, including an accounting of how city funds have been spent. From above, it can be seen that the seed of urban agriculture has been successfully planted in the city organism.

From above, it is also seen that the nexus is well-designed to work as a whole in San Francisco as well as to outreach other relative city departments to obtain valuable resources to support the nexus development.

11.4. Conclusion

Adaptation to climate change is essential and urgent, particularly for low-lying areas and island countries. Adaptation plans need to be designed appropriate to the needs and conditions of each country and area. But certain essentials are obvious.

Necessary requisites include adequate planning, involvement of those affected, collaboration among the departments of government, adequate training of responsible personnel, and involvement and education of the affected public.

The most important adaptation measures are those to protect populations from sea level rise and extreme weather events. These measures require

---

[173] San Francisco Administrative Code, Chapter 53.

early warning systems and relocation planning that can be very difficult, particularly if resettlement in another country is required. The problem of climate-related migrants needs priority attention.

Other adaptation measures can include natural protection of coastlines and protective barriers as with planting mangrove trees and creation of oyster shell barriers. Agriculture can be relocated and revised to include climate resistant crops. Electricity production can be decentralized and generation distributed with promotion of renewable energy. Where a grid is still necessary it can be strengthened. Water can be much better utilized through efficiency measures, the collection of rainwater, and the use of Gray Water.

Adaption plans should recognize the important nexus between water, food, and energy. Recognizing this interrelationship can greatly strengthen the effectiveness of adaptation measures.

Today there are examples of all these measures from which nations and municipalities can benefit, many of which are included in the case studies presented.

The most important requisite for adaptation is the will to act and international recognition of the urgency of action. As has been well demonstrated by the IPCC and many studies, it will be far cheaper to act now than to have to pay the incredible costs of inaction.

Chapter 12

Dealing with Climate Migrants: A New Challenge for Developing Nations

Asanga Gunawansa

12.1. Introduction

Climate change is one of the biggest environmental threats faced by the world today. Whilst mitigation of climate change by reducing the current levels of greenhouse gases will help future generations, adaptation initiatives are necessary to prepare the current generation for changing and unavoidable climatic conditions. Whilst the causes of climate change are being analyzed and their likely effects assessed, it is also important to anticipate people displacement scenarios due to climate change and improve the quality and readiness of communities to respond. However, whilst climate change and the efforts needed for its mitigation have been the subjects of intense debate and speculation within the scientific community and policy-makers, insufficient attention has been given to the humanitarian consequences of climate change.

It is pertinent to note that as early as in the First Assessment Report in 1990, the Intergovernmental Panel on Climate Change (IPCC) stated that the gravest effects of climate change may be those on human mobility (IPCC, 1990). However, not much progress has been made in addressing this issue. The outcome of the Copenhagen meeting of the COP15 and subsequent climate change negotiations to determine life after the Kyoto Protocol show that there is not much interest in the international community in the plight of poor countries, some of which are even showing signs of disappearance in the next few decades due to rising sea levels.

According to the International Organization for Migration (IOM), climate change and the environment are interrelated. Just as environmental degradation and disasters can cause migration, IOM concludes that movement of people can also entail significant effects on surrounding ecosystems. IOM stresses that whilst migration is often misperceived as a failure to adapt to a changing environment, instead, migration should be considered as an adaptation strategy to climate and environmental change (IOM, 2009).[1]

The IPCC estimated that 150 million people may be displaced by 2050 (IPCC, 2007),[2] whilst the Stern Review of the Economics of Climate Change cites estimates of 200 million people who might be displaced by 2050, due to adverse impacts of climate change, especially due to rising sea levels (Stern, 2007).[3] According to a study carried out by the UN Office for the Coordination of Humanitarian Affairs (OCHA) and the Internal Displacement Monitoring Centre (IDMC) of the Norwegian Refugee Council NRC, millions of people have already been displaced due to adverse climatic conditions (OCHA and IDMC/NRC, 2009).[4] In the same report, it was noted that in 2008 alone 36 million people were displaced as a result of sudden-onset natural disasters.

Although climate change is affecting the ecological system of the whole world and poses a threat to all countries, some regions are more vulnerable than the others due to their special geographic features, economic conditions, and technological disadvantages. According to a new report titled "In Search of Shelter", the areas worst affected are the world's great river deltas (largely situated in Asia), due to glacial melt; desert-dwellers in Africa subject to drought; islanders subject to rising

---

[1] International Organization for Migration (IOM), "Migration, Climate Change and the Environment," IOM Policy Brief, IOM, Geneva, May 2009.

[2] IPCC, "Climate Change 2007: Impacts, Adaptation and Vulnerability," Working Group II, Fourth Assessment Report, Cambridge University Press, Cambridge, 2007.

[3] United Kingdom Treasury, Stern Review on the Economics of Climate Change, 2005. Retrieved 10 January 2010, from http://webarchive.nationalarchives.gov.uk/+/http://www.hm-treasury.gov.uk/independent_reviews/stern_review_economics_climate_change/sternreview_index.cfm.

[4] OCHA and Internal Displacement Monitoring Centre (IDMC) of the Norwegian Refugee Council (NRC), "Monitoring Disaster Displacement in the Context of Climate Change," 2009. Retrieved 4 January 2010, from http://www.internal-displacement.org/8025708F004BE3B1/%28httpInfoFiles%29/12E8C7224C2A6A9EC125763900315AD4/$file/monitoring-disaster-displacement.pdf.

sea levels (mainly, low-lying islands in the Indian and Pacific Oceans); and the interior and coast of Mexico and the Caribbean (Warner *et al.*, 2009).[5] According to the same source, a mere one meter rise in sea levels could displace approximately 23.5 million people along the Ganges, Brahmaputra, Irrawaddy, Salween, Mekong, Yangtze, and Yellow rivers. A two meter rise could uproot 14 million people on the Mekong alone and swamp much of its farmlands. Thus, the countries in these regions are likely to be among the most vulnerable.

Migration into safer areas due to rising sea levels or extreme weather conditions such as cold waves or heat waves could be voluntary initiatives, sometimes encouraged by governments as an adaptation measure to climate change. However, unlike in such situations where migration can be planned, forced migration due to climate change-related disasters could be catastrophic due to lack of preparedness by the victims as well as the disaster struck states and their neighbors.

Given the projected scenarios, it is vital that evolving frameworks for climate change adaptation address displacement issues. Unfortunately, not only the developing countries, but also the developed countries currently lack appropriate mechanisms to deal with the threat of climate migration. For example, accelerating coastal erosion in Alaska is likely to force several communities to relocate; yet no process has been established to address this (Warner, 2009).[6] In the developing world, according to some current predictions, as a result of such sea level rise, the annual number of people flooded in coastal populations in Southeast Asia could increase by about 16 million, specifically from Thailand to Vietnam, and also Indonesia and the Philippines (WWF, 2007).[7] According to the Minister of Environment of Indonesia, of the approximately 17,000 islands of Indonesia, about 2,000

[5]K. Warner *et al.*, "In Search of Shelter," Center for International Earth Science Information Network, Columbia University, 2009. Retrieved 10 January 2010, from www.ciesin.columbia.edu/.../clim-migr-report-june09_final.pdf.

[6]K. Warner *et al.*, "Researching Environmental Change and Migration: Evaluation of EACH-FOR Methodology and Application in 23 Case Studies Worldwide, Conceptualising the Migration and Environment Nexus." IOM Blackwell Press, 2009.

[7]World Wildlife Federation (WWF), "Climate Change Impact on APEC Countries, WWF," 2007. Retrieved 4 January 2010, from http://assets.panda.org/downloads/cc_impacts_on_apec_countries.pdf.

islands may be submerged by 2030 due to sea level rise if the current trend of global warming continues unchecked (Reuters, 2007).[8] The Philippines, with approximately 7,100 islands and rocks, might not fare any better. Vietnam, which has approximately 3,000 islands with a total area of more than 1,600 square kilometers, and more than one million square kilometers of sea surface, is another country in the region that would face similar threats (World Bank, 2004).[9] There is also a fear that the Maldives might disappear in the next 75 years (Hoffman *et al.*, 1983; IPCC, 2001).[10] However, none of these countries appear to have yet put in place an effective relocation plan to safeguard the lives of the possible victims.

This chapter examines the issue of climate change-related migration and whether the existing international legal and policy framework is adequate to deal with the problem. Further, it proposes various initiatives that can be taken at the national, regional, and international levels to enable the nations and people of the world to be better prepared to deal with climate change-related displacement of people.

12.2. The Issue of Climate Migrants

12.2.1. *Most vulnerable areas*

Industrialization and overurbanization of our existing cities, amongst other mainly human-induced activities, have been the cause of environmental degradation over the years (IPCC, 2001).[11] Rising sea levels, glacier retreats, Arctic shrinkage, altered patterns of rain and agriculture, and extreme weather-related catastrophes such as hurricanes are often cited as consequences of climate change (IPCC, 2007), one of the main environmental problems faced by the world today. The Fourth Assessment Report of the IPCC (2007) authoritatively established that human-induced

[8]Reuters, "Indonesia May Lose 2,000 Islands to Climate Change," 29 January 2007. Retrieved 6 January 2010, from http://www.reuters.com/article/latestCrisis/idUSJAK299186.

[9]World Bank, "Modelled Observations on Development Scenarios in the Lower Mekong Basin," World Bank, Washington, DC, 2004.

[10]J.S. Hoffman, D. Keyes, and J.G. Titus. "Projecting Future Sea Level Rise," U.S. Environmental Protection Agency, Washington, DC, Gov. Print. Office, 1983, 266.

[11]Intergovernmental Panel on Climate Change (IPCC), "Working Group I Third Assessment Report," Cambridge University Press, Cambridge, 2001.

climate change is accelerating and already has severe impacts on the environment and human lives. One such significant impact is the increase in the frequency and severity of climate change-related natural disasters (OCHA and IDMC/NRC, 2009).

According to EM-DAT, the international disaster database, the recorded number of natural disasters has doubled from approximately 200 to over 400 per year over the past two decades, with the majority being climate-related disasters, i.e., disasters which climate change can influence both in terms of frequency and severity.[12] These include the meteorological (for example, storms), the hydrological (for example, floods), and the climatological (for example, droughts).[13]

Table 1 shows the 10 most extreme temperature disasters for the period 1900 to 2010 sorted by the total number of people affected at the country level, which shows that nine out of the 10 events have occurred during the past two decades.

Table 2 shows the top 10 most important climatological disasters for the period 1900 to 2010 sorted by total number of people affected at the

Table 1: Extreme Temperature-Related Disasters during 1900–2010.

Country	Date	Number of People Affected
China P Rep, Extreme winter conditions	January 2008	77,000,000
Australia, Heat wave	February 1993	3,000,500
Peru, Cold wave	June 2004	2,137,467
Tajikistan, Extreme winter conditions	January 2008	2,000,000
Peru, Cold wave	July 2003	1,839,888
Australia, Heat wave	December 1994	1,000,034
Liberia, Cold wave	1990	1,000,000
Peru, Extreme winter conditions	April 2007	884,572
Russia, Cold wave	January 1999	725,000
Kazakhstan, Cold wave	November 1997	600,000

Source: EM-DAT: The OFDA/CRED International Disaster Database.

[12]EM-DAT — The OFDA/CRED International Disaster Database. See website at http://www.emdat.be.

[13]Events caused by long-lived/meso to macroscale processes (in the spectrum from intraseasonal to multidecadal climate variability).

Table 2: Worst Climatological Disasters during 1900–2010.

Disaster	Date	Number of People Affected
India, Drought	May 1987	300,000,000
India, Drought	July 2002	300,000,000
India, Drought	1972	200,000,000
India, Drought	1965	100,000,000
India, Drought	June 1982	100,000,000
China P Rep, Drought	January 1994	82,000,000
China P Rep, Extreme temperature	January 2008	77,000,000
China P Rep, Drought	April 2002	60,000,000
India, Drought	April 2000	50,000,000
China P Rep, Drought	June 1988	49,000,000

Source: EM-DAT: The OFDA/CRED International Disaster Database.

Table 3: Worst Hydrological Disasters during 1900–2010.

Disaster	Date	Number of People Affected
China P Rep, Flood	July 1998	238,973,000
China P Rep, Flood	June 1991	210,232,227
China P Rep, Flood	June 1996	154,634,000
China P Rep, Flood	June 2003	150,146,000
India, Flood	July 1993	128,000,000
China P Rep, Flood	May 1995	114,470,249
China P Rep, Flood	June 2007	105,004,000
China P Rep, Flood	June 1999	101,024,000
China P Rep, Flood	July 1989	100,010,000
China P Rep, Flood	June 2002	80,035,257

Source: EM-DAT: The OFDA/CRED International Disaster Database.

country level which again shows that the greatest number of events has occurred during the last two decades.

Table 3 shows the top 10 most important hydrological disasters for the period 1900 to 2010 sorted by the total number of people affected at the country level, which also shows that the last two decades have witnessed the highest number of events.

The above figures show that in recent times, millions of people have been affected by natural disasters that could be linked to changing climatic conditions. The worst affected have been China and India, which together comprise approximately 40% of the world's population. Further, of the 20 disasters that caused most displacement in 2008, 17 were in this region (OCHA and IDMC/NRC, 2009). Although it may not be possible to establish without any doubt that any of the above incidents was caused by human-induced climate change, the affected countries need to be adequately prepared to deal with future disasters of similar nature. All the incidents recorded in the above tables are likely to, either temporarily or permanently, have displaced people.

The adverse effects of climate change, in particular rising sea levels, the phenomenon of desertification, floods, and heat waves, affect the living conditions of populations insofar as they can generate degradation of land, disappearances of territory, and increased pressure on natural resources. This could result in the displacement of large numbers of people. OCHA and IDMC/NRC (2009) noted that in 2008 alone, approximately 20 million people were displaced as a result of climate-related disasters. The International Organisation for Migration (IOM) predicts that by 2050, 200 million people will be displaced by environmental pressures (IOM, 2009).

According to the trends shown in above tables, most of the future displacement of people is likely to happen in China and India, going by the number of people affected. If we go by the number of countries, it is likely that the most vulnerable would be the small island states and coastal areas vulnerable to rising sea levels, and the heavily populated mega-deltas such as the Mekong and Red River Deltas which would be under threat from the flooding of the great rivers. According to the IPCC, more than one-sixth of the world's population lives in these areas affected by water sources from glaciers and snow pack that will "very likely" disappear (IPCC, 2007).

12.2.2. *Main displacement scenarios*

According to the Inter-Agency Standing Committee (IASC, 2008),[14] it is possible to distinguish between four main types of climate change-related

[14]IASC is a unique interagency forum for humanitarian coordination and a policy development and decision-making forum involving key UN and non-UN humanitarian actors. It was established in June 1992 in response to UN General Assembly Resolution 46/182 on the strengthening of humanitarian assistance. For more information, see http://www.humanitarianinfo.org/iasc.

displacement scenarios, where displacement is caused by:

- Weather-related disasters, such as hurricanes and flooding;
- Gradual environmental deterioration and slow onset disasters, such as desertification, sinking of coastal zones, and possible total submersion of low-lying island states;
- Increased disaster risks resulting in relocation of people from high-risk zones; and
- Social unrest and violence attributable to climate change-related factors.

Thus, in addition to the threat of climate change-related migration of people and the related economic consequences that the countries may face, another challenge is the possibility of conflicts between migrants and other sections of the population (Kolmannskog, 2008).[15] This possibility arises, for example, due to scarce resources, increasing competition, and cultural and ethnic differences between the migrants and the communities that absorb them. According to the United Nations Environmental Programme (UNEP, 2009),[16] the potential consequences of climate change for water availability, food security, prevalence of disease, coastal boundaries, and population distribution may aggravate existing tensions and generate new conflicts. According to one study, the effects of climate change interacting with economic, social, and political problems will create a high risk of violent conflict in 46 countries — home to 2.7 billion people (SIDA, 2008).[17] These countries, mainly in sub-Saharan Africa, Asia, and Latin America, are also the countries which are particularly exposed to projected negative impacts of climate change.

It is thus important that the affected and vulnerable countries take adequate risk mitigation and adaptation measures to deal with the threat of climate migrants.

[15] V.O. Kolmannskog, "Future Floods of Refugees," Norwegian Refugee Council, Oslo, 2008. Retrieved 6 January 2010, from http://www.nrc.no/arch/_img/9268480.pdf.

[16] United Nations Environmental Programme, "From Conflict to Peacebuilding: The Role of Natural Resources and the Environment," UNEP, Nairobi, 2009.

[17] International Alert and Swedish International Development Cooperation Agency, "A Climate of Conflict," SIDA, Stockholm, 2008, 7.

12.3. Current Legal and Policy Architecture to Deal with Climate Migrants

12.3.1. *Internally displaced persons*

People displaced from their homes as a result of climate change-related disasters or any other natural disasters are primarily the concern of the state in which they are citizens or residents. As most of the time, climate displacement due to disasters is temporary (until such time that the people return to their homes), or if the displacement is permanent, then relocation of the affected people takes effect within the jurisdiction of the affected country, and the displaced should be considered internally displaced persons (IDPs).

The 1998 United Nations Guiding Principles on Internal Displacement (GPID),[18] which has been recognized by states as an important international framework for the protection of internally displaced persons, although it does not have the same force as an international convention, has defined IDPs as:

> [P]ersons or groups of persons who have been forced or obliged to flee or to leave their homes or places of habitual residence, in particular as a result of, or in order to avoid, the effects of armed conflict, situations of generalized violence, violations of human rights or natural or human-made disasters, and who have not crossed an internationally recognized State border.

The inclusion of those affected by "disasters" in the above definition brings the climate refugees within the protection mechanisms of the existing legal regimes.

Normally, the basic human rights of a person such as the right to equal treatment, right to nationality, right to life, liberty and security of person, right to own property, to name a few, are guaranteed in most countries by

[18]Guiding Principles on Internal Displacement (GPID) (UN Doc. E/CN.4/1998/53/Add.2) was recognized as "an important international framework for the protection of internally displaced persons" in September 2005, by the heads of state and governments assembled at the World Summit in New York (G.A. Res. 60/L.1, 132, U.N. Doc. A/60/L.1).

way of the Constitution, the supreme law of the country.[19] Thus, persons affected by displacement within national borders are entitled to the full range of human rights guaranteed by the state concerned. The GPID, in fact, specifically states inter alia that IDPs "shall enjoy, in full equality, the same rights and freedom under international and domestic law as do other persons in their country. They shall not be discriminated against in the enjoyment of any right on the grounds that they are internally displaced". It provides further that, "Every human being shall have the right to be protected against being arbitrarily displaced from his or her home or place of habitual residence".[20]

Further, the Operational Guidelines on Human Rights and Natural Disasters developed by the Inter-Agency Standing Committee (IASC),[21] address the needs of all affected people, including the displaced. These guidelines aim inter alia to ensure that human rights principles and protection standards, including the fundamental principles of non-discrimination and accountability, are integrated into all disaster response, recovery, and reconstruction efforts from the earliest stage possible. Further, they aim to identify relevant measures to ensure that affected persons and their communities are fully consulted and can actively participate in all stages of the disaster response in accordance with their human rights (Brookings-Bern Project on Internal Displacement, 2008).[22]

Thus, as far as the legal rights of the displaced persons are concerned, there is no doubt that there is an obligation on the part of the state concerned to look after the welfare and interests of the affected persons. However, the real issue here is that, although the legal rights have been established and universally recognized, some countries, especially the

---

[19] For the full list of the human rights adopted in the Universal Declaration of Human Rights which was declared on December 10, 1948 by the General Assembly of the United Nations, see http://www.un.org/en/documents/udhr/.

[20] See GPID, principles 1 and 6.

[21] Supra note 3.

[22] Brookings-Bern Project on Internal Displacement, "Human Rights and Natural Disasters: Operational Guidelines and Field Manual on Human Rights Protection in Situations of Natural Disaster," Pilot Version, March 2008. Washington, DC, 2008. Retrieved 2 January 2010, from http://www.brookings.edu/~/media/Files/rc/reports/2008/spring_natural_disasters/spring_natural_disasters.pdf.

developing countries, may not have the financial and technical capacity to deal with climate-related disasters such as sea level rise, extreme heat or cold, or severe flooding, and protect their people against such disasters or effectively look after their interests in the aftermath of such disasters. In the circumstances, bilateral, regional, and global initiatives are important.

The guiding principles demand non-discrimination and fair treatment of those affected and consultation of those affected when relocating them. However, the possible non-observance of the principles is another issue. There have been allegations that in some of the countries affected by the Asian tsunami, buffer zones were being established in a discriminatory manner, allowing construction of tourism facilities while local residents were not allowed to return and reconstruct their homes (UNHRC, 2009).[23]

12.3.2. *Cross-border migrants*

Climate change-related disasters mentioned above could sometimes result in the affected and displaced people fleeing to other countries. One possible scenario of such displacement across national borders is the eventual total submergence of small island states due to rising sea levels. For example, if the current prediction that the Maldives will disappear within the next 75 years due to rising sea levels comes right and at the time it so happens that no solution has been found concerning the population that will be affected, it is likely that the neighboring countries in South Asia will have to deal with climate refugees trying to enter their borders. It should be noted here that migration is in fact one of the climate change adaptation strategies already being implemented in low-lying island states, such as Kiribati, the Maldives, and Tuvalu (IPCC, WGII, 2007).[24] It has also been reported that the government of the Maldives has made representations to

[23]The Office of the United Nations High Commissioner for Refugees, "Protection of Internally Displaced Persons in Situations of Natural Disasters," Report of the Representative of the Secretary-General on the Human Rights of Internally Displaced Persons, Walter Kälin, to the 10th Session of the Human Rights Council, 5 March 2009, A/HRC/10/13/Add.1.
[24]IPCC, "Climate Change 2007: Impacts, Adaptation and Vulnerability," Working Group II, Fourth Assessment Report, Cambridge University Press, Cambridge, 2007.

countries such as Sri Lanka and Australia regarding relocation of its people (Phillips, 2009).[25]

As far as legal protection is concerned, persons moving voluntarily or forcibly across an international border would be entitled to general human rights guarantees in a receiving state, but would often not have a right of entry to that state due to visa requirements and other regulations applicable to immigration. However, it should be noted that the Representative of the Secretary-General on human rights of internally displaced persons has suggested that a person who cannot be reasonably expected to return, for example, if the assistance and protection provided by the country of origin of such a person is far below international standards, should be considered a victim of forced displacement and be granted at least a temporary stay (OHCHR, 2008).[26]

Thus, it is clear that under the existing international legal regime, there is no permanent solution for the issue of cross-border climate migrants. Thus, the question arises whether the existing international law on human rights is suitable to deal with the issue and if not, whether a new law needs to be negotiated. According to Hampson (2008) the existing human rights law is not well-suited to address the issue of cross-border climate migration, as the cause of the problem is located outside the state of origin of the migrants as well as the recipient state.[27] She argues that a new law is not the most appropriate way forward, mainly because negotiating a new law would take a long time and would only offer a bottom-line level of protection. In the circumstances, she expresses the opinion that it would seem more

[25]M. Phillips, "Climate Change Threatens Maldives," *CBS News*, 7 December 2009. Retrieved 6 January 2010, from http://www.cbsnews.com/stories/2009/12/07/eveningnews/main5924789.shtml.

[26]Office of the High Commissioner for Human Rights (OHCHR), "Displacement Caused by the Effects of Climate Change: Who Will Be Affected and What Are the Gaps in the Normative Framework for Their Protection?," Background Paper, 2008. Retrieved 6 January 2010, from http://www2.ohchr.org/english/issues/climatechange/submissions.htm.

[27]F. Hampson, "Human Rights Law, Climate Change and the Question of Disappearing Territories." Presentation made to the OHCHR Study on the Relationship between Climate Change and Human Rights: Open-Ended Consultation Meeting, 22 October 2008. Retrieved 4 January 2010, from http://www2.ohchr.org/english/issues/climatechange/consultation.htm.

promising to examine existing laws and find long-term policy solutions to address the long-term problems of climate change ahead of time.

We agree with the view that the existing human rights laws do not provide clear answers as to the status of cross-border climate migrants and that, dealing with such an issue requires adequate long-term political solutions. However, it should be stated that having an international legal instrument that deals with the rights of cross-border climate migrants would add urgency for action against climate change by focusing on the impacts on vulnerable persons and help translate ethical and moral obligations into legal ones. Further, as argued by Humphreys (2008), such a move could help guide climate change policies, including by providing thresholds in the form of levels of protection of human rights regarded as minimum acceptable outcomes.[28] Further, in his view, in the area of adaptation, a human rights focus could help mobilize international funding to deal with the issue, while also guiding the use and prioritization of such funding.

So far, not only has the world failed to establish adequate legal principles to deal with cross-border climate migrants, but also, even appropriate policy initiatives have not been considered at the multilateral level. The outcome of the Conference of the Parties (COP15) of the United Nations Framework Convention on Climate Change (UNFCCC) in Copenhagen shows that despite the representations made by countries most vulnerable to the adverse impacts of climate change, their interests have been compromised to protect the economic interests of industrialized countries and a few fast growing developing economies. In summary, the Copenhagen Accord, agreed by major economies including the US and China, made a commitment to limit the rise in global temperatures to 2°C and raise US$100 billion annually by 2020 to help developing countries fight climate change. The Accord is criticized as it does not specify caps on emissions to achieve that objective. Further, according to the developing countries (such as their grouping, G77), the pledge to raise US$100 billion by the developed countries falls short of the US$400 billion estimated

[28] S. Humphreys, "Mapping the Human Rights Implications of Climate Change." Presentation made to the OHCHR Study on the Relationship between Climate Change and Human Rights (2008), *op. cit.*

annual requirement for helping developing countries (Vidal and Milimo, 2009).[29] Further, during the 2013 climate change negotiations (COP19 held in Warsaw), it has been observed that although the agreement in the Copenhagen round ostensibly formalized US$100 billion to the Clean Energy Fund by 2020 to assist developing countries in energy development, only US$7.5 million had been committed as of June 2013.

In these circumstances, there is no formal binding agreement has been reached to date that would take effect after the Kyoto Protocol which expired in 2012, the only multilateral instrument we had to deal with climate change. Although at the COP19 in Warsaw, member states agreed to work towards curbing emissions as soon as possible, with an idea date targeted at the first quarter of 2015, the survival of the current international consensus on the action required for mitigation of climate change and adapting to changing climate conditions is under severe threat in the absence of a formal binding agreement. In the absence of any formal international agreement, it is unlikely that rule of law will be applied to create and enforce any effective solution to deal with the issue of climate migrants and other climate change-related threats.

12.4. The Need to Provide Safe and Adequate Housing for Vulnerable Communities

The displaced people, both internal and external, have basic needs which will have to be provided. These needs include construction goods such as housing, schools and health facilities, as well as the physical infrastructure, which links these facilities together and facilitates the movement of persons, goods, and services. Moreover, the displaced people need premises in which to produce or provide services in order to earn a living. Thus, the construction industry is, or will be, required to play a crucial role in addressing some of the needs of the displaced people. However, in many developing countries, the provision of the basic requirements of constructed items is below what the governments themselves intend to attain. In this section, the case of housing is used to illustrate this point.

[29] J. Vidal and D. Milmo, "Copenhagen: Leaked Draft Deal Widens Rift between Rich and Poor Nations," *The Guardian*, 9 December 2009.

12.4.1. *The right to adequate housing*

The right to adequate housing is enshrined in several core international human rights instruments.

The Universal Declaration of Human Rights provides that:

> Everyone has the right to a standard of living adequate for the health and well-being of himself and of his family, including food, clothing, housing and medical care and necessary social services, and the right to security in the event of unemployment, sickness, disability, widowhood, old age or other lack of livelihood in circumstances beyond his control.[30]

Another such instrument is the International Covenant on Economic, Social and Cultural Rights (ICESCR) which provides inter alia that:

> "The States Parties to the present Covenant recognize the right of everyone to an adequate standard of living for himself and his family, including adequate food, clothing and housing, and to the continuous improvement of living conditions. The States Parties will take appropriate steps to ensure the realization of this right, recognizing to this effect the essential importance of international co-operation based on free consent.[31]

The right to adequate housing has been defined in the above conventions as "the right to live somewhere in security, peace and dignity".[32]

At the national level too, the right to adequate housing is recognized as a basic human right in many jurisdictions. For example, in the largest democracy in the world, India, this right is enshrined in the all-encompassing Right to Life under Article 21 of the Constitution. In China, although there is no Constitutional guarantee, the government has used a housing security policy to ensure the provision of housing to the people both during the planned economic period, and the period of reforms that began in the mid-1990s. According to Professor Chan Jian of the China Society for Human Rights Studies, during the long period under a planned economy in China, under the urban housing security system, the government allocated housing units to the people. With the transition from the planned economy

[30]Art. 25, para. 1.

[31]Art. 11.

[32]ICESCR general comment no. 12, para. 6.

to the market economy, the housing security system in China changed, with the allocation of houses as a benefit coming to an end. However, in 1998, the State Council introduced a three-level housing provision system under which the families with the lowest incomes rent the low-rent housing provided by the government, the families with the medium-to-low incomes buy the economically affordable housing, and other families with high incomes buy and rent houses at the market price (Chang and Goh, 2008).[33]

12.4.2. *The challenge for developing countries*

Whilst the developing countries need to recognize that the persons who need to be resettled or relocated have a right to adequate housing, providing that will be a major challenge for countries strapped with financial constraints. For example, despite the Constitutional recognition of the right to adequate housing as a basic human right, according to the UN Statistics Division (1995),[34] the number of homeless households in India was over 600,000. According to a 2001 census, India has about two million homeless people, a number that is unreported according various non-government agencies and autonomous branches of government bodies (Tulsyan, 2008).[35]

In these circumstances, countries which are already facing the problem of providing adequate housing to their people will find it extremely challenging to deal with the issue of climate change-related migration which will create a further demand for housing. This is because, whilst the right to housing is recognized under international law and in most national laws as a basic human right, the financial constraints make it extremely difficult for developing countries to honor this right.

According to the IPCC (2007) it is likely that climate change will affect the right to adequate housing in several ways, including sea level rise and

[33] J. Chang and W. Goh, "China's Housing Security System and Housing Rights Security for Citizens," China Society for Human Rights Studies, 2008. Retrieved 2 January 2010, from http://www.chinahumanrights.org/CSHRS/Magazine/Text/t20080730_363967.htm.

[34] United Nations, "Demographic and Social Statistics," Department of Economic and Social Affairs, 1995. Retrieved 12 January 2010, from http://unstats.un.org/unsd/Demographic/sconcerns/housing/comp1995/TABLE08.pdf.

[35] A. Tulsyan, "Street Side Story," India Together, 4 June 2008. Retrieved 6 January 2010, from http://www.indiatogether.org/2008/jun/pov-homeless.htm.

storm surges which will have a direct impact on many coastal settlements.[36] Such impacts have already led to the relocation of peoples and communities in the Arctic region and in low-lying islands.[37] According to the UNDP (2007/2008),[38] in particular, an estimated one billion people living in urban slums on fragile hillsides or flood-prone riverbanks in different parts of the world face acute vulnerability to extreme climate events.

In many developing countries, the capacity of the construction industry is insufficient to enable it to build what is necessary, even to meet current needs. Thus, the climate change scenario adds urgency to the need for measures to develop the construction industries of the poorer countries, and to enhance their performance. As it takes time for the capacity of the industry to be enhanced, these measures need to be taken now, in order to prepare the industries for the increase in needs and effective demand, with regard to constructed items such as housing, productive and social facilities, and infrastructure. Moreover, the reality of significant numbers of displaced persons puts the methods adopted in construction into sharp relief. For example, the construction of any item takes time so there is the need for early identification of the countries which will be affected, the physical planning process, design, and building.

12.4.3. *Need for national, bilateral, regional, and multilateral cooperation*

Given the disparity between states in terms of riches and technological know-how, no global problem can be solved without global cooperation. However, whilst the rich countries have come forward on many occasions to help developing countries, the world has not yet found permanent or long-term solutions to the key pressing issues such as poverty eradication and climate change.

The pre-conference publicity of the Copenhagen meeting of the COP15 provided hope for the entire world of a solution to deal more efficiently and effectively with climate change. However, the outcome of the meeting

---

[36]IPCC (2007), 332.

[37]*Ibid.*, 672.

[38]United Nations Development Programme, "Fighting Climate Change: Human Solidarity in a Divided World," Human Development Report, UNDP, Geneva, 2007/2008, 9.

disappointed many developing countries and small island states, as the developed countries and a few rising economies in the developing world, such as India and China, struck a deal that favored their economic aspirations and undermined the global effort to mitigate climate change and adapt to changing climatic conditions that the UN was seeking to achieve.

For instance, the World Bank reports that poor countries would need US$550 billion a year — including US$400 billion for technology development and US$150 billion — to cope with the effects of climate change. However, in Copenhagen, the secured pledges from developed countries amounted to less than US$100 billion a year. According to Lumumba Di-Aping, the Sudanese Chairman of the group of 132 developing countries known as G77 plus China: "This text destroys both the UN convention on climate change and the Kyoto protocol. This is aimed at producing a new treaty, a new legal initiative that throws away the basis of [differing] obligations between the poorest and the most wealthy nations in the world."

Thus, it is important that the countries which are vulnerable to adverse impacts of climate start putting in place appropriate local adaptation measures. Such measures could include creation of local adaptation funds in partnerships between public and private sector entities, education of vulnerable communities so that their early cooperation could be sought for necessary preparations and actions, and introduction of building regulations and construction standards that specify safe locations and specifications which will make buildings capable of withstanding adverse impacts of climate change. However, such measures will have to be complemented by support initiatives at bilateral, regional, and multilateral levels.

12.5. Proposed Policy Options

12.5.1. *National initiatives*

The NRC (2009) suggests that consistent application of a natural disaster displacement monitoring methodology would provide a baseline for informed estimates as to how current trends may be affected by climate change in the future, and would be a necessary element for any improvement in the response for the displaced. A further suggestion concerns the collection of data on related factors, including the duration of displacement and the needs of displaced populations, as such information would enable countries to be better prepared and make their reaction to climate disasters and resulting displacement of people rapid and efficient.

In addition to having the relevant data sets, appropriate financial, social, technological, and regulatory mechanisms should be developed. For example, the vulnerable states could create disaster funds to finance the welfare of and the temporary and permanent relocation of the displaced people depending on the level of the disaster and future threats. The funds should be complemented with social infrastructure that facilitates the acceptance and accommodation of migrants by communities in safe locations. Appropriate technology for relocation and reconstruction of cities is important to strengthen the adaptability to climate change and future threats. The regulatory mechanism should facilitate the financial, social, and technological initiatives in order to ensure smooth implementation and avoidance of conflict.

12.5.2. *Bilateral and regional initiatives*

As far as climate change is concerned, the type of impact and the capacity to adapt to such impacts vary according to regions. Countries in East Asia and South Asia are more vulnerable to rising sea levels and floods. Countries in Africa are more vulnerable to droughts. Those in the Americas may be more vulnerable to hurricanes. The countries in Europe may be more vulnerable to extreme weather conditions such as heat waves. If this is the case, some solutions for these problems can be found within regions.

Due to historical, cultural, ethnic, and geographical reasons, bilateral and regional cooperation can be more effective in reaching consensus on issues affecting the relevant countries, compared to solving issues at the global level. For example, if, as predicted, rising sea levels will result in the Maldives disappearing within the next 75 years, then the responsibility of finding alternative land for Maldivians can be considered within the region. Likewise, if rising sea levels will affect a large number of islands in Indonesia and the Philippines, then solutions to the problem can be considered within the region. As Dunpont and Pearman (2006) observe, Bangladeshis would seek refuge in India or Pakistan, Indonesians from

Sumatra would look to Malaysia, and so on.[39] According to Brown (2007) intercontinental migration is also a likely scenario following pre-existing paths and old colonial relationships.[40] For example, the United Kingdom might be an obvious destination for people from former colonies which are now under threat due to rising sea levels. Likewise, France would be an ideal destination for migrants from Francophone West Africa and Australia, and New Zealand for some groups in the South Pacific (Brown, 2007).

In addition, other solutions such as financial assistance, accommodation, sharing of technology, especially new building designs and construction materials more adaptable to extreme weather, and administrative support, could also be found at the bilateral and regional levels.

A variety of national and regional examples could be looked at to derive a roadmap or handbook on strategies for migration policy, asylum policy, and complementary and temporary protection in cases of environmental migration. For example, New Zealand has implemented a new labor migration policy called the Pacific Access Category (PAC), which allows 75 citizens of Kiribati, 75 citizens of Tuvalu, and 250 citizens of Tonga (including their partners and dependent children) to establish residency in New Zealand each year (Department of Labour, 2007).[41] While no mention is made in the PAC of any environmental issues, let alone climate change, accepting migrants from Pacific islands which are among the most vulnerable nations to climate change can be seen as one aspect of important new bilateral policies aimed at contributing solutions to these issues.[42]

---

[39]Dupont, A. and G. Pearman, "Heating up the Planet: Climate Change and Security," Lowry Institute for International Policy, Paper 12, Sydney, 2006, 59.

[40]Brown, O., "Climate Change and Forced Migration: Observations, Projections and Implications," Human Development Report 2007/2008, United Nations Development Programme. Retrieved 2 January 2010, from http://hdr.undp.org/en/reports/global/hdr2007-2008/papers/brown_oli.pdf.

[41]Department of Labour, "Migration Trends," Government of New Zealand. Retrieved 31 March 2015, from http://www.dol.govt.nz/PDFs/migration-trends-0607.pdf.

[42]New Zealand has received a great deal of popular press for supposedly agreeing to accept the entire population of Tuvalu should the islands becomes overwhelmed by sea level rise. However, according to the government of New Zealand, no such agreement exists. See New Zealand's immigration relationship with Tuvalu Statement from New Zealand Ministry of Foreign Affairs and Trade, July 2009.

Another example can be drawn from the current immigration and asylum policies of Sweden and Finland. Both these countries recognize "environmental migrants" as a category of "person in need of protection". In Sweden, for example, the Aliens Act[43] offers subsidiary protections for a "person otherwise in need of protection", who is "outside the country of the alien's nationality, because he or she . . . is unable to return to the country of origin because of an environmental disaster".[44] The Finnish Aliens Act,[45] provides that an alien may be granted humanitarian protection if they are, in their home country, under "threat of death penalty, torture or other inhuman treatment or treatment violating human dignity, or if they cannot return there because of an armed conflict or environmental catastrophe".[46] A key difference between these two statues is that, whilst the Swedish legislation does not go the extent to guarantee permanent protection to victims of environmental disasters, the Finnish legislation does not appear to have such limitations, offering complementary protection to foreign nationals who cannot return safely to their home country because of an environmental disaster (UNHCR, 2009).[47]

It should be pointed out that, although some countries might show willingness to accommodate the victims of environmental disasters, including climate change migrants, it is important to ensure that such initiatives have the local support of the citizens of the recipient country. If not, the migrants could meet with hostile treatment by the locals. Further, not every country in the world has the capacity to absorb a large number climate and other environmental migrants due to economic, cultural, and political reasons.

12.5.3. *Multilateral initiatives*

Climate change is a global problem and irrespective of the fact that the industrialized countries may have been more responsible than others to the

---

[43]Aliens Act (2005:716). Issued: 29 September 2005.

[44]*Ibid.*, Section 2.

[45]Aliens Act (301/2004).

[46]*Ibid.*, Sections 88A and 109.

[47]The Office of the United Nations High Commissioner for Refugees, Protection of Internally Displaced Persons in Situations of Natural Disasters, Report of the Representative of the Secretary-General on the Human Rights of Internally Displaced Persons, Walter Kälin, to the 10th Session of the Human Rights Council, 5 March 2009, A/HRC/10/13/Add.1.

current state of the environmental degradation of the globe, the adverse impacts of climate change could affect any one at any given time.

The immediate need thus is to find a solution to climate change that is acceptable to all nations. Any new post-Kyoto agreement should be more dedicated towards adaptation initiatives, whilst cutting down the levels of GHG emissions to save the world for the future generations. The current generation in the most vulnerable nations needs assistance from the developed nations to help them to adapt to irreversible adverse impacts of climate change.

Adaptation initiatives are important as climate change is likely to lead to increasing rates of displacement. Thus, it is vital that evolving frameworks for climate change adaptation address displacement issues. The initiatives that could be considered at the multilateral level include the creation of a sufficient adaptation fund, further research into mitigation and adaptation measures and the development of early warning systems, transfer of technology, and helping the developing countries achieve sustainable development. Whilst the funds will help the developing countries to relocate the vulnerable communities and victims of climate change, consistent application of natural disaster displacement monitoring methodologies and other early warning systems would provide a baseline for informed estimates as to how current trends may be affected by climate change in the future, thus providing opportunities to minimize risks and help prepare for future displacements.

What is currently lacking is effective information sharing and cooperation between the nations with financial and technical capacity to conduct research and the countries which lack such capacity. If such cooperation had existed, damages from disasters such as the Asian tsunami in 2004 which left approximately 225,000 dead and another 500,000 displaced could have been minimized. It is said that the estimated cost of a tsunami early warning system is only about US$20 million, whereas the cost of aid and reconstruction following the Asian tsunami exceeded US$7.5 billion (Pickrell, 2005).[48]

The multilateral initiatives suggested above are not new. They have been discussed at various international forums and have been campaigned

[48]Pickrell, J., "Figures and Facts: Asian Tsunami Disaster," *New Scientist*, 20 January 2005.

for by the developing countries. However, so far, only snail paced progress has been made. As far as research initiatives for data gathering is concerned, according to the United Nations University Institute of Environment and Human Security (UNU-EHS, 2008),[49] international entities such as itself, the IOM, UNEP, and the Munich Re Foundation have commenced collaboration on this matter, as it is crucial that the available data be improved to allow policy-makers to make the proper decisions. Thus, what is required is for countries to support the work programs of such institutions in addition to research that can be carried out at the national, bilateral, and regional levels.

12.6. Conclusion

The current scientific evidence indicates the vulnerability of millions to changing climate conditions and the possibility of environmentally-induced human migration on an unprecedented scale. Thus, what is needed is a complete reassessment of the vulnerabilities to climate change faced by different countries, and a refashioning of the vulnerable locations to adapt to increasing environmental volatility, before the exodus of people forced from their homes reaches biblical proportions.

This chapter has argued that whilst climate change mitigation has entered mainstream policy-making, adaptation still lags behind in many ways, especially in the case of finding effective solutions to deal with climate migrants. Further, this chapter has argued that although looking at climate change vulnerability and adaptive capacity in human rights terms could help nations to recognize the need to protect the vulnerable communities, the real answer to the issue of climate change and to climate-induced migration does not lie in entirely in a rights-based approach. This is because, within a country, within a region, and in the world, the capacity of people to adapt to disasters is not equal. In these circumstances, in the absence of cooperation among communities, cities, and nations, we cannot find an effective solution.

---

[49]United Nations University Institute for Environment and Human Security, "Policy Dialogue on Environmentally Induced Migration," Summary Report of Policy Dialog. 4 July 2008, Belgium, Brussels.

References

Brookings-Bern Project on Internal Displacement. "Human Rights and Natural Disasters: Operational Guidelines and Field Manual on Human Rights Protection in Situations of Natural Disaster." Pilot Version, Washington, DC, March 2008. Retrieved 2 January 2010, from http://www.brookings.edu/~/media/Files/rc/reports/2008/spring_natural_disasters/spring_natural_disasters.pdf.

Brown, O. "Climate Change and Forced Migration: Observations, Projections and Implications." Human Development Report 2007/2008, United Nations Development Programme. Retrieved 2 January 2010, from http://hdr.undp.org/en/reports/global/hdr2007-2008/papers/brown_oli.pdf.

Chang, J. and W. Goh. "China's Housing Security System and Housing Rights Security for Citizens." China Society for Human Rights Studies. Retrieved 2 January 2010, from http://www.chinahumanrights.org/CSHRS/Magazine/Text/t20080730_363967.htm.

Dupont, A. and G. Pearman. "Heating up the Planet: Climate Change and Security." Lowry Institute for International Policy, Paper 12, Sydney, 2006, 59.

Hampson, F. "Human Rights Law, Climate Change and the Question of Disappearing Territories." Presentation made to the OHCHR Study on the Relationship between Climate Change and Human Rights: Open-Ended Consultation Meeting, 22 October 2008. Retrieved 4 January 2010, from http://www2.ohchr.org/english/issues/climatechange/consultation.htm.

Hoffman, J.S., D. Keyes, and J.G. Titus. "Projecting Future Sea Level Rise." U.S. Environmental Protection Agency, Washington, DC Gov. Print. Office, 1983, 266.

Humphreys, S. "Mapping the Human Rights Implications of Climate Change." Presentation made to the OHCHR Study on the Relationship between Climate Change and Human Rights (2008), *op. cit.*

Intergovernmental Panel on Climate Change (IPCC). "Working Group I Third Assessment Report." Cambridge University Press, Cambridge, 2001.

Intergovernmental Panel on Climate Change (IPCC). "Climate Change 2007: Impacts, Adaptation and Vulnerability." Working Group II, Fourth Assessment Report, Cambridge University Press, Cambridge, 2007.

International Alert and Swedish International Development Cooperation Agency. "A Climate of Conflict." SIDA, Stockholm, 2008, 7.

International Organization for Migration (IOM). "Migration, Climate Change and the Environment." IOM Policy Brief, IOM, Geneva, May 2009.

Kolmannskog, V.O. "Future Floods of Refugees." Norwegian Refugee Council, Oslo, 2008. Retrieved 6 January 2010, from http://www.nrc.no/arch/_img/9268480.pdf.

OCHA and Internal Displacement Monitoring Centre (IDMC) of the Norwegian Refugee Council (NRC). "Monitoring Disaster Displacement in the Context of Climate Change," 2009. Retrieved 4 January 2010, from http://www.internal-displacement.org/8025708F004BE3B1/%28httpInfoFiles%29/12E8C7224C2A6A9EC125763900315AD4/$file/monitoring-disaster-displacement.pdf.

Office of the High Commissioner for Human Rights (OHCHR). "Displacement Caused by the Effects of Climate Change: Who Will Be Affected and What Are the Gaps in the Normative Framework for Their Protection?," Background Paper. Retrieved 6 January 2010, from http://www2.ohchr.org/english/issues/climatechange/submissions.htm.

Phillips, M. "Climate Change Threatens Maldives." *CBS News*, 7 December 2009. Retrieved 6 January 2010, from http://www.cbsnews.com/stories/2009/12/07/eveningnews/main5924789.shtml.

Pickrell, J. "Figures and Facts: Asian Tsunami Disaster." *New Scientist*, 20 January 2005.

Reuters. "Indonesia May Lose 2,000 Islands to Climate Change." 29 January 2007. Retrieved 6 January 2010, from http://www.reuters.com/article/latestCrisis/idUSJAK299186.

The Office of the United Nations High Commissioner for Refugees (UNHCR). "Protection of Internally Displaced Persons in Situations of Natural Disasters." Report of the Representative of the Secretary-General on the Human Rights of Internally Displaced Persons, Walter Kälin, to the 10th Session of the Human Rights Council, 5 March 2009, A/HRC/10/13/Add.1.

Tulsyan, A. "Street Side Story." India Together, 4 June 2008. Retrieved 6 January 2010, from http://www.indiatogether.org/2008/jun/pov-homeless.htm.

UNHCR. "Forced Displacement in the Context of Climate Change: Challenges for States under International Law." A paper submitted to the Sixth Session of the Ad Hoc Working Group on Long-Term Cooperative Action under the United Nations Framework Convention on Climate Change, Bonn, Germany, 2009.

United Nations (UN). "Demographic and Social Statistics." Department of Economic and Social Affairs, 1995. Retrieved 12 January 2010, from http://unstats.un.org/unsd/Demographic/sconcerns/housing/comp1995/TABLE08.pdf.

United Nations Development Programme (UNDP). "Fighting Climate Change: Human Solidarity in a Divided World." Human Development Report, UNDP, Geneva, 2007/2008, 9.

United Nations Environmental Programme (UNEP). "From Conflict to Peacebuilding: The Role of Natural Resources and the Environment." UNEP, Nairobi, 2009.

United Nations University Institute for Environment and Human Security (UNU-EHS). "Policy Dialogue on Environmentally Induced Migration." Summary Report of Policy Dialog. Belgium, Brussels, 4 July 2008.

United Kingdom Treasury. "Stern Review on the Economics of Climate Change." 2005. Retrieved 10 January 2010, from http://webarchive.nationalarchives.gov.uk/+/http://www.hm-treasury.gov.uk/independent_reviews/stern_review_economics_climate_change/sternreview_index.cfm.)

Vidal, J. and D. Milmo. "Copenhagen: Leaked Draft Deal Widens Rift between Rich and Poor Nations." *The Guardian*, 9 December 2009.

Warner, K., C. Ehrhart, A. Sherbinin and S. Adamo. "In Search of Shelter." Center for International Earth Science Information Network, Columbia University, 2009. Retrieved 10 January 2010, from http://www.ciesin.columbia.edu/.../climmigr-report-june09_final.pdf.

Warner, K., M. Stal, O. Dun, and T. Afifi. "Researching Environmental Change and Migration: Evaluation of EACH-FOR Methodology and Application in 23 Case Studies Worldwide." In Frank Laczko and Christine Aghazarm (eds.), *Migration Environment and Climate Change: Assessing the Evidence* (Geneva: IOM, 2009), Chapter IV.

World Wildlife Federation (WWF). "Climate Change Impact on APEC Countries." WWF, 2007. Retrieved 4 January 2010, from http://assets.panda.org/downloads/cc_impacts_on_apec_countries.pdf.

World Bank. "Modelled Observations on Development Scenarios in the Lower Mekong Basin." World Bank, Washington, DC, 2004.

Chapter 13

Climate Change, Migration, and International Law in Southeast Asia

*Benoît Mayer**

13.1. Introduction

Environmental factors have always constituted a strong, though often indirect, inducement to migration.[1] Climate change, through its impacts on local environments, is likely to result in an increase in human migration. Despite great scientific uncertainty as to the actual or future impact of climate change on migration, it is often referred to a controversial estimate according to which, by 2050, 250 million persons would have been displaced because of climate change.[2] This figure at least gives an idea of the projected scope of "climate migration".

The circumstances in which climate change may lead to migration are very diverse. Populations may move due to sudden-onset natural disasters (e.g., cyclones, drought) or because of slow-onset environmental changes (e.g., sea level rise, desertification). In most cases, however, migration results from the socio-economic phenomena that climate change exacerbates (e.g., famine, land degradation, lack of economic opportunities). Response measures to climate change may also cause migration, for instance through displacement and resettlement induced by large development projects carried out partly as a way of mitigating climate

*Associate Professor, Institute of International Law, Wuhan University, China, bnt.mayer@gmail.com.

[1]Etienne Piguet, "From 'Primitive Migration' to 'Climate Refugees': The Curious Fate of the Natural Environment in Migration Studies," *Annals of the Association of American Geographers*, 130 (148).

[2]Norman Myers, "Environmental Refugees: An Emergent Security Issue," presentation at the 13th Economic Forum, Prague, 23–27 May 2005.

change or of adapting to it (in particular hydroelectric projects), or possibly through the "push" and "pull" factors that the transformations toward a green economy induces.

Early normative studies on climate change-induced migration have submitted that no existing legal framework addresses the needs of environmental migrants in a satisfactory way.[3] To respond to what they identified as a new and dangerous phenomenon, these studies urged states to adopt an appropriate international regime for the protection of climate migrants. These studies have two main limitations that are highlighted in this chapter. On the one hand, they tend to assume that climate migration is a discrete phenomenon calling for "solutions", despite the irreducible variety of the causal links between climate change and migration. On the other hand, these studies apply a top-down approach that lacks adequate considerations for regional contexts. To contribute in resorbing these limitations of existing literature, this chapter is an attempt at downscaling the debate on possible responses to climate migration while giving due attention to the various forms of "climate migration". This chapter looks at climate migration and international law in the context of Southeast Asia, defined by reference to the 10 current member states to the ASEAN.

The chapter is constituted of three additional sections. Section 13.2 identifies several regional scenarios of climate migration. Section 13.3 exposes some difficulties with conceiving a governance of climate migration. Lastly, Section 13.4 analyzes some prospects for regional responses to climate migration in Southeast Asia.

13.2. Climate Change and Migration in Southeast Asia

Southeast Asia is particularly concerned by climate migration. On the one hand, migration (both within and across international borders) is "an established structural feature of the region".[4] Historically, migration

---

[3] See in particular Frank Biermann and Ingrid Boas, "Preparing for a Warmer World: Towards a Global Governance System to Protect Climate Refugees," *Global Environmental Politics*, 10 (2010), 60; Bonnie Docherty and Tyler Giannini, "Confronting a Rising Tide: A Proposal for a Convention on Climate Change Refugees," *Harvard Environmental Law Review*, 33 (2009), 349.

[4] Graeme Hugo, "The Future of Migration Policies in the Asia-Pacific Region, Background Paper for the World Migration Report," 5. Retrieved 7 April 2015, from http://publications.iom.int/bookstore/free/WMR2010_migration_policies_asia-pacific.pdf.

accounts for the ethnic diversity of each state of the region. To date, a third of the population of Singapore and Brunei was born in a country other than that in which they live, and the Philippines and Vietnam are major sources of migrant workers for many developed countries. On the other hand, climate change is having important consequences in Southeast Asia. In particular, sea level rise threatens coastal communities, while some extreme weather events (e.g., cyclones, storm surge, heavy precipitations, and droughts) may become more severe. Warming may also affect food production and vector-borne diseases such as dengue.[5] Rapid economic development and urbanization often add concomitant stress on fragile environments. The Internal Displacement Monitoring Centre estimated that 81% of the persons displaced by sudden natural disasters in the world between 2008 and 2012 were in Asia.[6]

In what follows, I identify four particular scenarios of climate migration in Southeast Asia, whereby migration follows from typhoons, floods, economic impacts, or response measures. These scenarios are not intended to present a comprehensive picture of climate migration in the region, but they are believed to reflect the diversity of climate migration and to provide a factual background for the more theoretical discussions of the following sections.

13.2.1. *Climate change, typhoons, and migration*

Several states in Southeast Asia are particularly exposed to typhoons, which may result in significant human displacement.[7] When typhoon Haiyan (locally called Yolanda) hit Tacloban and other regions of the Philippines on 8 November 2013, it killed more than 6,000 people and displaced an estimated 4 million. In the Warsaw Conference of the Parties to the UN Framework Convention on Climate Change (COP19) a few weeks

[5] See generally IPCC, AR5, Chapter 24 "Asia".

[6] Michelle Yonetani, *Global Estimates 2012: People Displaced by Disasters*, (Internal Displacement Monitoring Centre & Norwegian Refugee Council, 2013).

[7] For example, Graeme Hugo and Douglas K Bardsley, "Migration and Environmental Change in Asia," in Etienne Piguet and Frank Laczko (eds.), *People on the Move in a Changing Climate* (Netherlands: Springer, 2014), pp. 39–40. Retrieved 7 April 2015, from http://link.springer.com.libproxy1.nus.edu.sg/chapter/10.1007/978-94-007-6985-4_2.

later, the chief representative of the Philippines made an emotional speech associating the Haiyan with climate change.[8]

Yet, weather and climates are distinct concepts. A change in the climate is a change in the probability of certain weather patterns. Therefore, a single weather event can never be fully "attributed" to climate change, even when the probability of such event changes. This makes it difficult to determine whether or not a person displaced by a typhoon is a "climate migrant". Other typhoons stroked Tacloban before any significant anthropogenic interference in the climate system, for instance one, in 1912, which killed an estimated 15,000 people in a region that was then much less densely inhabited.[9] In fact, the fifth report of the IPCC is all but clear on the influence of climate change on the frequency or the severity of typhoons in Southeast Asia.[10]

13.2.2. *Climate change, floods, and migration*

The fourth IPCC report had considered that "[p]rojected sea level rise could flood the residence of millions of people living in the low lying areas of South, Southeast and East Asia such as in Vietnam, Bangladesh, India and China".[11] Graeme Hugo and Douglas Bardsley considered that, because of climate change, a third of the population of Southeast Asia is "at risk of coastal flooding and its associated impact", in particular "concentrated in Indonesia, the Philippines, Vietnam, Myanmar and Thailand, mainly in the megacities of Jakarta, Manila, Ho Chi Minh City and Bangkok which are located in low-lying coastal areas".[12]

Climate change is not an isolated cause of flooding, which may also have to do with the impacts of other anthropogenic activities, for instance through urbanization and deforestation (increasing water run-off) and the construction of dams (reducing sediment inputs in delta regions). Just like

---

[8]Henry Fountain and Justin Gillis, "Typhoon in Philippines Casts Long Shadow over U.N. Talks on Climate Treaty," *The New York Times*, 11 November 2013. Retrieved 19 May 2014, from http://www.nytimes.com/2013/11/12/world/asia/typhoon-in-philippines-casts-long-shadow-over-un-talks-on-climate-treaty.html.

[9]"15,000 Die in Philippine Storm," *The Washington Herald*, 30 November 1912, p. 1.

[10]IPCC, AR5, Chapter 24, "Asia," Section 24.3.1 (2014).

[11]IPCC AR4, WG2, Chapter 10, "Asia," Section 4.3.1.

[12]Hugo and Bardsley (2014), *op. cit.*, p. 39.

typhoons, floods could occur before climate change: not all floods are necessarily the consequence of climate change, and, in fact, the causal link is almost always very difficult to establish.

Not all floods cause migration, and not always in the same conditions. During temporary urban floods such as in Bangkok (2011) or Jakarta (2013), many remained in their homes, sometimes just moving on a higher floor of the same building. Households whose revenues depend from agriculture tend to migrate more in reaction to floods. Studies on the Mekong region (Vietnam and Cambodia) have suggested that seasonal labor migration to towns is a common practice during the flooding season, but "[s]uccessive flooding events can destroy crops and drive people to migrate [permanently] in search of alternative livelihoods".[13]

13.2.3. *Climate change, its economic consequences, and migration*

Rather than as a direct consequence of an environmental impact of climate change, many people migrate because of economic changes that can be related to the impacts of climate change. Thus, lower agricultural productivity or increased natural hazards contribute to incentivize rural households to seek alternative livelihood or to diversify their sources of revenues through migration.

This causal link between migration and climate change is indirect and relatively difficult to establish. The stories told by individual migrants almost never mention climate change, and rarely even mention environmental factors: migrants usually focus on what they perceive as lacking economic opportunities that affect them more directly. However, on a holistic perspective, it is likely that the impacts of climate change exacerbate economic factors of migration in many countries of Southeast Asia.

[13]Koko Warner, Olivia Dun, and Marc Stal, "Field Observations and Empirical Research," *Forced Migration Review*, 31 (2008), 13, 14. See also Olivia Dun, "Migration and Displacement Triggered by Floods in the Mekong Delta," *International Migration*, 49 (2011), 200; Paula Nuorteva, Marko Keskinen, and Olli Varis, "Water, Livelihoods and Climate Change Adaptation in the Tonle Sap Lake Area, Cambodia: Learning from the Past to Understand the Future," *Journal of Water and Climate Change*, 1 (2010); M. Keskinen *et al.*, "Climate Change and Water Resources in the Lower Mekong River Basin: Putting Adaptation into the Context," *Journal of Water and Climate Change*, 1 (2010), 103.

The economic consequences of climate change contribute not only to accelerate urbanization, but also to incentivize temporary or permanent international labor migration. Economic disparities within and among the countries of Southeast Asia multiply the economic incentive to migrate, for cities or neighboring countries offer significantly better economic opportunities than some of the rural areas affected by slow-onset environmental changes or recurrent sudden natural disasters.

13.2.4. *Climate change, response measures, and migration*

Often ignored, as if it could invariably be justified, the impacts of response measures may also include forced migration. With a rapidly growing demand for energy, countries in Southeast Asia are considering multiple hydroelectric projects, among others on the Mekong River. Measures that encourage the production of biofuel in Indonesia, Malaysia, Thailand, and the Philippines contribute to an increasing competition of land use, which, in a context of poor land titling, encourages land grabbing and the eviction of small landowners by biofuel companies.[14] Forest conservation measures may also contribute to this increasing competition for land use.

Through the Clean Development Mechanism (CDM) and other forms of international finance, developed countries fund emission limitation and reduction projects in developing countries, sometimes as a way to comply with their own emission limitation commitments.[15] Indonesia, Vietnam, Malaysia, Thailand, and the Philippines are important recipients of international funds on climate change mitigation. Yet, neither CDM nor REDD-plus come with social safeguards on involuntary displacement similar to those adopted by multilateral development banks since the 1990s.[16] From hydroelectric power to biofuel and forest conservation,

---

[14]See e.g., D. Diop *et al.*, "Assessing the Impact of Biofuels Production on Developing Countries from the Point of View of Policy Coherence for Development," Tanzania Field Mission. Draft Version, 2013; S. Kumar, Pujan Shrestha, and P Abdul Salam, "A Review of Biofuel Policies in the Major Biofuel Producing Countries of ASEAN: Production, Targets, Policy Drivers and Impacts," *Renewable and Sustainable Energy Reviews*, 26 (2013), 822.

[15]See in particular Kyoto Protocol to the United Nations Framework Convention on Climate Change, 1997, 12.

[16]See in particular, World Bank, OP 4.12 Involuntary Resettlement, December 2001.

climate finance may have grave consequences on the human rights of many affected groups, in particular through involuntary resettlement, land grabbing, eviction, and other forms of forced migration.

13.2.5. *The difficulty of conceiving of "climate migration"*

These four scenarios reflect the diversity of what can fall within the scope of "climate migration". They bear witness that climate change may exacerbate very different forms of migration. "Climate migration" may relate to temporary or permanent migration, to the migration of individual or groups, to voluntary or forced migration, etc. It is generally internal migration, as migrants only go as far as they need to go,[17] but climate change may sometimes have some statistical effect on international migration. None of these scenarios is exclusive to climate change: typhoons, floods, economic factors, or diverse policies could induce migration even in the absence of climate change.

Lastly, it is virtually impossible to distinguish individual "climate migrants" through any of these scenarios. In particular, a typhoon or a flood that causes migration can never be "attributed" to climate change, which is only about the likelihood of such events. Likewise, the economic consequences of climate change intermingle with a variety of other circumstances and they cannot be isolated as the unique cause of migration. Response measures are determined by decisions between multiple political options: they are not a proximate or foreseeable consequence of climate change. The influence of climate change on migration often appears clearly in statistical works, but not in individual cases.

[17]This is a significant difference between "climate migrants" and refugees. Refugees flee persecution or war, which generally affect them in the whole of their state of origin. Therefore, a refugee generally does not have any viable internal flight alternative. On the contrary, environmental phenomena rarely coincide with states borders. When they only affect part of a state's territory, resettlement within one's state of origin is in principle possible. In principle, the exception concerns small countries whose territory may be entirely affected by an environmental change (in particular small island developing states), or arguably countries whose population exceeds their carrying capacity (although this requires a delicate definition of a threshold of "carrying capacity"). In practice, the actual flows of international migrants tend to be the best indicator of the need for international migration.

13.3. Responding to Climate Migration

Climate migration has attracted growing public attention. This international debate, first focused on migration from small island developing states, turned to discussions on migration triggered by sudden disasters or slow-onset environmental changes, including through the economic consequences of climate change; less attention has yet been devoted to the impact of response measures on migration. It is of particular importance, when conceiving possible governance responses to climate migration, to keep in mind the conceptual difficulties highlighted above: the diversity of scenarios of climate migration, the fact that climate change exacerbates pre-existing migration scenarios but does not create new forms of migrations, and the indistinctiveness of individual "climate migrants".[18] This section contemplates possible responses to climate migration at a global scale in the perspective of Southeast Asia, whereas the next section discusses regional responses in Southeast Asia.

13.3.1. *Climate migration and vulnerability*

Climate change will exacerbate some forms of vulnerability triggered through migration, in particular through forced migration. This, however, does not suffice to justify specific responses to climate migration. Nations presumably have a duty to help those in need. Yet, it does not appear clearly that "climate migrants" (whomever this label designates) may be construed as a coherent category of individuals in need for protection.[19] Because of the variety of scenarios through which climate change exacerbates migration, not all "climate migrants" appear to be equally vulnerable or to have the same needs for protection. For instance, temporary and permanent migration calls for very different forms of assistance.

---

[18]Yet, they have often been overseen in the normative literature on climate migration. See Benoit Mayer, "Constructing 'Climate Migration' as a Global Governance Issue: Essential Flaws in the Contemporary Literature," *McGill Journal of International Sustainable Development Law and Policy*, 9 (2013), 87.

[19]Benoit Mayer, "Fraternity, Responsibility and Sustainability: The International Legal Protection of Climate (or Environmental) Migrants at the Crossroads," *Supreme Court Law Review* [*Canada*], 56 (2012), 723.

A central conceptual issue is that "climate migrants" do not appear to be necessarily more vulnerable than "climate non-migrants" (i.e., individuals affected by climate change but who do not migrate) or non-climate related migrants — a solid ground to question the relevance of the concept of "climate migration".[20] On the one hand, climate change will not only affect migrants, but also those who are not able or not willing to migrate. Inability or unwillingness to migrate may follow for instance from lacking the financial or social means necessary to migrate, or simply from a strong attachment to one's home. Thus, an important scientific report on "migration and global environmental change" highlighted that "millions of people will be unable to move away from locations in which they are extremely vulnerable to environmental change".[21] The climate-affected non-migrants, or individuals "trapped in place" might often be the most vulnerable.

On the other hand, from a migration perspective, the cause of migration does not matter as much as the conditions in which migration occurs, in particular the reception of migrants in the putative host society. In other words, whether or not a flood may be related to climate change is irrelevant in assessing the vulnerability of the persons affected. Environmental disasters unrelated to climate change cause similar types of human sufferings: hundreds of thousands were killed and several millions were displaced in Southeast Asia in the aftermath of the 2004 Indian Ocean tsunami, a disaster that had nothing to do with climate change. "Environmental migration" certainly extends beyond "climate migration". Economic deprivation may also force migration, whether or not it is exacerbated by climate change or other environmental phenomena. Some migration experts have consequently called for a protection of all "survival migrants" or responses addressing "crisis migration", notwithstanding the cause of migration.[22]

---

[20]Calum Nicholson, "Climate Change and the Politics of Causal Reasoning: The Case of Climate Change and Migration," *Geographical Journal*, 180 (2014), 151.

[21]Foresight, *Migration and Global Environmental Change: Final Project Report* (The Government Office for Science, United Kingdom, 2011), p. 9.

[22]Alexander Betts, *Survival Migration: Failed Governance and the Crisis of Displacement* (US: Cornell University Press, 2013); Susan F. Martin, Sanjula Weerasinghe, and Abbie Taylor, *Humanitarian Crises and Migration: Causes, Consequences and Responses* (UK: Routledge, 2014).

13.3.2. *Climate migration and common but differentiated responsibilities*

Rather than a question of international solidarity, it is possible to consider climate migration as a question of responsibility. It is the position taken by developing countries from the outset of climate negotiations that, "[s]ince developed countries account for the bulk of the production and consumption of environmentally damaging substances, they should bear the main responsibility in the search for long-term remedies for global environmental protection".[23] During the 1992 Earth Summit, Malaysia's PM Mahathir bin Mohamad was particularly prominent in calling for the responsibility of the West: "[t]he pittance they offer is much less than the loss of earnings by the poor countries and yet it is made out as a generous concession."[24] Despite its ambiguities, the principle of common but differentiated responsibilities may be construed as highlighting a special, causal responsibility of industrial states to repair the injury they have been causing to developing states through excessive emissions of greenhouse gases.[25]

Migration first entered climate negotiations as a form of adaptation,[26] but it was also under consideration within the ongoing work program on loss and damage associated with climate change impacts in developing countries that are particularly vulnerable to the adverse effects of climate change.[27] Assessing the damages resulting from climate change suggests an approach in terms of responsibilities, whereby responsible states would be called to indemnify the states disproportionately affected by climate change.

---

[23]G77, "Caracas Declaration," paras. II–34.

[24]Mahathir bin Mohamad, Prime Minister of Malaysia, Report of the United Nations Conference on Environment and Development, vol. III: Statements by the Heads of State or Government at the Summit Segment of the Conference, doc. A/CONF.151/26/Rev.1 (vol. III), at 233.

[25]See Benoit Mayer, "Conceiving the Rationale for International Climate Law." Retrieved 13 May 2014, from http://papers.ssrn.com/abstract=2432856; Lavanya Rajamani, *Differential Treatment in International Environmental Law* (UK: Oxford University Press, 2006), p. 155.

[26]Cancun Agreements: Outcome of the Work of the Ad Hoc Working Group on Long-Term Cooperative Action under the Convention (COP16), 2010, para. 14(f).

[27]UNFCCC, Decision 3/CP.18, "Approaches to Address Loss and Damage Associated with Climate Change Impacts in Developing Countries That Are Particularly Vulnerable to the Adverse Effects of Climate Change to Enhance Adaptive Capacity," para. 7.

However, responsibility fails to justify the definition of new obligations that affected states would have vis-à-vis their own citizens. The damages suffered by a state as a consequence of climate change are generally diffuse and relate to an increased exposure to weather hazards (increased likelihood of specific weather patterns) rather than to specific human or material damages: such damages can hardly be asserted in terms of damages suffered by specific individuals. Even if this could be done, the law on diplomatic protection does not define any obligation for a state to transfer any compensation received from another state to the injured person[28] — let alone to commit to specific policies. While climate change responsibility should support financial transfers between states, it does not justify any interference in the sovereignty of the injured states, including the right of such states to decide how to respond to climate change in accordance with unspecific human rights standards.

13.3.3. *Climate migration as a political momentum*

If "climate migrants" are not necessarily the most vulnerable individuals and climate change responsibility does not justify interference in the domestic affairs of injured states, the project of governing climate migration appears as somewhat arbitrary. Rather than a protection of climate migrants, one could be inclined to a protection of all vulnerable migrants and to a compensation of the states most affected by climate change.

This objection holds in a perfect world where ethical concepts guide international governance. In the real world, however, new norms are adopted through complex processes, where ethical ideas and states' interests are the object of contradictory arguments. Norm entrepreneurs and their organizational platforms play a particular role in pushing critical states to adopt and promote new norms,[29] and so do the intrinsic and extrinsic conditions in which political arguments are made and challenged.[30]

---

[28] ILC, "Draft Articles on Diplomatic Protection," 19, recognizing only a "[r]ecommended practice" to do so. See also PCIJ, Series A, no. 2, "Mavrommatis Palestine Concessions (Greece v. United Kingdom)," 12.

[29] See Martha Finnemore and Kathryn Sikkink, "International Norm Dynamics and Political Change," *International Organization*, 52 (1998), 88.

[30] Neta Crawford, "Homo Politicus and Argument (Nearly) All the Way down: Persuasion in Politics," *Perspectives on Politics*, 7 (2009), 103.

Despite its arbitrariness, climate migration may be an opportunity for a successful communication strategy just because it responds to public moral sentiments. The apparent newness of climate migration overcomes aid fatigue; it attracts considerable political attention and possibly resources because new problems require new solutions. Climate migration can thus be approached as essentially a marketing operation, the *rebranding* of forced migration and humanitarian assistance in the language of the time. By linking two of the main contemporary global concerns — climate change and migration — the concept of climate migration is a magic recipe for public attention.[31]

13.3.4. *Opportunities and limits for a global governance of climate migration*

While climate migration builds political momentum, the difficulty remains to define a workable solution that could benefit from this dynamic. Jane McAdam justly questioned "the utility — and, importantly, the policy consequences — of pinning 'solutions' to climate change-related displacement on a multilateral instrument, in light of the likely nature of movement, the desires of communities affected by it, and the fact that a treaty will not, without wide ratification and implementation, 'solve' the humanitarian issue".[32] Climate migration is not only a multifaceted phenomenon; it is also a contested political concept. Four trends of advocacy can be distinguished, promoting respectively humanitarian assistance, the rights of the migrants, climate change cooperation, and international security.[33] Each of these normative enterprises consider responses to climate migration as a first step, but each normative enterprise goes in different directions. Thus, the "solutions" they propose to climate migration vary significantly and cannot always be reconciled.

Possible responses range from an international treaty to informal bilateral cooperation. Several academic works proposed an international

---

[31] Benoit Mayer, "'Environmental Migration' as Advocacy: Is It Going to Work?," *Refuge: Canada's Journal on Refugees*, 29 (2014), 27.

[32] Jane Mcadam, "Swimming against the Tide: Why a Climate Change Displacement Treaty Is Not the Answer," *International Journal of Refugee Law*, 23 (2011), 2.

[33] Mayer, "'Environmental Migration' as Advocacy: Is It Going to Work?," (n 31).

convention that would be inspired by the international regime on refugee protection,[34] but climate change generally exacerbates migration scenarios that have little resemblance with flows of refugees (i.e., international, forced, long-term migration).[35] Despite the significant political impetus raised by climate migration, the ratification of a treaty would face significant obstacles.[36] Soft law instruments have sometimes been seen as practical alternatives, in particular on the ground of the success of the Guiding Principles on Internal Displacement. Thus, the ongoing "Nansen initiative", convened by the governments of Norway and Switzerland, may soon propose some guidelines on responses to international disaster-induced displacement.[37]

States may also make use of existing climate finance to carry on migration programs. However, the principle that international finance should cover the "incremental costs" of adaptation results in a significant hurdle for recipient states that must demonstrate the causal relation between climate change and migration. In addition, the funds made available for adaptation are relatively limited (about US$4.4 billion/year) and tightly controlled by donor states (less than US$1.5 billion/year is channelled through multilateral institutions).[38] While states have committed to "mobilize US$100 billion per year by 2020 to address the needs of developing countries",[39] these funds are most likely to focus on climate change mitigation projects with global benefits rather than on helping developing countries to bear the costs of adapting to climate change. Moreover, through conditions associated in particular with bilateral financial support, there is a risk that donor countries may impose their own policy priorities over developing states. With regard to migration, developed states may

---

[34]See e.g., Biermann and Boas (n 3); Docherty and Giannini (n 4).

[35]See supra note 18.

[36]Angela Williams, "Turning the Tide: Recognizing Climate Change Refugees in International Law," *Law & Policy*, 30 (2008), 502, 517.

[37]Walter Kälin, "From the Nansen Principles to the Nansen Initiative," *Forced Migration*, 41 (2012), 48.

[38]Barbara Buchner *et al.*, "The Landscape of Climate Finance," iv. Retrieved 9 April 2015, from http://climatepolicyinitiative.org/sgg/files/2011/10/The-Landscape-of-Climate-Finance-120120.pdf.

[39]Cancun Agreements (n 26).

be tempted to direct climate finance toward programs that support the management of international migration and the containment of migrants in the Global South, rather than programs that promote the rights or well-being of the migrants and other stakeholders.

There is also a danger that ill-founded yet powerful fears of migration (as causing political turmoil in the Global South or significantly exacerbating international migration to the Global North) could push some countries to further invest in control "solutions" such as fencing international borders and investing in military capacities while neglecting international development and further restricting international migration opportunities. Such policies are based on the false assumption that migration can be contained. Gregory White has shown that controlling migration is often perceived as having low costs, despite its high actual costs, and that it has a high political pay-off, despite its low long-term efficacy.[40]

13.4. Prospects for Regional Responses in Southeast Asia

The debate on the governance of climate migration has prevalently been conceived as a debate on global governance. This approach has its limits. In particular, the predetermined and asymmetrical interests of developed and developing states, as presumably donor and receiving states, hinder international cooperation. Another issue is that most of the debate originates from Western countries, which results in governance proposals that are ill-fitted to regional circumstances.[41] In what follows, I suggest some elements of reflection on the relevance of regional responses to climate migration in Southeast Asia.

13.4.1. *Specific regional circumstances*

Some specificities of Southeast Asia need to be taken into consideration. In particular, Southeast Asian states have not generally been keen to ratify international conventions on human rights law. Brunei, Malaysia, Myanmar, and Singapore are among the few states that have not ratified the 1966

---

[40]Gregory White, *Climate Change and Migration: Security and Borders in a Warming World* (UK: Oxford University Press, 2011).

[41]Benoit Mayer, "Environmental Migration in the Asia-Pacific Region: Could We Hang out Sometime?," *Asian Journal of International Law*, 3 (2012), 101.

Covenants on Economic, Social and Cultural Rights on Civil and Political Rights. Indonesia and the Philippines are the only ASEAN Member States to have ratified the 1990 Convention on the Protection of the Rights of All Migrant Workers and Members of their Families. The record of ASEAN Member States is poor with regard to the protection of refugees — let alone other forced migrants. This is important because the protection of "climate migrants" has sometimes been justified explicitly or implicitly by analogy with the protection of refugees. Amongst ASEAN Member States, Cambodia and the Philippines are the only ones that have ratified the 1951 Convention relating to the status of refugees. These considerations suggest that ASEAN Member States would be unlikely to access a treaty on the protection of climate migrants as has been proposed.

Moreover, the debate on the governance of environmental migration cannot ignore the region's perspective on international law. Theoretical propositions on "Asian values" have suggested regional specificities, such as a greater emphasis on economic and social rights and more attention to social harmony, traditional order, and economic growth.[42] In this debate, culture-centrism, political opportunism, and regional prides have perhaps led to extreme positions on both sides. As informed observers suggested, Southeast Asia's distinctiveness regarding human rights does not stem from an essential cultural particularism, but rather from the region's "will to differ" and to "evolve on its own path",[43] or even from a certain "public mood . . . of disgruntlement with the West".[44] One may or may not agree with the argument on "Asian values", and one would probably disagree with the way it might have been used to justify certain illiberal policies, but the development of this discourse and its relative success in regional political spheres reflect a particular defiance toward international governance. This

[42]Jack Donnelly, *Universal Human Rights in Theory and Practice* (US: Cornell University Press, 2003), p. 110; B. S. Chimni, "Third World Approaches to International Law: A Manifesto," *International Community Law Review*, 8 (2006), 3; Simon S.C. Tay, "Human Rights, Culture, and the Singapore Example," *McGill Law Journal*, 41 (1995), 743.

[43]Tay (n 42), 768. See also B.S. Chimni, "Asian Civilizations and International Law: Some Reflections," *Asian Journal of International Law*, 1 (2011), 39, 40–41.

[44]Joanne R. Bauer and Daniel A. Bell, "Introduction," in Joanne R. Bauer and Daniel A. Bell (eds.), *The East Asian Challenge for Human Rights* (Cambridge, UK: Cambridge University Press, 1999), pp. 3, 4.

defiance needs to be taken into consideration in approaching possible response to climate migration.

13.4.2. *Possible regional leadership*

Global governance is not necessarily the best approach for responses to climate migration. Angela Williams suggested that "an alternative system for addressing the plight of those displaced by climate change may be better coordinated by way of regional agreement, operating under an international umbrella framework".[45] I suggested elsewhere that a resolution of the UN General Assembly could give the tone of such a framework while encouraging the adoption of more substantive responses through bilateral or regional negotiations.[46] Francois Crepeau, the UN Special Rapporteur on the Human Rights of Migrants, recommended that "states should devise and implement regional migration policies and strategies to facilitate international climate-change-induced migration, where needed, including through the negotiation, conclusion and implementation of regional migration agreements".[47] Given the regional context of defiance vis-à-vis global institutions, regional governance is arguably a strong candidate in Southeast Asia.

The underlying assumption in support of regional response to climate migration is that states would have a greater incentive to cooperate on resettlement programs or domestic policies on migration that are more clearly defined and concern states with which they have tighter relations. Mutual support would not be purely altruistic: it would also aim at pursuing common interests between donor and receiving states, such as avoiding undocumented international migration, international criminality, political turmoil, or economic crises. Regional cooperation would also give more place to experimentation and to the identification of good practices.[48]

---

[45]Williams (n 36), 518.

[46]Benoit Mayer, "The International Legal Challenges of Climate-Induced Migration: Proposal for an International Legal Framework," *Colorado Journal of International Environmental Law and Policy*, 22 (2011), 357.

[47]François Crépeau, "Report of the Special Rapporteur on the Human Rights of Migrants," (United Nations General Assembly 2012) A/67/299 para. 93(f).

[48]Williams (2008), 521.

Following Europe, Southeast Asia is certainly the second most integrated region. As an "imagined community"[49] in the making, ASEAN has gone relatively far in dealing with the rights of migrants as well as with the protection of the environment. Although the limits of regional cooperation can easily be pointed out, Koh Kheng-Lian and Nicholas Robinson (2002) justly pointed out that ASEAN "has been remarkably successful in shaping a common regional environmental policy framework".[50]

13.4.3. *Possible regional responses to migration*

In 2007, the heads of states and government of the 10 ASEAN Member States adopted the Declaration on the Protection and Promotion of the Rights of Migrant Workers. If the declaration is relatively short and unsubstantial, and stops very short from international human rights standards, it does however call receiving states to "[i]ntensify efforts to protect the fundamental human rights, promote the welfare and uphold human dignity of migrant workers", and to "provide migrant workers... with adequate access to the legal and judicial system of the receiving state".[51] The following year, a Committee on the Implementation of the ASEAN Declaration was established. Lastly, the ASEAN Forum on Migrant Labour was established to convene a transnational dialogue between the different stakeholders.

While the Declaration stops very short from international human rights standards, its existence and that of the Committee and the Forum witnesses some important trends in Southeast Asia. ASEAN has been the forum for some discussions on the protection of migrants, in countries that have not otherwise accepted far-reaching human rights obligations. Intraregional international labor migration is an important phenomenon, and sending

---

[49]Simon Tay, Robert G. Patman, and Betty Mason-Parker, "Interdependence, States and Community: Ethical Concerns and Foreign Policy in ASEAN," in David B. MacDonald (ed.), *The Ethics of Foreign Policy* (UK: Ashgate, 2007).

[50]Koh Kheng-Lian and Nicholas A Robinson, "Strengthening Sustainable Development in Regional Inter-Governmental Governance: Lessons from the ASEAN Way," *Singapore Journal of International & Comparative Law*, 6 (2002), 640, 643.

[51]ASEAN Declaration on the Protection and Promotion of the Rights of Migrant Workers, signed on 13 January 2007 in Cebu, Philippines by the Heads of State/Government. Retrieved 9 April 2015, from http://www.asean.org/19264.htm, paras. 5 and 9.

countries — in particular Indonesia and the Philippines — have played an active role for the promotion of the rights of their nationals abroad. The very existence of the Declaration, the Committee, and the Forum reflects a strong will of ASEAN Member States to affirm their continued relevance, even as all member states are not able to agree on substantial protection standards.

ASEAN has also made some efforts in promoting human rights generally. In particular, the ASEAN Human Rights Declaration was adopted in November 2012. This Declaration recognizes migrant workers as "an inalienable, integral and indivisible part of human rights and fundamental freedoms".[52] It also highlights "the right to freedom of movement and residence within the borders of each State" and "the right to seek and receive asylum in another State in accordance with the laws of such State and applicable international agreements".[53] Implementation remains however largely desultory in several member states, in particular with regard to the latter provision on a right to asylum.

13.4.4. *Possible regional responses to climate change*

ASEAN has also been interested in regional environmental issues generally. One of the main concerns in this domain, in particular for Malaysia and Singapore, regards haze pollution. The 2002 ASEAN Agreement on Transboundary Haze Pollution, ratified by nine member states, has yet to be ratified by Indonesia, the main emitter of haze pollution. Many other instruments have been adopted on diverse environmental issues ranging from sea turtle conservation and protection to environmental education or from health to heritage parks.[54]

With regard more specifically to climate change, ASEAN has mostly served as a forum to establish a common position of its 10 member states before international negotiations. The positions that it has reflected do not

---

[52] ASEAN Human Rights Declaration, 19 November 2012, Phnom Penh, Cambodia, para. 4.

[53] *Ibid.*, paras. 15 and 16.

[54] See the documents compiled in Koh Kheng-Lian (ed.), *ASEAN Environmental Law, Policy and Governance: Selected Documents, Volume I* (Singapore: World Scientific, 2009); Koh Kheng-Lian (ed.), *ASEAN Environmental Law, Policy and Governance: Selected Documents, Volume II* (Singapore: World Scientific, 2013).

generally differ substantially from the position of the G77 (a coalition of developing countries), of which all ASEAN countries are members. In particular, ASEAN statements have emphasized the principle of common but differentiated responsibilities, insisted on the differentiated obligations of developed states in mitigating climate change mitigation and financing climate change adaptation, and called for efficient international cooperation with regard to the latter.[55] Several ASEAN countries have also strong interest in DCM and REDD projects as a source of substantial revenues.

13.4.5. *Possible regional responses to climate migration*

What role could international cooperation play in addressing climate migration in Southeast Asia? ASEAN appears as a non-negligible forum for regional cooperation, although the "ASEAN Way" is often characterized by incomplete achievements, multiple declarations without much implementation, and, even more, low standards of human rights protection. Unlike the European Union, ASEAN Member States have little in common; they feature a great level of cultural, economic, and political heterogeneity which results in substantial differences in what each member state conceives as its national interest.

On the other hand, climate migration may be a theme of interest for ASEAN Member States, as the territories of some of them include some of the most concerned regions of the world. While ASEAN Member States failed to define cooperative solutions to past forced migration (for instance with regard to Rohingya refugees), they may progressively realize the mutual benefits that can be drawn from regional cooperation with regard to migration. Most importantly, if ASEAN is ever to become a genuine imagined community, a common identity will come along with a sense of transboundary solidarity.

---

[55] See in particular ASEAN Declaration on the 13th Session of the Conference of the Parties (COP) to the UN Framework Convention on Climate Change (UNFCCC) and the third session of the Conference of the Parties serving as the Meeting of the Parties (CMP) to the Kyoto Protocol, Singapore, 20 November 2007, in Koh, *ASEAN Environmental Law, Policy and Governance* (n 54), 829; ASEAN Leaders' Statement on Climate Change to the 17th Session of the Conference of the Parties to the United Nations Framework Convention on Climate Change (UNFCCC) and the Seventh Session of the Conference of the Parties serving as the Meeting of Parties to the Kyoto Protocol, 2011, in *ibid.*, 1311.

Other relevant regional forums have played an important role in setting the issue of climate migration on national agendas. In particular, in 2010, the Asian Development Bank (ADB) launched a preliminary research to identify the "policy options to support climate-induced migration",[56] and a report was published two years later.[57] In partnership with the International Organization for Migration, the ADB established the Asia-Pacific Environment and Migration Network in 2012, as essentially a network of researchers. It has been suggested that the ADB could "provide financial expertise and capacity [and] use [its] convening power [to] support programs and projects".[58] It should not belong to multilateral development banks, which are largely controlled by donor states, to define responses to climate migration; such organizations may however play an important role "through political leadership, financial assistance and technical support".[59]

13.4.6. *Possible regional responses to concomitant factors of vulnerability*

If Southeast Asian states are to do something about climate migration, for instance through ASEAN, *what* exactly could they do? The question is a thorny one, given the diverse forms of migration exacerbated by climate change. For one, Southeast Asian states certainly need to push for international cooperation on climate change mitigation, as a way to reduce anthropogenic interference with the climate system. Southeast Asian states also need to promote adaptation to climate change, with a particular attention to the most vulnerable communities and individuals. Beyond this, they should foster human development as a way to alleviate social vulnerability. Opportunities need to be made available for those who

---

[56]Asian Development Bank, "Policy Options to Support Climate-Induced Migration," (December 2009). Retrieved 9 April 2015, from http://www2.adb.org/Documents/TARs/REG/43181-REG-TAR.pdf.

[57]Asian Development Bank, "Addressing Climate Change and Migration in Asia and the Pacific," (2012).

[58]Robert Dobias, "ADB's Role in Addressing Climate Change and Migration," Seminar on "Climate Change, Migration and Human Security in Southeast Asia," Singapore, 27 May 2011. Retrieved 9 April 2015, from http://www.rsis.edu.sg/nts/events/docs/Bob_Dobias.pdf.

[59]Crépeau (n 47), 93(g).

could benefit from individual migration, in particular through internal and international labor migration, and institutions need to be developed in order to ensure the effective protection of the rights of migrants. In a regional context where the rights of migrants are not systematically recognized or effectively protected, there is a real danger that some states could further take advantage of a worsening of the conditions of some populations within their own countries or in neighboring countries; this risk should be avoided by promoting a sense of common identity and solidarity within and among peoples.

Rarely has mankind been able to forecast future events as it is able to do — despite significant scientific uncertainty and the contingency to response measures — regarding the consequences of climate change. States need to take advantage of such exceptional foresight, including for what regards future migration scenarios. For instance, as recommended by the UN Special Rapporteur on the Human Rights of Migrants, "States should devise and implement local and national migration policies and strategies to facilitate internal climate-change-induced migration, where needed, including through building urban infrastructure that is sustainable, flexible and inclusive in order to accommodate internal migrants".[60] To ensure that the rights of the migrants are protected, "States should support civil society organizations that give a voice to such vulnerable populations, including migrants or potential migrants, with a view to ensuring their meaningful participation in all the deliberations regarding their future as citizens and/or migrants".[61] Education is also certainly an important strategy to take into consideration, not only to allow putative migrants to work elsewhere, but perhaps most importantly to facilitate the encounters between migrants and host communities. More generally, Southeast Asian states should put their laws and institutions in order, with specific concerns for the protection of the rights of migrants, in order to get ready for growing influxes of internal and international migrants in the coming years and decades.

All in all, the debate on climate migration calls for new attitudes toward migrants. It calls us to reconceive migration, not as a social anomaly, but as a perfectly normal phenomenon — migration happens in all human societies

[60]*Ibid.*, 93(e).

[61]*Ibid.*, 93(d).

and at all times. Each Southeast Asian country, by its ethnic and cultural diversity, testifies of the central role that migration played in our history. Remaining faithful to its history and traditions, Southeast Asia needs to build a collective identity, based on regional cooperation and humanity, which make it possible for migration to occur in the best conditions for the receiving and sending communities as well as for migrants themselves.

13.5. Conclusion

Climate change is likely to have significant consequences on migration within and perhaps across countries throughout Southeast Asia. Four representative scenarios were in particular discussed, whereby migration stems from floods, cyclones, economic impacts, or response measures to climate change. These scenarios evidence the irreducible diversity of the forms of migration that climate change exacerbates, and also shows that, though climate change increases the number of migrants, it does not create new forms of migration. "Climate migrants" can rarely be distinguished from other migrants, and they should arguably not be treated differently from other individuals with similar needs. Rather than genuinely being a new issue, climate migration is a wake-up call drawing attention on the shortcomings of the existing governance of migration. This wake-up call needs to be heard in Southeast Asia, where the protection of the human rights of migrants often remains insufficient.

There is no one-size-fits-all "solution" to climate migration. Climate change mitigation and adaptation, as well as regional cooperation on development, are necessary but insufficient responses. The creation of new opportunities for putative migrants should come along with stronger norms and mechanisms for the protection of the rights of all migrants in the region. A financial support from industrial states is required in application of the principle of common but differentiated responsibilities, and it could be channelled by multilateral development banks. Climate finance may help developing states in Southeast Asia to bear the incremental costs incurred because of climate change, but it should not impose excessive conditions or constraints on injured states.

Chapter 14

Achieving Human Rights in an Era of Climate Disruption: The Philippines

*Amado S. Tolentino, Jr.**

Developing and protecting rights as part of the efforts to safeguard the environment, attaining sustainable development, and alleviating poverty and inequalities are current concerns in many parts of the world. This was brought about by the grand-scale and unprecedented phenomenon of environmental refugees on account of, inter alia, natural disasters and climate disasters.

Refugees are people who seek asylum for fear of political, racial, or religious persecution. They leave their homes because of war or civil strife. This traditional notion of refugees, however, leaves out the new and growing phenomenon of people becoming environmental refugees to countries not their own due to natural occurrences like earthquakes, tsunamis, volcanic eruptions, floods, landslides, etc.[1] When people seek refuge within their own countries as environmental refugees they are commonly referred to as internally displaced persons (IDPs). In some Pacific island states, such persons prefer to be referred to as climate migrants.

It may be noted that climate change itself does not directly affect human rights. Rather, global warming causes environmental change which in turn affects human rights. Neither the UN Framework Convention

*Executive Governor (for developing countries), International Council of Environmental Law; Philippine Ambassador to Papua New Guinea and Qatar.

[1] "Disasters Creating Refugees," *Philippine Daily Inquirer*, 15 October 2006, p. 14.

on Climate Change[2] nor the Kyoto Protocol[3] includes any provision concerning specific assistance to those who will be directly affected by climate change and its effects. The Convention Relating to Refugees (Refugee Convention),[4] on the other hand, applies to those persecuted for reasons of race, religion, nationality, or membership in a particular racial or political group. As the refugee Convention is not applicable to climate change-displaced people at the moment, those persons deserve to get legal identity through the legal system of each country.

The number of internally displaced persons in the Philippines has tremendously increased during the last few years on account of conflicts and natural disasters. Early this year, it was reported that seven million Filipinos were displaced by calamities as a consequence of, among others, typhoons (Rammasun and Matmo) which caused floods, landslides, and monsoon rains that reached some parts of Luzon; the 7.2 magnitude earthquake that struck the islands of Bohol and part of Cebu; and Typhoon Haiyan in November 2013 which claimed lives, property, and crops.[5] It was also reported that the number of IDPs affected by the ongoing conflict between the Armed Forces of the Philippines and Muslim secessionist groups reached 500,000. Add to that the IDPs, mostly Muslim residents of Zamboanga including Badjaos, indigenous people dependent on the sea for livelihood, who fled their communities when armed men from a faction of the Moro National Liberation Front (MNLF) went into Zamboanga City in an attempt to occupy it. The most recent IDP-related reports concern some 500 displaced Manobo children suffering from ailments including diarrhea, tuberculosis, malnutrition, and gastritis living in temporary shelters in Davao City after military operations against the New People's Army in Davao del Norte.

Among basic human rights, the right to life and dignity are intrinsically tied to the right to an ecologically and humanely viable environment and

[2]United Nations Framework Convention on Climate Change, 1992. Retrieved 2 April 2015, from http://unfccc.int/resource/docs/convkp/conveng.pdf.

[3]Kyoto Protocol — Kyoto Protocol to the United Nations Framework, 1998. Retrieved 2 April 2015, from http://unfccc.int/resource/docs/convkp/kpeng.pdf.

[4]Convention Relating to the Status of Refugees, 1951. Retrieved 2 April 2015, from http://www.ohchr.org/Documents/ProfessionalInterest/refugees.pdf.

[5]"7M Filipinos Displaced by Calamities," *Manila Times*, 21 January 2014.

have been the subject of much discourse, vis-a-vis climate justice and the future generations of environmental refugees and internally displaced persons. Compared to their counterparts who crossed internationally recognized borders, IDPs do not enjoy adequate protection and assistance and instead experience suffering, neglect, and deprivation. This is so despite the duty of the government to utilize its resources and provide measures to address the situation.

Aside from the rights to life, liberty, and property (Sect. 5, Art. II) and a healthy environment (Sect. 16, Art. II), enshrined in the 1987 Philippine Constitution,[6] the fundamental law embodies as one of its state policies, "to value the dignity of every human life and to guarantee full respect for human rights" (Sect. 11, Art. II). This policy extends to environmental refugees and internally displaced persons as a consequence of natural calamities and armed conflicts. In furtherance thereof, the Constitution mandated the Philippine Commission on Human Rights (CHR) to provide "appropriate legal measures for the protection of human rights of all persons within the Philippines . . . and provide for preventive measures and legal aid services to the underprivileged whose human rights have been violated or need protection" (Sect. 18, Art. XIII).

Towards that end, the Commission on Human Rights and the UN High Commissioner for Refugees[7] forged an agreement to strengthen the monitoring of IDPs in the Philippines as early as 2011. At the same time, Bills were filed before the 15th Congress of the Philippines (legislature) to protect the rights of internally displaced persons.[8]

Indeed, the Philippine government as well as its international partners like the UN High Commissioner for Refugees, UN Commission on Human

---

[6]Philippines Constitution, 1987. Retrieved 2 April 2015, from http://www.ilo.org/wcmsp5/groups/public/---ed_protect/---protrav/---ilo_aids/documents/legaldocument/wcms_126176.pdf.

[7]United Nations High Commissioner for Refugees. Retrieved 2 April 2015, from http://www.unhcr.org/cgi-bin/texis/vtx/home.

[8]15th Congress of the Philippines, Senate Bill No. 00530, "Improving Philippine Commitment to Human Rights Promotion and Protection by Providing the Necessary Mechanisms for the Prevention of the Occurrence and Protection from the Adverse Effects of International Displacement and for Other Purposes," 2010. House Bill no. 00239, "An Act Protecting the Rights of Internally Displaced Persons Providing Penalties for Violations Thereof and for Other Purposes," 2010.

Rights,[9] International Committee of the Red Cross, and the Red Crescent among many others, made significant efforts to address the protection, assistance, and needs of IDPs.

14.1. Problems, Issues, and Concerns about Internally Displaced Persons

In Southeast Asia, the Philippines ranks third after Myanmar and Indonesia as a country with the most number of IDPs and among the top 40 countries, all over the world where internal displacement is considered as a significant phenomenon.[10]

When people are displaced, they are vulnerable and susceptible to being victims of human rights violations. They are deprived of their ordinary and normal environment and their ability to meet their most basic needs in peril.

In the recent natural disasters which hit the Philippines, humanitarian aid including relief goods distribution was a big problem. Many goods did not reach the intended beneficiaries. There were allegations of hoarding and even stealing for profit. It also took time for humanitarian aid groups to access the areas that were extensively damaged by the disaster.

Displaced people's rights to basic needs should be paramount and must be safeguarded. The state should exert effort to likewise ensure that all necessary assistance to humanitarian aid givers is provided. Even the restrictions on delivery of aid such as excessive delays to obtain the permits needed must be looked into by the government.

Discrimination is another problem encountered by IDPs. In the Bohol earthquake and the storm surge in Leyte, there were reports that victims were subjected to unequal treatment ascribed to political factors. It was claimed that those who did not vote for the incumbent official were not given the relief goods and assistance they needed. In the Mindanao region where the major cause of displacement is armed conflict between the government and non-state armed groups, there were reports of discrimination on account of religious beliefs. Efforts to eradicate ill-treatment and discrimination

---

[9] "CHR, UN Boost IDP Monitoring," *Manila Bulletin*, 15 July 2011.

[10] "Protection for Displaced Persons Urged," *Manila Bulletin*, 2 November 2010.

ascribed to political and religious factors must be looked into and remedied.

Protection of rights of women and children must also be given priority by the state because they experience heightened vulnerability to sexual and gender-based violence especially in evacuation camps and in their makeshift shelters. In Leyte, there were reports of rape and sexual abuse against women and children, days after the storm surge. Child trafficking is another serious risk when people are displaced; children who have lost their families are vulnerable to this kind of human rights violation. The state should secure and intensify their security against sexual predators and human traffickers.

Important too is the protection of property of IDPs. In Tacloban City (Leyte), there are families who found their homes looted of contents upon their return. One who went to the Tacloban airport to collect relief goods found his car stolen from his garage along with other valuables. When disasters strike, government authorities should afford and increase security as soon as possible and military visibility must be present to prevent loss of what was left of victims' valuables.

IDPs should likewise be protected from the combined impact of conflicts and natural hazards. A natural hazard often forces IDPs to flee again either from the place where they had taken refuge from conflict or from the places to which they had returned or relocated. In the Philippines, those living in poorly equipped evacuation camps and makeshift shelters in Central Mindanao were exposed to flooding more than their counterparts in the general population. Natural hazards and environmental degradation can also create apprehensions over scarce resources.

14.2. Adopted but Vetoed Proposals for IDP Rights Protection

The 15th Philippine Congress (lawmaking body) (2010–2013) adopted three legislations on internal displacement in 2012. House Bill 00239 entitled "An Act Protecting the Rights of Internally Displaced Persons Providing Penalties for Violations Thereof and for Other Purposes" was a response to the need to enact a law that adopts human rights-based approach to disaster risk reduction and management to ensure the continued protection of those who are internally displaced in a situation of natural or manmade disasters. In consonance with human rights treaties and conventions like the Universal Declaration on Human Rights[11] and the UN Guiding Principles on Internal Displacement,[12] the approved Bill sets forth the rights and guarantees for resettlement, return, and reintegration of IDPs in all phases of displacement. The Bill likewise recognizes the responsibility of national authorities to provide protection and assume accountability in cases of violations of rights of IDPs.

On the other hand, Senate Bill 00530 of the 15th Congress entitled "Improving Philippine Commitment to Human Rights Promotion and Protection by Providing the Necessary Mechanisms for the Prevention of the Occurrence and Protection from the Adverse Effects of Internal Displacement and for Other Purposes" is a measure that seeks to localize the guiding principles on internal displacement set forth by the United Nations. In this proposal which was adopted, the government is mandated to establish a mechanism to support victims of internal displacement in armed conflict, to be able to ensure that the rights and lives of its citizens shall not be placed at risk.

The 15th Philippine Congress likewise approved "An Act Protecting the Rights of Internally Displaced Persons, Providing Penalties for Violations Thereof and for Other Purposes" or the "Rights of Internally Displaced Persons Act of 2013". In accordance with the principles of the 1987 Philippine Constitution and the standards set by international humanitarian law and human rights laws, international treaties and conventions adhered to by the Philippines, this measure seeks to promote and protect the rights of IDPs in situations of armed conflict, generalized violence, tribal or clan wars, violations of human rights, aggressive implementation of development projects, and natural and man-made disasters.

However, on 23 May 2013, President Benigno Aquino III vetoed the consolidated enrolled Bill (Senate Bill 3317 and House Bill 5627) entitled "An Act Protecting the Rights of Internally Displaced Persons, Providing Penalties Thereof and for Other Purposes" stating that while the objectives of the Bill on IDPs is laudable, some of the provisions are unconstitutional

[11]The Universal Declaration of Human Rights, 1948. Retrieved 2 April 2015, from http://www.un.org/en/documents/udhr/.

[12]Guiding Principles on Internal Displacement, 1998. Retrieved 2 April 2015, from http://www.unhcr.org/43ce1cff2.html.

on the following grounds: (1) The Bill's provision on damages unlawfully differentiates between displacement caused by security agents of the state and other entities; (2) the power granted to the Commission on Human Rights (CHR) to determine damages incurred against IDPs and facilitate the award of such claims is a power that belongs exclusively to the judiciary; (3) the additional powers it grants to the CHR exceed those which the Constitution intended to give it, as it was conceived as an investigative and recommendatory agency; and (4) the provision allowing individuals to claim financial assistance and compensation from the government opens the door to a slew of claims or cases against the government and goes against the non-suability of the state principle.[13]

14.3. "An Act Providing for the Rights of Internally Displaced Persons" — A Law in the Making

Pending before the 16th Congress of the Philippines is a refiled Bill which covers the rights of internally displaced persons in situations of natural or man-made disasters.[14] Formulation of the Bill was preceded by effectivity of some international legal instruments of which the Philippines is a party. Among them are: the Hyogo Framework of Action (HFA)[15] and the ASEAN Agreement on Disaster Management and Emergency Response (AADMER). The HFA outlines five priorities for action and offers guiding principles and practical means for achieving resilience. Its goal is to substantially reduce disaster losses by 2015 by building the resilience of countries and communities to disasters. It means reducing loss of lives and social, economic, and environmental assets when disaster strikes. On the other hand, AADMER is expected to provide a framework for the development of operational procedures to respond collectively and expeditiously to disasters. The Agreement has provisions for movement of relief, of relief assistance, expedited customs, and immigration clearance,

---

[13]Veto Message on Senate Bill no. 3317 and House Bill no. 5627 of President Aquino, 2013.
[14]16th Congress of the Philippines, House Bill no. 239, "An Act Protecting the Rights of Internally Displaced Persons, Providing Penalties for Violations Thereof and for Other Purposes," 2013.
[15]Hyogo Framework for Action, 2005–2015. Retrieved 2 April 2015, from http://www.unisdr.org/we/inform/publications/1037.

etc. The utilization of military and civilian personnel in disaster relief and the establishment of a center to coordinate regional disaster response are part of the Agreement. It has a provision also for the setting up of an ASEAN disaster relief fund as well as simulation exercises to test emergency responses on a regular basis.

It should be known and noted that on account of the recognized vulnerability of the Philippine archipelago to natural calamities, a Climate Change Law (Republic Act or RA 10175) was enacted in 2009 with a mandate to address the vulnerability of local communities and which adopts a gender sensitive, pro-children, and pro-poor perspective. It also emphasizes the participation of all stakeholders including local government units, NGOs, and others in responding to the adverse impacts of climate change. Likewise, it seeks to integrate disaster risk reduction into climate change programs and initiatives.[16] Additionally, due to the increasing yearly occurrences, commonly earthquakes, strong typhoons, floods, and landslides/mudslides, a National Disaster Risk Reduction Management Coordinating Council (RA 10121) was created in 2010 to ensure that unforeseen disastrous events would be responded to quickly, especially by local governments prepared with their contingency plans. It adopts a disaster risk reduction and management approach that is holistic, comprehensive, integrated, and proactive in lessening the socio-economic and environmental impact of disasters. It also aims at mainstreaming disaster risk reduction and climate change in national and local plans and development processes. RA 10121 was amended by the People's Survival Fund Act (RA 10174) (2012) which aims to provide direct access funding to local government units and communities for their own planned adaptation activities. This can be in the form of provision of information and medicines against vector-borne diseases triggered by climate change such as dengue and malaria; forecasting and early warning systems; support for contingency planning in the most vulnerable areas; and guarantee for farmers' risk insurance needs.[17]

---

[16] Legal Mandate, Climate Change Commission, 20 January 2014. Retrieved 2 April 2015, from http://climate.gov.ph/legal_mandate.

[17] Republic Act no. 10174, "An Act Establishing the People's Survival Fund to Provide Long-Term Finance Streams to Enable the Government to Effectively Address the Problems of Climate Change," 16 August 2012.

The proposed law for IDPs provides for their protection during and after displacement as well as their resettlement, relocation, and reintegration. It defines "internal displacement" as the "forced or obliged movement of a person or groups of persons to flee or leave their homes or places of habitual residence without crossing an internationally recognized border as a result of or in order to avoid the effects of natural or human-made disaster". The proposal sees to it that displacement should not be carried out in a manner that violates the rights to life, liberty, dignity, and security of those affected. It contains terms on protection against crimes, discriminatory practices, and curtailment of human freedoms during the period of displacement; extension of humanitarian assistance and provision of support during the return, resettlement, or local integration. The draft law is also about provision and access to basic necessities; guarantees freedom of movement, family unity; health and education; property and possessions; and protection from all forms of harm and abuse. Likewise being considered is integration, specifically of the needs of pregnant women, mothers, and newborns into the emergency and relief response.

In addition, IDPs shall not be discriminated against in the right to: (1) enjoy freedom of thought, conscience, religion or belief, opinion, and expression; (2) to seek freely opportunities for employment and to participate in economic activities; (3) to vote and participate in governmental and public affairs including the right to have access to the means necessary to exercise these rights; and (4) to communicate in a language they understand. Other rights contemplated by the proposed law include, but are not limited to, the following: the right to appropriate clothing; the right to essential medicine and sanitation; the right of persons with disabilities; the right to family reunification; the freedom to choose residence; and the right to seek freely opportunities for employment.[18]

If approved by Congress and, hopefully signed into law by the President, the law to protect the rights of IDPs will be the first in Asia and a model for countries facing similar issues of displacement and violation of people's rights, as the government is unable to serve to the general welfare needs of its constituents. Human rights law clearly imposes obligations on states that are relevant to the enjoyment of a safe, clean, healthy, and

[18] Supra note 15.

sustainable environment. In pursuance thereof, states should adopt legal and institutional frameworks protective of those rights. This was recognized by the UN Human Rights Council in its March 2014[19] Resolution on Human Rights and the Environment.

It was recently announced that the Philippine national government is crafting a disaster preparedness and risk reduction roadmap that includes a storm surge advisory protocol and flood advisory system in highly vulnerable areas in the country as well as a disaster response master plan. From local government units' Disaster Risk Reduction and Management Councils, recent reports reveal a variety of disaster preparedness and mitigation plans encompassing rehabilitation, recovery, and quick response ranging from purchase of motor boats, rubber boats, life-saver equipment, rescue vehicles including ambulances, and other rescue needs to clearing and upgrading of drainage, canals, river banks, to acquisition of lands for the construction of evacuation centers and relocation sites. Noted are also plans to acquire early warning devices such as rain gauges and flood early warning devices and stepping up tree planting activities as part of the disaster mitigation programs.[20]

Overall, the foregoing would be in furtherance of the promotion of the rights of IDPs. All this, however, needs effective implementation of the series of climate change-related Philippine policies, legislations, and directives aimed at reducing disaster risk, improving clean energy production, and adapting to shifts in climate change. Concomitant therewith is the of budget/funds for enhancing geo-hazard maps to include multiple hazards, early warning systems, and other infrastructure to improve the capacity to reduce risk. Truly, it is most advantageous if concrete steps are taken towards management of weather and geological-related risks; minimization of vulnerability of people and improvement in the preparedness capacity of institutions and communities; and building resilience so that there will

[19]The UN Human Rights Council (HRC) Resolution on "Human Rights and the Environment," 2014. Retrieved 2 April 2015, from http://daccess-dds-ny.un.org/doc/UNDOC/LTD/G14/123/55/PDF/G1412355.pdf?OpenElement.

[20]"Government Crafting Protocols for Warning Public on Storm Surges, Floods," *Manila Bulletin*, 1 February 2014.

be lesser need for recue, evacuation, and rehabilitation and improvement in the capacity to get back on one's feet immediately.

14.4. Conclusion

The growing population of internally displaced persons alarms both the local and global community. States and international organizations must work hand in hand in order to address the matter of protection, assistance, and needs of IDPs.

Activities for the IDPs, like those performed by the International Committee for the Red Cross, must be supported. These include, but are not limited to, the following: (1) interventions to persuade authorities and armed groups to fulfill their obligations to protect civilian populations; (2) efforts to restore family links; (3) distribution of relief material such as food, water, and essential household items, and seed and farming tools; (4) first aid and surgery, and health and healthcare programs; and (5) livelihood support programs such as microeconomic initiatives, agricultural, and livestock support. As articulated by ICRC, "promoting self reliance to the affected communities is the main aim to ensure that their basic needs are met and programs that are helpful to avoid displacement are not and should not be necessary, and improve the capacity of the host to provide for the IDP needs".[21]

[21]"Internally Displaced People," International Committee of the Red Cross, Geneva, 1997.

Theme VI

Adaptation — Disaster Management, Risk Reduction, and Humanitarian Assistance

Chapter 15

The Legal Regime of Climate Change Adaptation and Disaster Management in Taiwan: Focus on the Major Amendment Following the Devastating Typhoon Morakot of 2009

Gao Ming-Zhi Anton

15.1. Introduction

The United Nations Framework Convention on Climate Change (UNFCCC) requires each party which has ratified it to include policies to reduce emissions and implement measures to mitigate it at its source. However, such a mainstream approach has faced challenges, due to rising temperatures and the need to deal with disasters caused by extreme weather. Also, the accumulation over the years of the inability to reduce emissions and the current bottleneck in the climate negotiations have exacerbated the situation. The emphasis therefore has moved from mitigation to both mitigation and adaptation in national policies. This can be seen in the adaptation policies that have emerged around the world in the past decade.[1]

At first glance, Taiwan's approach to climate change seems to follow this global trend: the idea of "adaptation" has been embedded in several policy documents and discussions since the 1992 UNFCCC (such as the

[1]For example, see Germany's "German Strategy for Adaptation to Climate Change," 2008. Retrieved 9 April 2015, from http://www.bmub.bund.de/fileadmin/bmu-import/files/english/pdf/application/pdf/das_gesamt_en_bf.pdf; France's "National Adaptation Strategy," 2006. Retrieved 9 April 2015, from http://www.developpement-durable.gouv.fr/La-strategie-nationale-d,14477.html; and for complete details on the climate adaptation policies of European countries: http://climate-adapt.eea.europa.eu/web/guest/countries (Retrieved 9 April 2015).

important 1998 "Energy Policy White Paper"[2]). However, the role of "comprehensive" adaptation came relatively late, despite the announcement of the UN's "Adaptation Policy Framework" in 2006.[3] Up to 2010, there still were no comprehensive adaptation measures, and no central and coherent "policy", only a "dispersed" approach adopted by different competent authorities. For instance, the Ministry of Economic Affairs was in charge of the water strategy, while the National Science Council was in charge of the development in climate adaptation technology.[4] Finally, in 2012, there was a breakthrough with the completion of a nationwide, single adaptation policy, the "National Climate Change Adaptation Policy Framework".[5]

The role of law seems very important in this policy document: for each of the eight policy objectives, the "establishment of the legal framework and government institute" is listed as the first objective and first priority.[6] This policy framework also mentions the role of adaptation-related laws, which are summarized in Table 1.

From the policy statements, it appears that a legal regime to deal with climate change adaptation issues is lacking. Thus, not only are new legal regimes needed for but also the existing ones need to be "adapted" to climate change adaptation issues. However, it should be pointed out that there has long been in existence a very important but usually ignored "adaptation" legal regime in Taiwan.

Taiwan is unfortunate in being located in an earthquake and typhoon zone that has suffered considerably from national disasters in the past.

[2]Energy Commission, Ministry of Economic Affairs, "1998 Energy Policy White Paper," 1998.

[3]UNFCCC, "UNDP Adaptation Policy Framework." Retrieved 9 April 2015, from https://unfccc.int/adaptation/nairobi_work_programme/knowledge_resources_and_publications/items/5501.php.

[4]Kuo, "Taiwan Climate Adaptation Policy Framework," 6 July 2010. Retrieved 9 June 2015, from http://tinyurl.com/13w7mm3.

[5]"National Climate Change Adaptation Policy Framework of Taiwan," 2012. Retrieved 9 April 2015, from http://www.ndc.gov.tw/dn.aspx?uid=11505.

[6]See page 29 of the "National Climate Change Adaptation Policy Framework of Taiwan," 2012.

Table 1: The Role of Law in the "National Climate Change Adaptation Policy Framework" of Taiwan, 2012.

Key Legislative Directions	Policy Statements
The promulgation of new legislation	(1) Promulgating *Framework Climate Change laws*, in which the principles, objectives, and policy measures of climate mitigation and adaptation should be included.[7] (2) Promulgating *new laws* to regulate the behavior of businesses, civil groups, and individuals, etc.[8] (3) Facilitating the review process of important Bills in parliament, including the *National Land Planning Bill, Coast Bill*, and *Wetland Bill*.[9]
The revision or review of existing legislation	(1) Reviewing existing legislation and *epidemic-related laws*.[10] (2) Biodiversity: Reviewing *existing regulations on dealing with alien species*, in order to control and eradicate the intrusion of such.[11] (3) Revising legislation related to *land, tax, finance, insurance, labor, environmental safety, energy and resource utilization, building, cross-strait relations, and international trade*.[12] (4) Reviewing the existing legal regime and institutional design with a view to create a *sound energy supply and business environment* to enable the adaptation of the business and industrial sectors.[13] (5) Reviewing the *legal regime related to the public infrastructure*.[14] (6) Reviewing *the existing Urban Planning Act and related land use and building codes* to facilitate the construction of green infrastructure.[15]

(*Continued*)

[7] *Ibid.*, p. 49.
[8] *Ibid.*, p. 46.
[9] *Ibid.*, p. 38.
[10] *Ibid.*, p. 46.
[11] *Ibid.*, p. 45.
[12] *Ibid.*, p. 42.
[13] *Ibid.*
[14] *Ibid.*, p. 40.
[15] *Ibid.*, p. 39.

Table 1: (*Continued*)

Key Legislative Directions	Policy Statements
	(7) Revising *existing laws relating to national land planning*.[16] (8) Reviewing *the existing building code to facilitate the deployment of water-saving devices* in public buildings or the public infrastructure.[17]

Before suffering with climate change issues, Taiwan had already tried very hard to "adapt" to or live with the extreme weather conditions, while climate change just made such disasters worse. Thus, a disaster management approach to deal with climate issues has become a tradition in Taiwan, but climate change policies have tended to ignore the role the Disaster Prevention and Protection Act (DPPA) could play. The DPPA seeks to tackle several important natural disasters relating to climate adaptation, including windstorms, floods, earthquakes, droughts, frosts, debris flow, etc.[18]

This chapter aims to raise *awareness of the close connection between climate change adaptation and the DPPA*, and to provide an overview of the DPPA and focus on the recent legal reforms following the devastating typhoon of 2009, which was more than likely caused by climate change. The reformed elements during this wave of reform illustrate the key elements in combating climate disasters.

To understand the DPPA, this chapter will try to provide a simple model (Figure 1) to illustrate the relationship between the different chapters of the Act. Briefly, different work is distributed (Section 15.3) across

[16] *Ibid.*, p. 38.

[17] *Ibid.*

[18] Article 2 of the DPPA: The Act provided the following definitions:

1. Disaster: hazards caused by any of the following:

   (1) Natural disasters including windstorms, floods, earthquakes, droughts, frosts, and debris flow; and

   (2) Fire, explosion, gas leaks, fuel pipe and power transmission line failures, mining disasters, air crashes, shipwrecks, motor vehicle accidents, forest fires, and toxic chemical materials disaster.

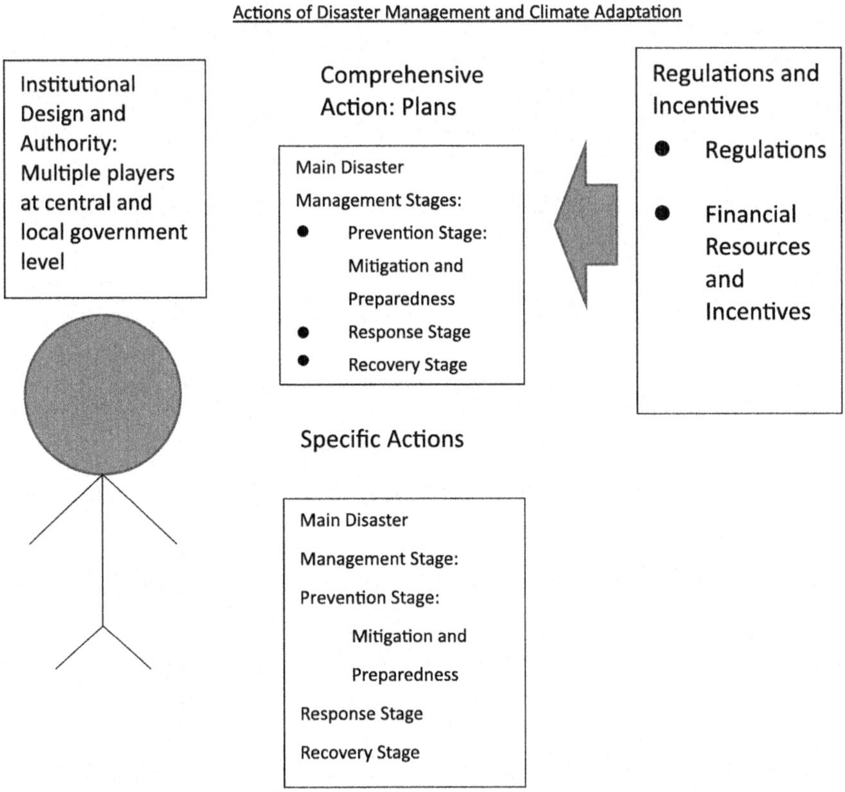

Figure 1. Brief Model of the Key Regime under the DPPA.

different institutions and authorities (Section 15.2), while certain "carrots" and "sticks" (Section 15.4) are provided to support the functions of the institutions and implementation of actions. Furthermore, this chapter will elaborate on the transformation of key schemes before and after Typhoon Morakot in order to identify the most important scheme for perfecting the climate adaptation and disaster management regime.

15.2. Institutional Design and Authority

Institutional design usually acts as the driving engine for further policy, rules, and actions to tackle disaster management and adaptation issues. It also frequently guides further actions through the key disaster management stages, including prevention, emergency response, and recovery and reconstruction, and assesses the success of the disaster management and adaptation policy. Thus, the Taiwan disaster management and adaptation regime (TDMAR) places great emphasis on the role of institutional design, which features two key aspects: first, there is a very delicate and sometimes complex *horizontal distribution of power* across different authorities at both the central and local government level; and second, there is usually a high degree of *vertical distribution of power* with the local government bearing the greater load under the TDMAR — a unique situation under the current centralization of Taiwan Constitutional law. However, whether these multiple players have functioned well in the past is what this article will elaborate on.

15.2.1. *Competent authorities under the Disaster Prevention and Protection Act: For all disasters?*

According to Article 4 of the Disaster Prevention and Protection Act (DPPA), the main *competent authorities* include the Ministry of the Interior (MOI) at the central level, as well as the municipal, county (city), and township (city) governments at the local level. There are further specialized and professional offices under the MOI and local government, which will be explained. However, an interesting aspect of the DPPA is that these competent authorities are not necessarily responsible for "all disasters".

This type of "*two-layer*" authority is similar to that of the Taiwanese administrative legal system, and can be found in most legislation.[19] Even though it appears to be an "equal" distribution of power and duties between central and local governments at first glance, the actual implementation usually relies on the resources from the "central" government, in this case the MOI. Thus, the competent authority regime is relatively inclined towards a *centralized* system, reflecting the above-mentioned centralized government scheme under Taiwanese Constitutional law.

This scheme operated very well for the first few years after the implementation of the DPPA; because Taiwan is a small island, perhaps a centralized system could deal with disasters efficiently. However, such

[19]For example, see Article 2 of the *Environmental Education Act* and Article 3 of the *Gender Non-Discrimination Education Act.*

a centralized system faced many challenges and much criticism after the unusually violent Typhoon Morakot of 2009. Due to the over-reliance on the central government to intervene, the emphasis on the role of the local government on the front line became a legislative issue. It was argued that instead of asking for help from the central government and waiting to see whether and what resources were provided, it would make more sense and be much more efficient for the local government to take more responsibility in the first instance. Consequently, under the new revisions, the role of municipal, county (city), and township (city) governments are re-emphasized first in the second paragraph of the authority clause, and further institutions will be developed later on.[20]

15.2.2. *Authorities at central government level for specific disasters*

Apart from the MOI at the central government level, there are also a number of other authorities in charge of different disasters under Article 3(1) of the DPPA. The main functions of these authorities are further stipulated in Article 3(2):[21]

(1) Windstorms, earthquakes, fire, and explosion: *Ministry of the Interior.*
(2) Floods, droughts, gas leaks, fuel pipe and power transmission line failures, and mining disasters: *Ministry of Economic Affairs.*

---

[20]Article 2(2) of the DPPA: "The municipal or county (city) government and township (city) office handle the affairs of disaster prevention and protection according to Article 18, paragraph 11, subparagraph 2, and Article 19, paragraph 11, subparagraph 2, and Article 20, paragraph 7, subparagraph 1 as prescribed in the Local Government Act and the Act accordingly."

[21]Article 3(2) of the DPPA: "The preceding paragraph prescribing the central regulating authorities for disaster prevention and protection shall be in charge of disaster prevention and protection business in the following areas:

1. Commanding, supervising and coordinating the central regulating authorities and municipal and county (city) government for disaster prevention and protection.
2. Drafting and implementing the amended disaster prevention and protection operation plan.
3. Supporting and dealing with disaster prevention and protection works.
4. Dealing with the command or coordination of the affairs of disaster prevention and protection, and non-compliance with the Act in non-local administrative areas.

(3) Frost, debris flow, and forest fire: *Council of Agriculture, Executive Yuan.*
(4) Air crashes, shipwrecks, and motor vehicle accidents: *Ministry of Transportation and Communications.*
(5) Toxic chemical materials: *Environmental Protection Administration (EPA), Executive Yuan.*

The rationale behind such a dispersed scheme is related to the *professional* issues for certain disasters. For instance, EPA is the main authority under the *Toxic Chemical Substances Control Act*[22] and is more capable of dealing with similar disasters. Similar reasoning can also be applied to other authorities for specific disasters.

Yet, there are still many challenges for such a dispersed scheme, particularly when there is a "compound" disaster involving different authorities for specific disasters. For instance, a typhoon may lead to not only windstorms but also floods, power transmission line failure, and shipwrecks, under which circumstances, the phenomenon of "everybody's business is nobody's business" causes problems at the central government level. It may also lead to these authorities being unwilling to deal with cross-local government disasters.

Following Typhoon Morakot, one of the reforms did address this problem: authorities at the central government level for specific disasters are required to "deal with the command or coordination of the affairs of disaster prevention and protection . . . in non-local administrative areas". It is hoped that such an approach can balance both the professional nature of these authorities and the collaboration issues, the latter of which is important for facing the challenge of extreme weather conditions brought about by climate change.

---

5. Where the disaster region involves a sea area and crosses over two municipal or county (city) governments, or where the disaster situation is significant and the municipal or the county (city) government is unable to coordinate and process in due time."

[22] *Toxic Chemical Substances Control Act*, amended 11 December 2013. Retrieved 9 April 2015, from http://law.moj.gov.tw/Eng/LawClass/LawContent.aspx?PCODE=O0060012.

15.2.3. *Authorities planning for disaster at all levels: Before the emergency response stage*

15.2.3.1. *Central government*

A. Disaster Prevention and Protection Council

In order to plan for the disasters that might happen each year, a high-level planning platform and regular meeting is provided for by the DPPA in the form of the Disaster Prevention and Protection Council, established by the Executive Yuan.[23] This platform can be seen as a *cross-ministerial* platform[24] demonstrating that the government is paying attention to the issue of disaster management and planning. Nevertheless, further *soldiers* are needed for dealing with disaster management issues, but this scheme has remained unchanged despite major disasters, including Typhoon Morakot.

---

[23]Article 6 of the DPPA: "(Responsibilities of the Central Disaster Prevention and Protection Council).

The Executive Yuan established the Central Disaster Prevention and Protection Council to be in charge of the following areas:

1. To define the basic disaster prevention and protection guidelines.
2. To authorize the Disaster Prevention and Protection Basic Plan and the disaster prevention and protection operation plans of the central regulating authorities.
3. To authorize important disaster prevention and protection policies and countermeasures.
4. To authorize emergency response and countermeasures in case of a nationwide disaster.
5. To supervise and evaluate the performance of disaster prevention and protection of the central administration and municipal and county(city) governments.
6. To carry out any other duties as stipulated by the law or regulations."

[24]The membership of this Council is detailed in Article 7(1) of the DPPA: "(Organization of the Central Disaster Prevention and Protection Council).

There shall be one each of a convener and a deputy convener to be respectively chaired by the Premier and the Vice Premier of the Executive Yuan; and the secretary general and several commissioners will be assigned or employed on a part-time basis from the incumbent ministers without portfolios, governors of related agencies, disaster prevention and protection experts, and scholars by the Premier of the Executive Yuan for the Central Disaster Prevention and Protection Council."

B. Disaster Prevention and Protection "Commission": The establishment of the Disaster Prevention and Protection "Office" in the post-Morakot era

To implement the disaster prevention and protection policy authorized by the Central Disaster Prevention and Protection Council, and to promote major disaster prevention and protection duties and actions, the Disaster Prevention and Protection "Commission" was established.[25]

With a view to dealing with the after-effects of Typhoon Morakot and reinforcing the function of this Commission in supporting the Council, the Commission was renamed the "Central" Disaster Prevention and Protection Commission, and a new permanent institution, the Office of Disaster Management, was created to provide aides and advice.[26] It is hoped that a permanent agency will prove beneficial in assisting the planning and decision-making by the Council and Commission for tackling large disasters.

C. The Disaster Prevention and Protection Expert Consultation Committee and National Science and Technology Centre

Taking into account the professional and technology-oriented nature of disaster management, the input from expert groups is essential. In order to provide disaster prevention and protection advice; facilitate the research, development, and implementation of disaster prevention and protection technology; and strengthen disaster prevention and protection policies and evaluations, before Typhoon Morakot, the DPPA stated that a consultation committee and technology center "may" be established.

Following the typhoon, though, the main legal change was to make such a committee and center "mandatory".[27]

D. Fire Fighting Agency or National Disaster Prevention and Protection Agency under the MOI

To implement the disaster prevention and protection policy and actions, the original DPPA, before Typhoon Morakot, stipulated that the MOI should transform its Fire Fighting Agency into the Fire Fighting and

---

[25] See Article 7(2) of the DPPA.

[26] *Ibid.*

[27] See Article 7(3) of the DPPA.

Disaster Prevention and Protection Agency.[28] Although the Fire Fighting Agency tried to expand its competence into the disaster management area, the nominal title of the agency remained unchanged, the "National Fire Fighting Agency", which appears very unprofessional in terms of disaster management.

With a view to emphasizing the importance of disaster management, the DPPA in the post-Morakot era renamed the Agency as the "National Disaster Prevention and Protection Agency".[29] It remains to be seen, however, whether this gives more impetus to the function of the agency or simply carries a symbolic meaning.

15.2.3.2. *Local government*

A. Municipal, county, and city levels

Similar to that at the central level, a Municipal or County (City) Disaster Prevention and Protection Council[30] was created under the DPPA to undertake disaster management functions.[31] Similarly, these regulations did not change after Typhoon Morakot either.

With a view to supporting the function and dealing with the affairs of the Municipal or County (City) Disaster Prevention and Protection Council, a

---

[28] See Article 7(4) of the original DPPA.

[29] See Article 7(5) of the DPPA.

[30] Article 9(1) of the DPPA: "There shall be one convener and one or two deputy conveners to be respectively chaired by the administrator and deputy administrator of the municipal or county (city) government; and several commissioners will be assigned or employed on a part-time basis from the military representatives, governors of related agencies and divisions, disaster prevention and protection experts, and scholars by the municipal or county (city) mayor for the Municipal or County (City) Disaster Prevention and Protection Council."

[31] Article 8(1) of the DPPA: "(Responsibilities of the Municipal or County (City) Disaster Prevention and Protection Council).

The municipal or county (city) governments will establish the Municipal or County (City) Disaster Prevention and Protection Council to be in charge of the following areas:

1. To authorize the regional disaster prevention and protection plan of the municipality or county (city).
2. To authorize the essential disaster prevention and protection policy and countermeasures.
3. To authorize the emergency response and countermeasures in case of a disaster within its jurisdiction.
4. To supervise and evaluate works related to disaster prevention and protection within its jurisdiction.
5. To carry out any other duties as stipulated by the law or regulations."

"specialized competent authority" was made mandatory under the original DPPA,[32] as well as an optional "Disaster Prevention and Protection Experts Consultation Committee" being recommended.[33]

However, in order to address the unclear designation of this specialized competent authority and the need for professional advice, the revised DPPA, post-Morakot, stipulated that local governments should create "The Municipal or County (City) Office of Disaster Management".[34] The Disaster Prevention and Protection Experts Consultation Committee also became mandatory under the revised DPPA.[35]

B. Township and district levels

The institutional design at township level also shows how the DPPA treats the vertical separation of powers and the autonomy of local government very seriously in legal terms. Similar to the central and municipal or county (city) governments, the creation of Township Disaster Prevention and Protection Councils was also authorized.[36] The responsibilities[37] and composition[38]

---

[32] See Article 9(2) of the original DPPA.

[33] See Article 9(3) of the original DPPA.

[34] See Article 9(2) of the revised DPPA.

[35] See Article 9(3) of the revised DPPA.

[36] See Article 10 of the DPPA.

[37] Article 10 of the DPPA: "(Responsibilities of the Township (City) Disaster Prevention and Protection Council).

The township (city) offices will establish the Township (City) Disaster Prevention and Protection Councils to be in charge of the following areas:

1. To authorize the regional disaster prevention and protection plan developed for the township (city).
2. To authorize the essential disaster prevention and protection policy and countermeasures.
3. To promote and command refugee accommodation and settlement; organize, report on and evaluate disaster situations; execute emergency response measures; remove barriers at the site of the disaster; and implement disaster rescue countermeasures.
4. To promote community disaster prevention and protection affairs.
5. To carry out any other duties as stipulated by the law or regulations."

[38] Article 11 of the DPPA: "There shall be one each of a convener and a deputy convener and several commissioners for the Township (City) Disaster Prevention and Protection Council. The mayor of the township (city) shall be the convener; the chief secretary or secretary of the township (city) office shall be the deputy convener; and representatives from the

of such councils can be regarded as a scaled-down version of the central and municipal councils.

The scope of the Council was widened further after Typhoon Morakot, to include: "3. To promote and command refugee accommodation and settlement; organize, report on and evaluate disaster situations; execute emergency response measures; remove barriers at the site of the disaster; and implement disaster rescue countermeasures." It is apparent that the role of the Township Council is much enhanced after a large disaster.

At the township level, though, the competent authority for assisting the council with disaster management becomes much less clear, although under Article 11 of the original DPPA, a "designated agency" should be assigned this role.

However, it is this designated power that leads to the confusion and delays. Different townships have different units or personnel dealing with disaster management, and those personnel may not be professionally trained or are only part-time employees. In order to improve the professional ability of the township offices, the revised DPPA requires a "Township (City) Office of Disaster Management" to be established,[39] with regard to both office space and human resources, to avoid any ambiguity.

Finally, it is optional for districts of special municipalities and cities to establish Disaster Prevention and Protection Councils and Offices of Disaster Management in the post-Morakot revision of the DPPA. This rule can be seen as a much more advanced idea of the *localization of disaster management*, which is hardly found in other individual administrative laws in Taiwan. This new provision may also be related to the situation in which a relatively large number of municipalities (five) exist on the small island of Taiwan, some of which cover a large territory.

15.2.3.3. *The relationship between councils at all levels and the all-out defence system*

Article 15 of the original DPPA urged councils at all levels to consider dealing with disaster prevention, protection, emergency response, and

divisions designated in the regional disaster prevention and protection plan shall be assigned or employed as the commissioners on a part-time basis."

[39] See Article 11(2) of the DPPA.

convocation issues with the system of all-out defence mobilization and preparation.

After Typhoon Morakot, two changes were brought about: first, the joint scheme expanded to include the "civil defence system" in the all-out defence mobilization and preparation system; second, the issues were now regarded as "disaster preparatory and emergency response issues".[40]

15.2.4. *Special authority: At the emergency response stage*

15.2.4.1. *Local Disaster Response Centre & Central Emergency Operation Centre*

Article 12 of the DPPA stipulated that a Local Disaster Response Centre should be established at the municipal, county (city), or township (city) level to either prevent disaster or effectively promote disaster response countermeasures.[41] This can be seen as a *bottom-up* approach to responding to disasters.

However, for major disasters, a *top-down* response strategy is adopted. In case of a major disaster or the risk of such a disaster presenting itself, authorities at the central government level for specific disasters will, based on the severity, nature, conditions, and influential factors, establish a Central Emergency Operation Centre, and immediately report to the convener of the Central Disaster Prevention and Protection Council who will appoint a Central Commander.[42]

Following the establishment of the Central Emergency Operation Centre, the municipal or county (city) government may be requested to establish a Disaster Response Centre.[43]

These top-down and bottom-up approaches have remained unchanged following Typhoon Morakot.

---

[40]Article 15 of the revised DPPA.

[41]Article 12(1) of the DPPA: "In case of a disaster taking place or the risk of a disaster presenting itself, the convener of the Municipal, County (City) or Township (City) Disaster Prevention and Protection Council shall, depending on the severity of the disaster, create, and direct a Disaster Response Centre to prevent disaster or effectively promote disaster response countermeasures, and be the commander."

[42]See Article 13(1) of the DPPA.

[43]See Article 13(2) of the DPPA.

15.2.4.2. *Emergency response team to assist the center*

To implement all the response measures and handle disaster prevention and protection affairs, or coordinate with the Disaster Response Centre in executing the disaster response countermeasures, any agency, division, or public enterprise designated in the Disaster Prevention and Protection Operation and Regional Disaster Prevention and Protection Plans will establish *emergency response teams.*[44]

15.2.4.3. *Rescue and rescue command*

After the major earthquake of 7.3 degrees on 21 September 1999, the weaknesses in the human and financial resources available to deal with such major disasters became evident. Therefore, following similar schemes in Japan and the US, Article 16 obliges the Fire and Disaster Management Agency of the MOI to create a *Special Search and Rescue Team and Training Centre at the central government level*, and the municipal or county (city) governments to establish a *Search and Rescue Organization.*

In response to the above-mentioned change of name for the Fire Fighting Agency by the DPPA, the Fire and Disaster Management Agency was also renamed as the "National Disaster Prevention and Protection Agency". However, this is only a symbolic response to a minor revision of the DPPA. However, a major revision to the DPPA does exist in response to the needs and challenges of the post-Morakot era. To integrate and facilitate the implementation of disaster prevention and protection, the Central Disaster Prevention and Protection Commission will establish the *National Rescue Command Centre* to command, supervise, and coordinate the personnel from the various functional authorities and government-related agencies in carrying out disaster rescue missions.[45]

15.3. Actions of Disaster Management and Adaptation

15.3.1. *Comprehensive actions: Plans*

15.3.1.1. *Plans at central level*

In order to provide a comprehensive framework for disaster management and planning, the *Central Disaster Prevention and Protection Commission*

---

[44] See Article 14 of the DPPA.

[45] See Article 7(4) of the revised DPPA.

is obliged to draw up **a** Disaster Prevention and Protection "Basic" Plan, the contents of which are also specified by the DPPA.[46] After approval by the Central Disaster Prevention and Protection Council, this plan will become the basis for competence authorities at all levels and public utilities to prepare more detailed plans.[47] This plan should also be updated and reviewed regularly.[48]

In accordance with the Disaster Prevention and Protection Basic Plan, authorities for specific disasters at central government level are required to draw up the Disaster Prevention and Protection Operation Plan, which is subject to the approval of the Central Disaster Prevention and Protection Council.[49]

Following Typhoon Morakot, though, a change was made to the existing rules. With a view to increasing scrutiny by Parliament, Article 17(3) of the revised DPPA stipulates that the Executive Yuan will submit *The White Paper for Disaster Prevention and Protection* to the Legislative Yuan on an annual basis for review and approval.

---

[46]Article 18(1-2) of the DPPA: "The Disaster Prevention and Protection Basic Plan shall include:

1. An overall long-term Disaster Prevention and Protection Plan.
2. Essential items for the Disaster Prevention and Protection Operation Plan and Regional Disaster Prevention and Protection Plan.
3. Any other items deemed necessary by the Central Disaster Prevention and Protection Council.

The Disaster Prevention and Protection Basic Plan, Disaster Prevention and Protection Operation Plan and the Regional Disaster Prevention and Protection Plan referred to in the first paragraph shall include the following contents:

1. Related affairs of disaster prevention.
2. Related affairs of disaster emergency response and countermeasures.
3. Related affairs of recovery and reconstruction after a disaster.
4. Any other items deemed necessary by the other administrative agencies, public enterprises, and municipal, county (city) and township (city) councils."

[47]See Article 17(1) of the DPPA.
[48]See Article 17(2) of the DPPA.
[49]See Article 19(2) of the DPPA.

15.3.1.2. *Plans at local level and of public utilities*

Similarly, the executive units of Municipal or County (City) Disaster Prevention and Protection Councils have a duty to draw up a "Local Prevention and Protection Plan" in accordance with both the Disaster Prevention and Protection Basic Plan and related Disaster Prevention and Protection Operation Plans, as well as potential characteristics of regional disasters. This local plan should be approved by the respective Disaster Prevention and Protection Councils, and the Central Disaster Prevention and Protection Council notified.[50]

At the township level, the authorities are also required to take into account the Disaster Prevention and Protection Plan drawn up by its governing authority, to consider the potential characteristics of regional disasters (volubility) specific to its own area and prepare its own "Local Disaster Prevention and Protection Plan".[51] This plan should then be approved by its own Disaster Prevention and Protection Councils, and the superior competent authorities notified.

Not only government authorities but also public utilities are required to draw up disaster management plans. Public utilities will draw up their Disaster Prevention and Protection Operation Plans in accordance with the Disaster Prevention and Protection Basic Plan, and seek authorization from their central regulating authorities.[52]

15.3.1.3. *Summary: The hierarchy for different plans and conflict resolution between those plans*

Of all the plans related to disaster management, the Disaster Prevention and Protection Basic Plan is supreme: all other plans, including the local municipal, county,[53] and township[54] plans, cannot contradict this basic

---

[50]See Article 20(1) of the DPPA.

[51]See Article 20(3) of the DPPA.

[52]See Article 19(1) of the DPPA.

[53]Article 20(2) of the DPPA: "No Local Disaster Prevention and Protection Plan as set forth in the first paragraph herein shall ever contradict the Disaster Prevention and Protection Basic Plan and related Disaster Prevention and Protection Operation Plans."

[54]Article 20(4) of the DPPA: "No Local Disaster Prevention and Protection Plan as set forth in the preceding paragraph shall ever contradict the superior Disaster Prevention and Protection Plan."

plan. Furthermore, plans proposed at lower government levels cannot contradict those of the upper government levels either. Finally, none of the basic, local, or operation plans should contradict the DPPA.[55]

It is obvious that there are a lot of plans in place that are very likely to conflict with each other. With a view to avoiding or resolving potential conflict, therefore, a dispute resolution scheme is also detailed.

Before Typhoon Morakot, the Disaster Prevention Commission of the Executive Yuan undertook this function. After the typhoon, however, due to the change in name for the Commission, the current conflict resolution mechanism is called the "Central" Disaster Prevention Commission of the Executive Yuan.[56]

15.3.2. *Actions at the main disaster management stage*

15.3.2.1. *Prevention stage: Mitigation and preparedness*

The prevention stage can be further divided into two sub-stages. The first is the mitigation stage, for which Article 22 provides a wide range of functions to be undertaken by "competent authorities at all levels"[57]

---

[55]Article 18(3) of the DPPA: "Any Disaster Prevention and Protection Plan and disaster prevention and protection-related regulations enacted by administrative agencies according to any other laws should not be contradictory to the Act."

[56]Article 21 of the DPPA: "Any pending conflict between Disaster Prevention and Protection Operation Plans or Regional Disaster Prevention and Protection Plans shall be reported to the Central Disaster Prevention Commission of the Executive Yuan for settlement."

[57]Article 22(1) of the DPPA: "To reduce the occurrence or prevent further expansion of disaster, various tiers of government, depending on their functional authorities, shall implement the following mitigation:

1. Draft a budget and prepare, execute and review the Disaster Prevention and Protection Plans.
2. Provide education, training and information on disaster prevention and protection.
3. Support, and apply the results of, research and development into disaster prevention and protection technology.
4. Provide mountain maintenance, flood control and other conservation works for their homeland.
5. Inspect, reinforce and maintain old buildings, important public buildings, and disaster prevention and protection facilities and equipment, and improve urban disaster prevention and protection mechanisms.

and "public utilities",[58] both of whom are required by law to appoint responsible personnel to be in charge.[59] Further regulations concerning "the disclosure of information of disaster potentials" will be provided in the Ordinance.[60]

During Typhoon Morakot, the main issue with the mitigation measures was the problems in microwave communication, particularly the disruption or interference to the signal transmission by tall buildings in the city. These measures include the power of the MOI to designate a "barrier-free zone for microwave transmission",[61] the duty of collaboration on the builders

6. Observe, collect, analyse and establish essential meteorological, geological, hydraulic and other related information required for disaster prevention and protection.
7. Investigate potentials for disaster, degree of hazard, and situation simulation and risk analysis, and publish the results as applicable.
8. Produce a signed agreement of mutual support related to disaster prevention and protection between local governments and public enterprises.
9. Promote, guide, assist and encourage disaster prevention and protection groups and community voluntary organizations.
10. Plan and promote disaster insurance.
11. Provide disaster prevention and protection support for the disadvantaged minority.
12. Establish engage in and cooperate internationally with disaster prevention and protection information networks.
13. Undertake any other affairs related to disaster prevention."

[58]Article 22(3) of the DPPA: "Public enterprises shall depend on their Disaster Prevention and Protection Operation Plan to implement mitigation activities."

[59]Article 26 of the DPPA: "Various tiers of government and related public enterprises shall assign full-time staff to execute disaster prevention works; while township (city) offices could assign part-time, before assigning full-time, staff to execute disaster prevention works."

[60]Article 22(4) of the DPPA: "Pursuant to Item 7 of Para. 1, documents to be published, regions, operational procedure and other related items of disaster potentials shall be reimbursed with the procedure to be enacted by the Central Disaster Prevention and Protection Operation regulating authorities and the municipal or county (city) governments."

[61]Article 23(4) of the DPPA: "To ensure the transmission of microwave communication for a successful response to disaster rescue, an efficient wireless transmission distance range shall be enacted and duly publicized by the Ministry of the Interior."

of tall buildings in such zones,[62] and the compensation scheme for these builders.[63]

The second stage is called the "preparedness" stage. Unlike the first stage, this is closer to the actual occurrence of a disaster and certain, more advanced, measures should be taken by both "competent authorities at all levels"[64]

---

[62]Article 23(5) of the DPPA: "Before the enactment of Enforcement Rules for the Regional Planning Act, if the builder has had new construction or extension construction, a permit shall be obtained from the municipal or county (city) government, provided one of the following requirements is met:

1. An improvement program has been approved by the Ministry of the Interior.
2. The builder has agreed to allow the Ministry of the Interior to adopt a revised minimum transmission loss method in which a microwave radio or a satellite digital broadcast transmission is installed on the top of the said building to ensure uninterrupted communication."

[63]Article 23(6–7) of the DPPA: "The Ministry of the Interior shall compensate the builder for any liquidated damages incurred from the improvement program adopted by, or for the microwave radio or satellite digital broadcast transmission installed on, the top of the said building. The builder shall obtain a nominal reimbursement of the liquidated damages provided an agreement of such is entered into between the two parties. The procedures, requirements, application deadline and amount of liquidated damages, and other relevant settlement affairs, shall be reviewed and approved by the Ministry of the Interior."

[64]Article 23(1) of the DPPA: "To effectively execute emergency response measures, various tiers of government and related public enterprises, depending on their functional authorities, shall implement the following preparedness actions:

1. Preparedness of disaster prevention and protection organizations.
2. Training and drills for disaster prevention and protection.
3. Disaster monitoring, forecasting, warning announcements and facility reinforcements.
4. Collection and notification of details concerning the disaster situation, and the establishment, maintenance and reinforcement of communication facilities for command purposes.
5. Storage and inspection of supplies and facilities for disaster prevention and protection.
6. Preparedness and inspection of facilities and equipment for disaster prevention and protection.
7. Improvement, reinforcement or detachment of barriers to disaster response counter-measures.
8. Coordination in international disaster rescue support.
9. Any other preparedness actions required for emergency response."

and "public utilities",[65] both of which are required by law to appoint *full-time responsible personnel* to be in charge.[66] Also, at this stage, there is the more bothersome requirement of providing regular *training and drills* for disaster prevention and protection. Thus, there are further requirements concerning implementers[67] and joint participants.[68] Finally, there is even very specific provisions on the "Measures of Emergency Refuge", because of its controversial nature and its significant effect on the willingness of the affected citizens.[69] This provision gives the necessary power to the competent authorities of second tier and township level governments to tackle such a "hot potato" issue.

Although full-time employees are vital for the execution of preventive actions, there are sometimes budget or vacancy issues for township governments. Thus, the revised version of the DPPA following Typhoon Morakot gives more flexibility to the original full-time clause: in principle, full-time employees are necessary,[70] but there is some leeway for township

---

[65]Article 23(3) of the DPPA: "Public enterprises shall, depending on their Disaster Prevention and Protection Operation Plan, implement preparedness actions."

[66]Article 26 of the DPPA: "Various tiers of government and related public enterprises shall assign full-time staff to execute disaster prevention works; while township (city) offices could assign part-time, before assigning full-time, staff to execute disaster prevention works."

[67]Article 25(1) of the DPPA: "Government at various tiers and related public enterprises shall implement training and drills for disaster prevention and protection."

[68]Article 25(2–3) of the DPPA: "To implement such training and drills for disaster prevention and protection as specified in the first paragraph herein, all the employees from the various government tiers and public enterprises, residents, and other public and private schools, groups, companies and factories are obliged to participate or provide assistance. Anyone who participates in the training and drills as specified in the preceding paragraph shall be granted official leave from their organization, school, group, company and factory."

[69]Article 24 of the DPPA: "In a case of a disaster or where there is concern for a disaster, the municipal and county (city) governments, or the township (city) office, shall advise or force people to evacuate and provide proper refuge to safeguard people's lives and properties or prevent further expansion of the disaster. Within the scope of the necessary response to the disaster, the municipal or county (city) governments and township (city) office shall advise or force the owners of objects or items that may expand the disaster or interrupt the rescue to remove the object or item and take applicable actions."

[70]Article 26 of the DPPA: "Various tiers of government and related public enterprises shall assign full-time staff to execute disaster prevention works..."

governments where part-time is allowed.[71] This part-time clause may seem reasonable, but it remains to be seen whether it will jeopardize the prevention actions.

15.3.2.2. *Response stage*

A. Main institutions and actions

Similar to the prevention stage, Article 27 also distributes the response actions across "competent authorities at all levels"[72] and "public

[71]"...while township (city) offices could assign part-time, before assigning full-time, staff to execute disaster prevention works."

[72]Article 27(1) of the DPPA: "To execute emergency response measures, various tiers of government shall, depending on their functional authorities, implement the following subparagraphs:

1. Disaster alarm announcement and response, resident evacuation, rescue and refuge advice, assessment of disaster situation and survey for reporting loss.
2. Setting limits for a precaution zone, providing traffic control, maintaining civil order and ensuring crime prevention.
3. Fire fighting, flood prevention and other response measures.
4. Providing temporary accommodation, social rescue and special protection measures for the disadvantaged minority.
5. Providing emergency care for child, youth and student victims of the disaster.
6. Emergency handling of hazardous materials, facilities and equipment.
7. Infectious disease prevention and control, reject processing, environmental disinfection, inspection of food sanitation and any other public health affairs.
8. Search and rescue, emergency medical service and transportation of the injured.
9. Assist with the inspection and disposal of the bodies of victims and their belongings.
10. Supply and allocate essential living items and portable water.
11. Disaster prevention and emergency repair to conserve water and agriculture.
12. Emergency repair of public facilities, including railways, highways, bridges, public transportation, rapid transit system, airports, seaports, public gas and fuel pipelines, power transmission lines, telecommunications, tap water, and agricultural and fisheries.
13. Emergency evaluation of hazardous buildings.
14. Preservation and disposal of drift, sunken objects and other objects from the salvage operation.
15. Complete documentation concerning the course of the disaster response.
16. Any other affairs taken in response to disaster and preventing its expansion."

utilities".[73] Subsequently, the Ordinance was amended to include "emergency evaluation of hazardous buildings",[74] and additional powers of Injunctive Orders or inspection powers are given to the competent authorities at all levels to execute their response duties.[75] The DPPA remained unchanged in this respect after Typhoon Morakot.

The Emergency or Disaster Response Centre plays a crucial role at the response stage. Thus, the DPPA elaborates further its functions and detailed regulations were included: first, there should be a permanent venue for such a center;[76] second, once the center is established, there are regulations governing the distribution of works among related personnel and the Response Centre Commander.[77]

The Commander is authorized by law to undertake a range of functions and execute a number of response-important decisions, such as the announcement and execution of emergency response measures.[78] One of the key issues is integrating the role of "the civil defence organization,

---

[73]Article 27(3) of the DPPA: "Public enterprises shall depend on their Disaster Prevention and Protection Operation Plan to implement response actions."

[74]Article 27(4) of the DPPA: "Pursuant to Item 13 of Para. 1, applicable disaster types, timing of implementation, operational personnel, procedures and the posting of danger signs in the case of an emergency evaluation of hazardous buildings shall be enacted by the Ministry of the Interior."

[75]Article 32 of the DPPA: "(Injunctions Issued to the Essential Supplies' Dealers for the Implementation of Disaster Response Countermeasures).

For the purpose of implementing those countermeasures set forth in Article 26.1 and 27.1, governments at each tier may issue injunctive orders to dealers who manufacture, transport, sell and are in custody of those supplies needed for the implementation, to requisition or take those supplies into custody.

To execute any penalty under the preceding paragraph, the government may dispatch personnel, with ID, to enter the business premises of any of those dealers, or the place where the materials are stored, for inspection."

[76]Article 28(2) of the DPPA: "Each tier in the Disaster Response Centre shall maintain a permanent venue for its operation, keep sufficient equipment for disaster prevention and protection, and hold periodic drills."

[77]Article 28(1) of the DPPA: "After creating the Disaster Response centre tiers, the governor of every group shall either personally be or designate personnel of the functional authority to be stationed in the centre and execute the disaster response actions. Then, the Response Centre Commander shall be in charge of leading, coordinating and integrating the works."

[78]Article 31(1): "After the establishment of the Disaster Response Centres, the Commander may, subject to the extent necessary for the disaster response, undertake the following actions

community disaster prevention and protection groups, and civil voluntary organizations for disaster prevention and protection" in the emergency response team.[79] Similar to the competent authorities at all levels, the additional power of Injunctive Orders or inspection powers are given to the Commander to execute their response duties.[80]

In response to Typhoon Morakot, and to reinforce the function of the Emergency Response Centre and avoid the Central Disaster Prevention

---

in the name of various tiers of government:

1. Announcement and execution of emergency response measures.
2. Set limits of the precaution zone, restricting or preventing people from entering, or ordering them to leave, by issuing temporary passes.
3. Specify an interval within a road, water area and/or air altitude to limit or forbid the passage of vehicles, sea vessels and aircrafts.
4. Requisition related professionals, technicians and operational personnel of purchased resources for their assistance in rescue operations.
5. Requisition and purchase private search dogs, salvage tools, vehicles, sea vessels, aircrafts, land, water rights, buildings and work objects.
6. Command, supervise and coordinate the national army, fire department, police department, related government agencies, public enterprises, reserve force, civil defence organization, community disaster prevention and protection groups, and civil voluntary organizations for disaster prevention and protection to carry out disaster rescue operations.
7. Demolish hazardous buildings and work objects, and remove barriers at the site of the disaster.
8. Have top priority in using the media and communications equipment to collect and broadcast the information on the disaster situation and emergency response.
9. File, attend, supervise and coordinate overseas disaster relief organizations in Taiwan.
10. Organize, evaluate and report on the disaster situation.
11. Any other required response."

[79]See Article 31(3) of the DPPA.

[80]Article 32 of the DPPA: "(Injunctions Issued to the Essential Supplies Dealers for the Implementation of Disaster Response Countermeasures).

For the purpose of implementing those countermeasures set forth in Article 26.1 and 27.1, governments at each tier may issue injunctive orders to dealers who manufacture, transport, sell and are in custody of those supplies needed for the implementation, to requisition or take those supplies into custody.

To execute any penalty under the preceding paragraph, the government may dispatch personnel, with ID, to enter the business premises of any of those dealers, or the place where the materials are stored, for inspection."

and Protection Centre from failing to cope with the disaster or implement response measures in the case of cross-municipality or cross-county (city) disasters, the *Remote (or Backup) Disaster Backup and Recovery Centre* was established.[81]

B. Disaster notification and disaster alarm signal

The above-mentioned institutions are usually triggered into action by a disaster notification, the responsibility for which is given, under Article 30, to any individual,[82] as well as competent authorities at all levels and public utilities.[83] The alarm signals needed during the emergency response are also regulated by the DPPA,[84] and use of disaster alarm signals without permission is prohibited.[85]

C. Vertical cooperation and the participation of the national army

The upper levels of government, including city, county, and central governments, are required to *take the initiative* in assisting the lower levels of government, including township, city, and county governments, should they fail to cope with a disaster.[86]

---

[81]See Article 28(3) of the DPPA.

[82]Article 30(1-2) of the DPPA: "Anyone shall take the initiative to report any disaster or the risk of a disaster to the fire or police department, the village master or secretary.

Upon receiving the report, the department or the personnel as set forth in the preceding paragraph shall promptly take essential measures."

[83]Article 30(3) of the DPPA: "Upon discovering or being notified of a disaster situation or the risk of a disaster, the governments at each tier and the public enterprises should take the initiative to collect and communicate related information on the disaster situation and promptly make the necessary arrangements."

[84]Article 35(1) of the DPPA: "(Classification, Content and Announcement of the Alarm Signal Required in the Emergency Response).

Unless otherwise provided by other laws, the classification, content, method and announcement timing of alarm signals needed during the emergency response shall be drawn up by each central disaster prevention and protection operation regulating authorities and approval sought from the Central Disaster Prevention and Protection Council for publication."

[85]Article 35(2) of the DPPA: "Alarm signals referred to in the first paragraph, and any similar signals, shall not be used unless otherwise permitted."

[86]Article 34(1-2) of the DPPA: "If the township (city) office fails to cope with a disaster, the county (city) government shall take the initiative to send its employees to provide support,

If the municipal or county (city) governments, or the central disaster prevention and protection operation regulating authorities, fail to cope with the disaster, then they may request that the *national army* participate in disaster response activities.[87]

However, in order to respond to the needs of a *major disaster*, such as Typhoon Morakot in 2009, the national army is authorized to *take the initiative* in assisting with disaster management and actively support response measures for a major disaster rescue.[88] The detailed rules concerning the intervention of the national army will be further specified.[89]

15.3.2.3. *Recovery stage*

To facilitate reconstruction and recovery after disasters, *competent authorities at all levels* are required to implement recovery and reconstruction works,[90] while encouraging the civil organizations and enterprises to provide support. In addition, similar to their role during the other

---

or assign coordinators to provide support and assistance as requested by the township (city) office. If the municipal or county (city) governments fail to cope with a disaster, the central disaster prevention and protection operation regulating authorities shall take the initiative to send its employees to provide support, or assign coordinators to provide support and assistance as requested by the municipal or county (city) governments."

[87] See Article 34(4) of the DPPA.

[88] See Article 34(4) of the revised DPPA.

[89] Article 34(5) of the new DPPA: "The procedures for: requesting rescue support or active support for response measures for disaster rescue from the national army; the arrangement and dispatch of the national defence force; commanding and coordinating negotiations and communications, and education and training, for disaster rescue affairs, shall depend on the Ministry of National Defence and the Ministry of the Interior jointly enacting the schedule of duty operation and other relevant operational regulations as prescribed in the fourth paragraph therein."

[90] Article 36(1) of the DPPA:

"1. Investigate, study, evaluate and analyse disaster situations and the needs of victims.
2. Authorize and implement programmes and proposals for recovery and reconstruction after a disaster.
3. Record and allocate volunteers.
4. Distribute and handle donated resources and money, and allocate relief.
5. Attend to casualties, resettle people and order maintenance in disaster areas.
6. Provide medical care, epidemic prevention and psychological counselling.
7. Rebuild school dormitories and supplementary public facilities.

stages, public utilities shall follow their Disaster Prevention and Protection Operation Plan and implement recovery and reconstruction actions.[91]

However, in order to facilitate the recovery process, the creation of a new institutional design is sometimes needed. Therefore, at the discretion of the competent authorities at all levels, a tasking commission, that is, a Reconstruction Promotion Commission, whose members are assigned from all the agencies concerned, may be created to promote reconstruction.[92] The Reconstruction Promotion Commission may only be dissolved upon completion of the recovery and reconstruction after a disaster.[93]

Finally, in order to speed up the reconstruction process and alleviate the burden of a wide range of existing administrative laws, such as the Urban Planning Law, Water and Soil Conservation Law, or Slopland Conservation and Utilization Act, the DPPA permits exemptions from the legal requirements under two situations. First, in the case of *interruption to traffic* or where damage to public utilities affects the *safety of people*, due to the disaster, competent authorities at all levels may *simplify*

8. Ensure schooling and boarding for affected students.
9. Approve or assist with the drafting of urgent repairs to and recovery of heritage, historical sites and buildings.
10. Ensure emergency responses, including investigation of damage to heritage, historical sites and buildings, and their urgent repair and reinforcement.
11. Undertake a safety evaluation and handling of damaged structures.
12. Ensure the recovery of houses, public buildings, urban renewal and handling of land ownership.
13. Repair utilities, including conserving the water supply, soil and water conservation, environmental protection, telecommunications, electricity, water, oil and gas, and regulate resource supplies.
14. Repair tracks, roads, bridges, public transportation, airports, seaports, and agriculture and fisheries.
15. Undertake environmental disinfection and reject processing.
16. Provide employment services for people and the reconstruction of industry.
17. Deal with other issues related to recovery and reconstruction."

[91] See Article 36(3) of the DPPA.

[92] See Article 37(1) of the DPPA.

[93] See Article 37(2) of the DPPA.

administrative procedures and not be subject to the red tape involved in certain legislation.[94] Second, where *normal living conditions* are affected by a natural disaster, competent authorities at all levels may, in terms of delimiting, acquiring, altering, evaluating, managing, and protecting land and buildings, again simplify administrative procedures and not be subject to certain legal requirements.[95]

These exemptions appear quite reasonable and straightforward at first glance: the main priority for reconstruction being "the quicker, the better". However, the author takes a different view, and there should be a balance between the two contrasting approaches to reconstruction: the need for it to be completed quickly, and the need to take care to avoid the *reappearance* of any disaster damage. With regard to the latter, it seems that although the legal exemptions, in line with many environmental, health, and safety (EHS) concerns, may be helpful in speeding up reconstruction, the lack of consideration for the existing legal requirements may also pose a threat to and undermine capability in facing the challenges of future disasters. Existing rules, though bothersome, may be of benefit for "future" disaster prevention and emergency responses, even though they may slow down the

[94]Article 37(1) of the DPPA: "In the case of interruption to traffic or where damage to public utilities affects the safety of people, due to the disaster, various tiers of government shall simplify the administrative procedures and shall not be subject to the Regional Planning Act, Urban Planning Law, Water and Soil Conservation Law, Slopland Conservation and Utilization Act, Forest Law, Water Act, National Park Law and other restrictions in related laws or acts, in order to enable urgency repairs or reconstruction to be completed.

Simplified administrative procedures and measures that are not subject to the related laws or acts under the preceding paragraph shall be replaced with procedures enacted by the central disaster prevention and protection operation regulating authorities and the municipal or county (city) governments."

[95]Article 37(2) of the DPPA: "Where normal living conditions are affected by a natural disaster, various tiers of government shall, in terms of delimiting, acquiring, altering, evaluating, managing, and protecting land and buildings, simplify the administrative procedures and shall not be subject to the Regional Planning Act, Urban Planning Law, Construction Law, Urban Renewal Regulation, Environmental Impact Assessment Law, Water and Soil Conservation Law and other restrictions in related laws or acts, in order to provide proper settlement for the disadvantaged minority or to adopt recovery measures.

Simplified administrative procedures and measures that are not subject to the related laws or acts under the preceding paragraph shall be replaced with procedures enacted by the central disaster prevention and protection operation regulating authorities and the municipal government or county (city) governments."

current reconstruction. However, finding the balance is a real difficulty in disaster and adaptation issues.

To conclude, when faced with a large unpredictable typhoon like Morakot, it is inevitable that there will be numerous casualties. In order to facilitate the issuing of "death certificates" in such circumstances, Article 47-1 was promulgated,[96] whereby certificates should be issued within *one year* of the disaster;[97] of course, there is a rectification clause should the missing person have survived and be found.[98]

15.4. Regulations and Incentives to Facilitate the Adaptation Works

15.4.1. *Regulations*

15.4.1.1. *Penalties*

Under Articles 38–42 in Chapter 7 of the DPPA, there are penalties in place to ensure compliance with the aforementioned rules. For instance, according to Article 32(2), the government may dispatch personnel, with ID, to enter the business premises of any dealers in essential supplies, or the place where such materials are stored, for inspection; yet according to Article 40, the offences of avoidance of, interference with, or refusal of this inspection will be subject to a penalty of between NT$30,000 and NT$150,000, which remained unchanged after Typhoon Morakot.

---

[96]Article 47-1(1) of the revised DPPA: "When a person is missing due to a disaster, the relevant institution, after detailed investigations, shall issue the death certificate based on its authority or the application made by the inheritor, if there is sufficient evidence that the person died due to the disaster despite the body having not yet been found."

Article 47-1(3) of the revised DPPA: "The missing person in the first paragraph is presumed to be dead at the time written on the death certificate."

[97]See Article 47-1(2) of the revised DPPA.

[98]Article 47-1(4–6) of the revised DPPA: "If the missing person survives, the prosecution institution may void the death certificate based on its authority or the application made by either the missing person themselves or the applicant in the first paragraph, or an interested party. When the body of the missing person is found after the issue of a death certificate, according to the law, the body shall undergo an autopsy and an autopsy certificate issued, with the death certificate issued previously being made void.

Authentication for revocation of the death certificate in the last two paragraphs shall apply mutatis mutandis for Article 640 of The Code of Civil Procedure."

15.4.1.2. *Reimbursement of salvage expenses*

During the emergency response, everyone should comply with the Commander's orders or decisions relating to: "2. Delimiting a precaution zone to restrict or prevent people from entering, or to order them to leave, by issuing temporary passes; 3. Specifying an interval within a road, water area and/or air altitude to limit or forbid the passage of vehicles, sea vessels and aircrafts."[99] For those violating these orders, even though they may not be faced with a penalty, they may facc a financial claim from the government. Moreover, in the event that anyone in distress, due to violating the orders, has been rescued by the Disaster Response Centre, they, or the responsible enterprise for which they acted, may receive a written request for the reimbursement of the rescue expenses.[100]

15.4.2. *Financial resources and incentives*

15.4.2.1. *Budget for the DPPA*

The DPPA requires competent authorities at all levels to duly prepare budgets to implement disaster prevention and protection.[101] There is a certain flexibility to the rules in the Budget Act relating to expenditure on disaster management.[102] Moreover, the central government is responsible for providing budget support to those municipal and county (city) governments facing a "major disaster".[103]

[99] See Article 31(1) of the DPPA.

[100] See Article 31(2) of the DPPA.

[101] See Article 43(1) of the DPPA.

[102] Article 43(2) of the DPPA: "Receipts and expenditures for the current fiscal year may be adjusted as applicable, without being subject to Articles 62 and 63 of the Budget Act, when such adjustment is required in the event that the budget prepared for the expenditure on disaster prevention and protection is insufficient to pay the costs of the response measures during the disaster and the recovery and reconstruction after the disaster."

[103] Article 43-1 of the DPPA: "Municipal and county (city) governments unable to afford the expenditure arising from major natural disasters shall request a subsidy from central government.

Pursuant to the preceding paragraph, the timing of the application for the subsidy, qualifications, guidelines and request procedures shall be replaced with a procedure to be enacted by the Executive Yuan."

These budget clauses provide financial resources particularly for "major disasters", which appear appropriate for tackling adaptation to extreme weather conditions. However, in reality, the vague references to "the competent authorities at all levels" and non-specific definition of "major disaster" lead to political conflict and the problem of "everybody's business becomes nobody's business".

15.4.2.2. *Low-interest loans and tax credits*

To provide the funds for reconstruction purposes, the Disaster Prevention and Protection Council of the Executive Yuan shall promptly oblige the financial institutions to provide loans at lower interest rates:[104] further rules concerning the details of the loan scheme will also be provided.[105] Moreover, the Council should consider drawing up rules on tax exemptions, or deferring or postponing tax payments.[106]

The loan and tax credit arrangements remained the same after the revisions to the DPPA following Typhoon Morakot. Only two minor revisions were made: first, since the provision for tax exemptions and a postponed payment scheme was included in the Tax Collection Act[107] and other related tax laws, the power of the Council in this respect was removed;

[104] See Article 44(1) of the original DPPA.

[105] Article 44(2) of the original DPPA: "The Disaster Prevention and Protection Council of the Executive Yuan shall approve the ceiling for the amount of loans and interest subsides, and the loan operating procedures for the purpose of the first paragraph herein. The ceiling for the amount of interest subsides shall be paid from the budget prepared by the government at each tier. In deciding the range of interest subsides, the extent of damage that people suffered as a result of the disaster and their capability to undertake independent reconstruction shall be taken into consideration."

[106] See Article 44(3) of the original DPPA.

[107] For example, see Article 10 of the Tax Collection Act: "Where a taxpayer fails to pay taxes within the statutory payment period due to a natural disaster or incident, the competent tax collection authorities may, based on the situation, extend the payment period and make a public announcement to that effect."

Also, see Article 26 of the Tax Collection Act: "In the event that a taxpayer is unable to pay taxes in full within the statutory tax payment period due to a natural disaster or incident, or suffering material property loss, an application may be filed with the competent tax collection authorities within the said statutory period for the deferral of the tax payment or for payment by instalments; the period for such deferral or instalment scheme shall not exceed three (3) years."

and second, the name of the Disaster Prevention and Protection Commission of the Executive Yuan was changed to the "Central" Disaster Prevention and Protection Commission.

15.4.2.3. *Donation*

The government is allowed to use civil donations for disaster relief purposes: such donations must be exclusively applied to matters directly related to disaster relief, and any use for the government's own administrative or operational purposes is strictly prohibited. Detailed accounts of payments made from such donations will be published.[108]

15.4.2.4. *Compensation*

A. Compensation for those who undertake disaster prevention and protection activities under the DPPA

Should anyone be injured or disabled, become ill, or die in the course of carrying out disaster prevention and protection activities under the Act, they will be eligible for compensation payment(s).[109] For

[108]See Article 45 of the DPPA.

[109]Article 47(1) of the revised DPPA: "Anyone who is injured or disabled, becomes ill, or dies in the course of carrying out disaster prevention and protection activities under the Act, may claim compensation according to the requirements related to their regular job."

For the detailed calculation method, see Article 47(3):

"1. Injury or illness: a certificate for such injury or illness issued by the government shall be produced to a public hospital or a contracted hospital in order to receive medical care: provided, however, that medical care may be sought from any hospital in the case of an emergency.
2. Disabled: the disabled payment shall be paid in a lump sum depending on the severity of the injury, as follows:
   (1) Extremely severe and severe degree: 36 cardinals.
   (2) Intermediate degree: 18 cardinals.
   (3) Mild degree: 8 cardinals.
3. Death: 90 cardinals of the pension is payable in a lump sum.
4. Disabled and dies within one year after the event: payment of a lump sum to make up the deficiency as provided in the preceding subparagraph."

See also, Article 47(3-6): "For the purpose of this Article, the amount of 1 cardinal shall not be less than that of a monthly remuneration of the annual amount payable to a civil servant in a fifth-rank post.

those who are not eligible, certain compensation schemes may still apply.[110]

The compensation rules remained very similar before and after Typhoon Morakot. There was only a small revision to the Chinese meaning of being disabled, from the relatively negative term of "handicapped" to a more neutral term of "disabled".

B. Compensation according to the expropriation clause under the DPPA

Anyone may suffer a loss as a result of the legal actions, disposals, Injunctive Orders, or other orders made pursuant to Articles 24(2),[111] 31(1),[112] or 32(1)[113] of this Act.

Article 33 gives the right to claim for damages unless such a loss is attributable to a civilian.[114] The compensation shall be paid in money and

Authentication of the degree of disability set forth in the second paragraph shall be determined pursuant to the People with Disabilities Rights Protection Act and related regulations.

When any payment already claimed under the first paragraph herein is lower than that specified in subparagraphs 2–4 of the second paragraph, the deficiency shall be made up to the claimant.

Payments to be made under the second paragraph, and the deficiency to be made up as provided in the preceding paragraph of this Article, shall be issued by the government."

[110]Article 47(2) of the revised DPPA: "In the event that the claimant set forth in the first paragraph herein is prevented from claiming payments, the payment(s) shall be paid according to the following requirements, with reference to that allowed for the voluntary fire-fighter; and the government shall prepare the budget for the payment of the required expenditures."

[111]"Within the scope of the necessary response to the disaster, the municipal or county (city) governments and the township (city) office shall advise or force the owner of objects or items that may expand the disaster or interrupt the rescue to remove the object or item and take applicable actions."

[112]For example: "4. Requisition related professionals, technicians and operational personnel of purchased resources for their assistance in rescue operations."

[113](Implementation of Disaster Response Countermeasures). For the purpose of implementing those countermeasures set forth in Article 26(1) and 27(1), governments at each tier may issue Injunctive Orders to dealers who manufacture, transport, sell, and are in possession of those supplies needed for implementation, in order to requisition or take those supplies into custody.

[114]See Article 33(1) of the DPPA.

at a sum no greater than the actual loss incurred,[115] but there is a time limitation on making such a claim.[116]

Moreover, in order to enforce the requisition clause of Article 31(1) ("5. Requisition and purchase of private search dogs, salvage tools, vehicles, sea vessels, aircrafts, land, water rights, buildings and work objects"), the DPPA requests that the MOI promulgate a further Ordinance.[117]

15.4.2.5. *Commendations of merit for disaster prevention*

Any prevention and protection group, voluntary organization, or individual who has made extraordinary contributions to disaster prevention and protection works shall be duly commended by the government.[118]

15.4.2.6. *Assistance for civil voluntary organizations*

The government shall pay *accident insurance* and help provide *salvage equipment* to prevention and protection groups or voluntary organizations.[119]

15.4.2.7. *Disaster relief for victims*

The victims of the disaster may apply for disaster relief under the conditions defined by further rules.[120] For instance, the Ordinance on Windstorm, Earthquake, Fire and Explosion Disasters (Categories and Standards of Assistance for Windstorm, Earthquake, Fire and Explosion Disasters)[121]

---

[115]See Article 33(2) of the DPPA.

[116]Article 33(3) of the DPPA: "The claim shall be honoured by the government, once investigated and confirmed, within two years, but no claim for damages will be accepted if filed five years or more after the date that the loss was sustained."

[117]Article 49 of the DPPA: "Any requisition or purchase made under the Act shall be reimbursed or compensated for, with the guidelines, procedures, payment methods and other related matters to be enacted by the Ministry of the Interior."

[118]See Article 46 of the DPPA.

[119]See Article 50(2) of the DPPA.

[120]Article 48 of the DPPA: "Categories and standards for the purpose of the support to be provided in the event of a disaster shall be agreed and specified jointly by the central disaster prevention and protection operation regulating authorities and the municipal or county (city) governments."

[121]*Categories and Standards of Assistance for Windstorm, Earthquake, Fire and Explosion Disasters*, 20 January 2011. Retrieved 9 April 2015, from http://law.moj.gov.tw/Eng/LawClass/LawContent.aspx?PCODE=D0120017.

provides: (1) Assistance for families of deceased victims; (2) Assistance for families of missing victims; (3) Assistance for severely injured victims; and (4) Assistance for victims needing relocation.[122] Usually, relief takes the form of an amount of money: for example, the Standard of Assistance for: families of deceased victims is NT$200,000 per victim; families of missing victims is NT$200,000 per victim; and severely injured victims is NT$100,000 per victim.[123]

15.5. Conclusion

This chapter sets out the main reforms following Typhoon Morakot, especially the intensive reforms concerning "institutional design and authority". In fact, most of the reforms focused on this area, even some of the very minor revisions such as the names and titles of the authorities. However, the long-standing, controversial structure of "dispersed" authorities for specific disasters at the central government level remains unchanged, while it remains to be seen whether the effort to localize competent authorities comes to fruition. Overall, the effectiveness of such reforms in fighting extreme weather conditions is still to be proved.

During the legal reforms to the DPPA, there was discussion of whether it was the scale of a disaster or unusually extreme weather conditions that was to be blamed for the losses, rather than any flaws in the DPPA itself. Thus, some might say that it is pure chance whether the authorities and their actions will succeed or fail in the face of certain extreme weather conditions. Such an assumption would imply that laws such as the DPPA play no decisive role in climate adaptation, and so there is no need to revise it, or that any revision only has *symbolic* meaning.

In light of this argument, the author would like to make the following comment. It is true that luck or the act of God, that is the scale of the extreme weather conditions, does play its part; however, the "actions of man" may also play a role in preventing and protecting against large disasters. A well-structured and functioning legal regime could contribute to the capability

[122] See Article 3 of the *Categories and Standards of Assistance for Windstorm, Earthquake, Fire and Explosion Disasters, op. cit.*

[123] See Article 5 of the *Categories and Standards of Assistance for Windstorm, Earthquake, Fire and Explosion Disasters, op. cit.*

of institutions (Section 15.2 of this chapter) in taking appropriate action (Sections 15.3 and 15.4 of this chapter) to avoid making the situation worse. In this respect, adaptation and disaster management laws can be seen as a way of avoiding or alleviating "man-made" chaos or disaster after an actual natural disaster — the issue of "everybody's business is nobody's business", for instance, or of offloading one's responsibility onto another. By properly distributing the *rights and responsibilities* for activities across multiple authorities, it is hoped that the DPPA can contribute, at least partially, to combating unforeseen extreme weather conditions and disasters.

Chapter 16

Legal Options for Mainstreaming Climate Change Disaster Risk Reduction in Governance for Kenya

Robert Kibugi

16.1. Introduction

Climate change has resulted in complex adverse impacts in Kenya. In addition to changes in rainfall patterns and temperature variability, the country is experiencing a high rate of extreme weather events that are attributable to climate change. These events, including droughts and floods, are now occurring on an average annual cycle, and quite often magnified to natural disaster levels. Climate change-induced natural disasters bring out adverse effects on the Kenyan economy, population, and environment — and are therefore a threat to the sustainable development of the country. People are displaced from homes, social, and economic systems that are destroyed and in some cases completely obliterated. Extreme floods have resulted in massive mudslides that damage the landscape and make the land unusable for various economic activities including agriculture.

The frequent occurrence of climate change-induced extreme weather events is a threat to the sustainable development of the country. The national plans to attain middle-income status by the year 2030, or the Constitutional obligation to pursue ecologically sustainable development are under threat. Ongoing investments in infrastructure, including new highways, bridges, airport, and seaports, are likely to be harmed unless climate proofing is integrated into national law. This is mainly because the climate-induced natural disasters are a threat to building resilience and adaptive capacity to climate change for the Kenyan economy, population, and environment. It is important that national climate change governance mechanisms mainstream disaster risk reduction — into sectoral and subnational government functions. Equally, the role of citizens in building and adopting a culture of safety, in line with the Hyogo Framework of Action 2005–2015: Building the Resilience of Nations and Communities to Disasters[1] is critical. This chapter will examine how Kenya's evolving governance system for climate change can be modified to mainstream disaster risk reduction, with the aim of building resilience and adaptive capacity to climate change.

16.2. Climate Change Risks to the Sustainable Development of Kenya

In 2010, Kenya became one of the first countries in Africa to provide legal character to the concept of sustainable development through its recognition in the Constitution. This happens in two ways. First, the Constitution sets out principles and values of national governance, that are binding during the making of, interpretation, and implementation of national laws and public policies.[2] To this end, sustainable development is set out as one of the binding values of national governance, in addition to other critical ones such as participation of people in government and devolution of governmental functions.[3] Second, the Constitution sets out an obligation on the state and the people to cooperate in order to attain ecologically sustainable development in the use of natural resources.[4] This means that sustainable development has found a basis for legal implementation, through the supreme law of the land in Kenya.

That notwithstanding, actual normative content of how to design and implement sustainable development in Kenya is still elusive, with climate change and its adverse impacts, including disasters, complicating this further. It is notable that extreme events in Kenya form the very core of

[1]UNISDR, "Hyogo Framework for Action," The United Nations Office for Disaster Risk Deduction. Retrieved 10 April 2015, from http://www.unisdr.org/we/coordinate/hfa.

[2]See, Article 10, Constitution of Kenya, 2010.

[3]*Ibid.*

[4]See, Article 69(2), Constitution of Kenya, 2010.

evidence of climate change. For instance, in recent years, the frequency, intensity, spatial extent, duration, and timing of extreme weather and climate events has gained tempo, with the country's drought cycles reducing from 20 years (1964–1984), to 12 years (1984–1996), to two years (2004–2006), to a yearly occurrence of drought recorded in the period between 2007 and 2012.[5]

The impacts of these disasters, especially droughts and floods, exacerbate the levels of poverty, compromise adaptation efforts, and act at cross-purpose to the socio-economic and environmental stability that attainment of sustainable development could bring. In fact, according to the 2014 Draft National Climate Change Policy:[6]

> . . . the realization of sustainable development in Kenya, despite significant progress to date, is threatened by climate change and its resultant impacts, as the country has in the recent past seen increased evidence of climate change such as rising temperatures and changing rainfall patterns, and has experienced extensive climate related impacts through the increased frequency and intensity of extreme weather events such as droughts and flooding. These manifestations of climate change constitute a serious threat to Kenya's natural, built economic and physical systems, on which the country's sustainable development and future prosperity depends.[7]

In reality, according to the National Climate Change Action Plan (NCCAP 2013–2017),[8] the Government of Kenya has recognized the growing threat of climate-related risk to the achievement of its development goals; and although various measures are in place, it is more urgent to implement more effective measures because climate change, for instance, increases drought vulnerability. Much of the adverse impacts of climate change that Kenya faces today require implementation of adaptation measures,

---

[5] Kenya (2014) Draft National Framework Climate Change Policy, Final Review October 2014, paragraph 2.1.6.

[6] Kenya (2014) Draft National Framework Climate Change Policy, Final Review October 2014, paragraph 1.1.2.

[7] Supra note 5, para. 1.1.2.

[8] "Kenya (2013) National Climate Change Action Plan," Ministry of Environment and Mineral Resources, Nairobi, at 39.

in order to build resilience and enhance adaptive capacity of the society and economy. It is important to address this because, as pointed out earlier, climate change and the resulting disasters pose a major challenge to the desired attainment of sustainable development (which has Constitutional basis), and the socio-economic and environmental plans put in place to facilitate pursuit of sustainable national development. Vision 2030,[9] which is the national development policy blueprint for the 2008–2030 period, aims to create "a globally competitive and prosperous country with a high quality of life by 2030", and to transform Kenya into "a newly-industrializing, middle-income country providing a high quality of life to all its citizens in a clean and secure environment".[10]

The current 2013–2017 Medium Term Plan (MTP)[11] for implementing Vision 2030 puts the risk posed by disasters in perspective, noting that "Kenya is susceptible to natural disasters such as drought and flooding which are likely to increase as a result of climate change". The MTP is also cognizant that the management of these disasters is a cross-cutting issue that requires collaborative action by public and private sector agencies at national, county, and community levels.[12] With the background of the objectives sought by Vision 2030 for Kenya, and the obligation of sustainable development imposed by the Constitution, the NCCAP adopted a philosophical approach to guide the conceptualization, design, and implementation of climate change interventions.[13] This approach, focusing on attainment of low-carbon climate resilience development pathway, presents helpful conceptual tools to link climate-induced disasters with sustainable development. This chapter focuses on the linkage between adaptation and sustainable development, with the argument that building resilience should include disaster risk reduction in order to align adaptation to sustainability outcomes.

---

[9]"Kenya (2012) Sessional Paper No. 12 of 2012," Government Printer, Nairobi, ii.

[10]*Ibid.*

[11]"Kenya (2013) Second Medium Term Plan to Vision 2030: 2013–2017," Ministry of Devolution and Planning.

[12]*Ibid.*

[13]"Kenya (2013) National Climate Change Action Plan," Ministry of Environment and Mineral Resources, Nairobi, at 41.

16.3. Conceptualizing the Link between Adaptation and Disaster Risk Reduction for Sustainable Development

Focusing on sustainable development and adaptation, the NCCAP attempts to debunk the implications of an integrated low-carbon climate resilient development pathway by breaking down the two key elements, as follows:[14]

1. Sustainable Development — Achieving sustainable development should be at the forefront of all climate actions; climate change and development are intricately linked.
2. Adaptation — Reducing vulnerability to avoid or cushion the impacts of climate change, and enable people to respond to climate risks by moving toward a climate-resilient society.

In its Fifth Assessment Report in 2014, the Intergovernmental Panel on Climate Change (IPCC) addressed itself to the question and challenges arising in the link between climate resilient development pathways and sustainable development. Climate resilient pathways, the report argues in Chapter 20,[15] bring together sustainable development, and the effects of climate change as threats/opportunities to sustainable development. Climate resilience is deemed critical and defined as the ability of a social, ecological, or socio-ecological system and its components to anticipate, reduce, accommodate, or recover from the effects of a hazardous event or trend in a timely and efficient manner.[16] This approach is somewhat consistent with the Kenyan approach in the NCCAP, which equally makes an attempt to link climate change and sustainable development, and particularly brings out adaptation as the functional principle governing

---

[14] *Ibid.*

[15] F. Denton, T. J. Wilbanks, A. C. Abeysinghe, I. Burton, Q. Gao, M. C. Lemos, T. Masui, K. L. O'Brien, and K. Warner, "Climate-Resilient Pathways: Adaptation, Mitigation, and Sustainable Development." In C. B. Field, V. R. Barros, D. J. Dokken, K. J. Mach, M. D. Mastrandrea, T. E. Bilir, M. Chatterjee, K. L. Ebi, Y. O. Estrada, R. C. Genova, B. Girma, E. S. Kissel, A. N. Levy, S. MacCracken, P. R. Mastrandrea, and L. L. White (eds.), *Climate Change 2014: Impacts, Adaptation, and Vulnerability. Part A: Global and Sectoral Aspects. Contribution of Working Group II to the Fifth Assessment Report of the Intergovernmental Panel on Climate Change* (UK and NY: Cambridge University Press, 2014), pp. 1101–1131, at 1108.

[16] *Ibid.*

climate resilience. The importance of this approach is succinctly brought out by this paragraph in the NCCAP:[17]

> The country's economy is highly dependent on climate sensitive sectors including agriculture, tourism, and energy. Climate change potentially poses one of the greatest challenges for Kenya to realise its vision to become a prosperous country. The World Bank affirms that poverty and vulnerability to climate change remain the most critical development challenges facing Kenya.
>
> Agriculture is the backbone of the Kenyan economy directly contributing 24% of the GDP valued at KSh342 billion in 2009 and another 27% indirectly, valued at KSh385 billion. The sector accounts for 65% of informal employment in rural areas. Kenya faces major food security challenges due to the over dependence on rain-fed agriculture for food production. The number of Kenyans requiring food assistance rose from 650,000 in 2007 to almost 3.8 million in 2009/2010. Pastoral and marginal agricultural areas are particularly vulnerable to the impacts of climate change. Extended periods of drought erode livelihood opportunities and community resilience in these areas; leading to undesirable coping strategies that damage the environment and impair household nutritional status, further undermining long-term food security.

The foregoing paragraph, together with the earlier illustration of increased drought and flood frequencies, demonstrates the extent of Kenya's vulnerability to climate variability. Adaptation, as a climate change intervention tool incorporates both building climate resilience and enhancing adaptive capacity. Building climate resilience requires Kenyan governance systems, ecosystems, and society to have the capability to maintain competent functions in the face of climate change and to return to some normal range of function even when faced with adverse impacts of climate change.[18] This is because climate change is altering the face of disaster risk, not only through increased weather-related risks and sea level and temperature rises, but also through increases in societal vulnerabilities, for example,

[17] "Kenya (2013) National Climate Change Action Plan," Ministry of Environment and Mineral Resources, Nairobi, at 4.

[18] "Kenya (2014) Draft National Framework Climate Change Policy," Final Review October 2014, paragraph 4.1.2.

from stresses on water availability, agriculture, and ecosystems.[19] Indeed, disaster risk reduction and climate change adaptation share a common space of concern: reducing the vulnerability of communities and achieving sustainable development.[20]

For this reason, adaptive capacity is key to improving socio-economic characteristics of communities, households, and industry as it includes adjustments in both behavior and in resources and technologies. Adaptive capacity is a necessary condition for the design and implementation of effective adaptation strategies so as to reduce the likelihood and magnitude of harmful outcomes resulting from climate change.[21] An enhanced understanding of the adaptive capacity of Kenya can be derived from vulnerability assessments.[22] Vulnerability is a common feature between climate change and disaster risk, since both seek to reinforce the resilience and adaptive capacity of various people. According to the Hyogo Framework of Action, a definition adoption by the Kenyan national disaster management policy, disaster risk reduction is: "The concept and practice of reducing disaster risks through systematic efforts to analyze and manage the causal factors of disasters, including through reduced exposure to hazards, lessened vulnerability of people and property, wise management of land and the environment, and improved preparedness for adverse events."

Using this definition, the two key elements of disaster risk: vulnerability and hazard — are also key aspects of climate change adverse impacts resulting from extreme weather events, such as droughts and floods. In essence, deployment of adaptation interventions through building resilience and adaptive capacity will have the effect of moderating the adverse impacts of hazards, by reducing the vulnerability of society, environment, and economy. This outcome could reinforce sustainable development activities of public policy in Kenya, so long as the legal framework in Kenya

---

[19] Marilise Turnbull, Charlotte L. Sterrett, and Amy Hillebo, "Toward Resilience: A Guide to Disaster Risk Reduction and Climate Change Adaptation," Practical Action, Warwickshire, UK, at 6.

[20] *Ibid.*

[21] "Kenya (2014) Draft National Framework Climate Change Policy," Final Review October 2014, paragraph 4.1.3

[22] *Ibid.*

approaches climate change-induced disaster risk reduction, with an explicit understanding of the link to adaptation and sustainable development.

16.4. Assessment of the Current Regulatory Framework on Disaster Risk Reduction and Climate Change Risks

In this section, the chapter examines whether Kenyan law contains an explicit linkage between disaster risk reduction and climate change adaptation. Currently, there is no single comprehensive legislation on disaster risk reduction in Kenya. However, elements of the legal and policy regime on disasters in Kenya can be drawn from the 2010 Constitution, as well as a number of sectoral laws and policies. With regard to policies, key recent policy formulation exercises include the National Policy on Disaster Management of 2009 and the Kenya Vision 2030. The sectoral laws do not cover disasters in a comprehensive manner but only disaster areas relevant to each particular law. These include the Physical Planning Act, the County Governments Act, the Public Health Act, land use laws, laws on food security, and others discussed below. Therefore, there is in place legal and institutional frameworks on disaster management (including non-climate-induced) — although, as the discussion below shows, there is hardly any mainstreaming of functions to integrate the whole disaster risk reduction and management continuum from planning and budgeting to implementation.

16.4.1. *The Constitution of Kenya 2010*

The Constitution of Kenya 2010 contains provisions that are relevant to disaster risk reduction. In particular, the Bill of Rights is of key importance given the protection of the right to life,[23] right to clean and healthy environment,[24] right to human dignity,[25] right to be free from discrimination on prohibited grounds,[26] right to freedom of movement and

[23]Article 26.
[24]Article 42.
[25]Article 28.
[26]Article 27.

security of the person,[27] and economic and social rights.[28] All these rights reflect the social contract between the Kenyan state and its citizens, under which the state is the guarantor of the rights. Indeed, Article 19 of the Constitution, explaining the purpose of the fundamental rights, is rather clear that:

> The purpose of recognising and protecting human rights and fundamental freedoms is to preserve the dignity of individuals and communities and to promote social justice and the realisation of the potential of all human beings.

For this reason, the Kenyan state has a duty to ensure these rights are realized, and to eliminate or mitigate circumstances, such as climate-induced disasters, that could compromise realization of the basic rights in question. Design and implementation of disaster risk reduction strategies constitutes a key element in upholding rights to life, dignity, and environmental health. Thus, where disasters are prevented or vulnerability is reduced, these fundamental rights will, in return, be promoted. In addition, efforts to prevent or respond to disasters must comply with all human rights such that although human rights maybe limited during a state of emergency, this can only be within the scope permitted by the Constitution.[29] Finally, it should also be noted that a Contingencies Fund and County Emergencies Funds have been established to facilitate urgent public expenditure, which includes funds emergencies such as response and management of disasters.[30]

16.4.2. *The National Policy for Disaster Management, 2009*[31]

This National Policy for Disaster Management was formulated in 2009. The purpose of this policy is to provide a sound policy framework for disaster management of both natural- and human-induced disasters

---

[27]Article 29.
[28]Article 43.
[29]Article 58.
[30]Article 208 of the Constitution; see also the Contingencies Fund and County Emergencies Fund, 2011.
[31]Supra 9.

in Kenya. The policy recognizes that disaster risks pose a significant threat to economic development as the planned use of scarce resources is diverted from their intended use, economic activities are interrupted, and overall sustainable development is impeded. In response to the aforesaid problems, the policy was formulated to reduce vulnerabilities to disaster risks by creating a strong legal and institutional framework. It envisages both disaster risk management as well as disaster risk reduction. Disaster management is a reaction to the occurrence of a disaster; hence, it is designed to deal with losses caused by the disaster rather than its prevention.[32] The policy conceptualizes disaster risk reduction as occuring where there is focus on early warning, reduced vulnerability, effective management of land and the environment, and public participation.[33]

Although the policy has not been approved by parliament as a Sessional Paper, it sets out fundamental principles of disaster management. These principles are similar to the best practices set out by the globally accepted Hyogo Framework for Action — as the foundation of a good disaster risk reduction legal framework:[34]

- Establishment of an effective Disaster Early Warning, Information and Prediction System;
- Mainstreaming disaster management in development programming;
- Strong link between Early Warning and Disaster Response through harmonized and standardized rapid response interventions to disasters;
- A government-led institutional framework that allows effective coordination of the relevant stakeholders and enhances trust-building and collaboration among partners;
- Provision for and differentiation between procedures and mechanisms of response to both slow- and rapid-onset disasters;
- Established and functional semi-autonomous agency responsible for disaster management with oversight from the Ministry of State for Special Programmes;

[32] Refer to the general interpretation section in the 2009 policy.
[33] Refer to the Draft Policy on definitions of disaster management terms.
[34] See Chapter 3 of the 2009 Draft Policy on Disaster Management.

- Provision for flexible financial and response procedures to facilitate appropriate, effective, and timely planning and management of disaster control initiatives; and
- Reinforcement of the disaster management strategies for vulnerable and affected communities.

Further, the policy provides for goals and objectives of disaster risk reduction and management. The main goal is a safer, more resilient, and more sustainable Kenyan society. The place of public participation is also recognized through a provision for capacity building and involvement of communities likely to be affected by disasters in designing, planning, and implementation, as well as monitoring and evaluation of disaster reduction projects. Gender equity is also prioritized to the extent that all stakeholders in disaster risk reduction will take gender equity into consideration in all programmes.[35] The policy provides a constructive foundation that is sufficiently broad to integrate climate-induced disasters. Nonetheless, it does not provide how DRR responses will be mainstreamed into functions of various government functions and mandates.

16.4.3. *The Arid and Semi-Arid Lands (ASALs) Policy*

The National Policy for the Sustainable Development of Northern Kenya and other Arid Lands (2010) is a broader strategy premised on the Kenya Vision 2030. The aim of this Policy is to facilitate and fast-track sustainable development in Northern Kenya and other arid lands by increasing investment in the region and by ensuring that the use of those resources is fully reconciled with the realities of people's lives.[36] The implementation of the policy will contribute towards effective disaster risk reduction in arid and semi-arid areas through increased resilience and preparedness. It will further reduce dependence on relief interventions and

[35] See Part 3.2 of Chapter 3, "Draft Disaster Management Policy," 2009.

[36] "Draft National Policy for the Sustainable Development of Northern Kenya and other Arid Lands," 2010, 13.

the heavy financial burden of emergency response.[37] The Policy commits the government to:

- Gazette and manage emergency drought reserve areas, and encourage the development of buffer areas of crop and forage production as part of contingency planning;
- Mainstream climate foresight and climate adaptation into planning at all levels;
- Systematically strengthen the strategies used by communities to adapt to climate variability and to reduce and manage the risks from natural disasters; and
- Explore opportunities and develop appropriate mechanisms through which communities can benefit from bio-carbon initiatives such as carbon sequestration, whereby the resulting carbon credits are sold in the carbon markets to generate payment for environmental services.[38]

The National Policy acknowledges that drought response and contingency plans to prevent widespread loss of livestock and other assets in ASALs are activated too late when very little can be salvaged. In order to reduce the effects of drought and climate change on vulnerable communities in the ASALs and help them to adapt, the Policy set the basis for the creation of the National Drought Management Authority to ensure timely activation of contingency plans to protect assets. The Policy's area of action is limited to drought, whereas climate change-induced disasters affecting ASAL lands extend beyond drought. With the advent of devolved government in Kenya since 2013, the Policy should factor in the roles of county governments in dealing with drought since counties possess the relevant Constitutional mandate on implementing disaster risk reduction, such as agriculture, livestock husbandry, etc.

16.4.4. *Sectoral laws and institutions on disaster risk reduction*

A significant amount of functions that impact adaptation, disaster risk reduction, and sustainable development are undertaken through sectoral laws that

---

[37] *Ibid.*

[38] "National Policy for the Sustainable Development of Northern Kenya and Other Arid Lands," 2010, 20.

confer authority on various agencies and levels of government. These laws may not be explicitly concerned with climate change adaptation — or with disaster risk reduction, but could be deployed to that effect. Nonetheless, the discussion in this section and below points to a need for some modifications to enhance legal options available for mainstreaming disaster risk reduction into governance systems.

16.4.4.1. *The Physical Planning Act*[39]

This legislation regulates physical developments in all sectors of the economy. It regulates developments at the national, regional, and local levels. In terms of planning, this law requires development planning with respect to urban and rural land areas. With respect to rural land, the physical planning legislation empowers the Director of Physical Planning to prepare a regional physical development plan over any government land, trust land, or private land for the purpose of improving the land and providing for the proper physical development of such land, and securing suitable provision for transportation, public purposes, utilities and services, commercial, industrial, residential, and recreational areas, including parks, open spaces, and reserves, and also the making of suitable provision for the use of land for building or other purposes.[40] In the context of urban land, the law provides for preparation of a local physical development plan for the general purpose of guiding and coordinating development of infrastructural facilities and services, and for the specific control of the use and development of land or for the provision of any land for public purposes. Although the preparation of these plans is not mandatory, they provide useful planning tools, which could put into place critical mechanisms for disaster risk reduction.[41]

Under the provisions for development control, the Act provides that every person (including corporate bodies) anticipating to initiate a development project is required to apply for the development control approval of a county government within the area where the proposed project is to be situated. The county government is mandated to issue or withhold approval.[42] In giving or withholding approval, the county

[39]Chapter 286 of Laws of Kenya.

[40]Section 16.

[41]Section 24.

[42]See Part V of Cap. 286, from Section 29 to Section 40.

government must consider the impacts of the project to the environment as well as human life. Indeed, development control permits are subject to the provisions on environmental impact assessment (EIA) under the framework Environmental Management and Coordination Act (EMCA), and Section 36 of the Physical Planning Act. The provisions on EIA provide a firm and useful basis for incorporation of disaster risk reduction and preparedness mechanisms during the grant of development control permission.

16.4.4.2. *Environmental Management and Coordination Act (EMCA) 1999*[43]

This 1999 legislation provides the law of general application on environmental matters. For its implementation, compliance, and enforcement, the law established the National Management Authority (NEMA) as the principal agency of government for oversight on environmental management.[44] As regards disaster risk reduction, NEMA is mandated to require and assess environmental impact assessment study reports for any projects that are set out in the Second Schedule to the EMCA. This second schedule is extensive, and includes a broad range of development activities, including construction of roads, houses, towns, as well as mining. These EIA studies are critical for evaluating the possible and extent of damage that development activities might cause to the environment. In addition, the application of development control together with EIA procedures could play an important role in identifying potential disaster risks, and either stopping the project altogether, or require mitigation measures as part of the Environmental Management Plans. This in turn, could prevent the establishment of environmentally destructive projects. Also, by conducting regular environmental audits, as required by EMCA, on existing projects, NEMA ensures that any project impacting the environment negatively is stopped through forfeiture of the license. Through these two functions, NEMA can analyze environment-related disaster risks, such as likelihood

---

[43]Legislation no. 8 of 1999.

[44]Refer to Section 7 of EMCA which establishes NEMA, Section 8 provides for the Headquarters of NEMA which is at Nairobi. The management/administration is under a Board.

of drought and flooding, and whether the proposed development project has disaster mitigation and response elements. This can then be factored into the approval process for development process.

16.4.4.3. *The Public Health Act*

Under this Act, local authorities (now county governments) are mandated to carry out extensive assessment of areas under their jurisdiction to ensure public sanitation is at an acceptable standard. This has clear implications for the health of residents, because where sanitation is poor the outbreak of pests and disease will likely be significantly higher. Ensuring public spaces have effective sanitation and public health risks are minimized provides a valuable contribution to overall disaster risk reduction.

16.4.4.4. *Forests Act of 2005*

Section 52 of the forest legislation prohibits lighting of fires, smoking, or carrying inflammable substances in forests. Provisions to prevent deforestation and provide substantial forest cover also mitigate the potential risk of landslides and flooding due to run-off. The Grass Fires Act[45] provides a framework to regulate and issue permits to any person who intends to set fire to any vegetation. It also makes it a criminal offence to negligently set fire to vegetation. These legislative provisions are important because in addition to prohibiting actions that increase greenhouse gas emissions, they play an important disaster risk (fire) reduction role.

16.4.4.5. *The Drought Management Authority Order 2011*

This Order, [46] issued by the President as subsidiary legislation under powers derived from Section 3(1) of the State Corporations Act, effectively created the National Drought Management Authority. The Authority is required to establish mechanisms to ensure that drought does not become famine and the impacts of climate change are sufficiently mitigated. Specifically, the Authority's functions require it to establish, institutionalize, and coordinate structures for drought management; operate an efficient drought early

---

[45]Cap 327 Laws of Kenya.

[46]National Drought Management Authority Order, 2011 Legal Notice no. 171.

warning system; support drought-related policy formulation; coordinate the preparation of risk reduction plans; undertake risk reduction awareness and education; and coordinate the implementation of risk reduction activities. This Authority, albeit recently created, has the potential to play an important role in reducing the disaster risk attributed to drought and famine. However, it will be necessary for the Authority to work with sectoral agencies such as the national and county government ministries that are responsible for environment, water, and agriculture in order to ensure disaster risk reduction plans for drought are mainstreamed into the sectoral operations.

It is worthy to note that this Order embraces the concept of sustainable development of Arid and Semi Arid Lands (ASALs) in Kenya. This is important as the area is vulnerable to disasters, and therefore necessary for the government to take a more preventive approach towards curbing drought and related risks through early warning, preparedness, and enhancing resilience of communities living in ASALs. Under the Order, development funds should be effectively devolved to local levels in ASALs through locally established institutions under the Drought Management Authority.

16.4.4.6. *The Kenya Red Cross Society*

The Kenya Red Cross Society is established by the Kenya Red Cross Act,[47] as a body corporate, recognized as a voluntary aid society. The Red Cross has many objectives, including the provision of relief to victims of catastrophes and disasters of a public nature.[48] Kenya Red Cross, in collaboration with the government of Kenya has prioritized disaster risk reduction (DRR) and disaster mitigation as a key strategy for achieving its long-term strategic objective; and its DRR strategy will contribute to community safety and resilience to disasters through a strengthened capacity of communities to cope with the impacts.[49] The Red Cross continues to play an instrumental role in DRR and disaster management,

[47] Cap 256 Laws of Kenya.

[48] Section 5.

[49] Reference: Retrieved 22 June 2015, from http://www.kenyaredcross.org/index.php?option=com_content&view=article&id=221&Itemid=61.

and has variously been appointed by the government as the lead agency to respond to various disaster emergencies.

The Kenya Red Cross is one of the most active humanitarian agencies in Kenya with regard to prevention and managing disaster risks. In general terms, the major role of this organization is providing relief and rescue operations to disaster victims. In effect, the Kenya Red Cross provides relief, rescue, rehabilitation, reconstruction, and recovery to the affected people especially in the provision of food, housing, and medical care to displaced and affected families.

In order to strengthen the resilience of communities, the Act provides that the organization shall participate in activities aimed at disaster risk reduction, for example, settlement of communities in areas less prone to disasters. Further, the Act encourages partnering with local communities and dissemination of information so as to create awareness.

16.4.4.7. *Kenya National Disaster Operations Centre*

The Kenya National Disaster Operations Centre was established in 1998 in response to challenges arising from the El Niño rains and the US Embassy bombing of August 1998. The Centre is based in the Presidency, and is staffed by officers drawn from various ministries/departments of the government on a 24-hour basis. These include the Ministry of Defence, National Coordination, Health, Ministry of Roads and Public Works, Devolution, and National Youth Service.[50] The Centre performs various functions, including mobilizing national resources to combat rapid onset disasters; coordinating disaster management activities; collaborations with other stakeholders; national level coordination of all disaster management activities; ensuring agencies are informed of the activation of disaster contingency plans; and translating the decisions of the National Disaster Coordination Committee (NDCC) into action.[51]

From the foregoing discussion, it is apparent that Kenya has in place a variety of laws and institutions whose mandates impact some aspects of

[50]Reference: "National Disaster Operation, Kenya Office of the President — Regional Consultation." Retrieved 10 April 2015, from http://www.slideshare.net/SCRpresentations/national-disaster-operation-kenya-office-of-the-president-regional-consultation.

[51]*Ibid.*

disaster risk reduction and management, but there is no direct linkage to climate change. In addition, the legal framework is rather fragmented with little appreciation of the efficiencies brought by mainstreaming of functions.

16.5. Towards a Mainstreaming Approach in Governance, and Opportunities for Climate Change-Induced Disaster Risk Reduction

Since there is a commonality between disaster risk and climate change risks, and a further linkage between disaster risk reduction and adaptation, it is necessary to examine legal options that ensure harmonious responses that promote sustainable development. A useful approach in public policy and administration is implementing the concept of mainstreaming of functions, from planning and budgeting, to implementation and oversight.[52]

16.5.1. *The concept of mainstreaming through horizontal and vertical integration of functions*

Mainstreaming is a new approach that provides the framework to achieve coordinated action on climate change across government functions. Mainstreaming in this sense requires cross-sectoral policy integration that operates both horizontally, by providing an overarching guide for all sectors; and vertically, by requiring all sectors and levels of government to implement climate change responses in their core functions. Mainstreaming provides a helpful tool in framing responses to climate change-induced disaster risk reduction. This is because climate change vulnerability and disasters, as earlier mentioned, bear commonalities — such that a harmonized legal approach will ensure efficiencies in framing, financing, and implementing the sufficient responses. Climate change and disaster risk revolve around vulnerability of the people, economy, and environment, and both require building resilience in order to avoid or manage risks. Therefore,

[52]The mainstreaming approach has already been adopted for climate change legislative and policy documents in Kenya, to which this author was a drafter. The arguments made here are very similar to the arguments on mainstreaming, vertical, and horizontal integration, that are set out in the 2014 Draft Climate Change Policy, fully referenced elsewhere in this chapter.

an assessment of mainstreaming through the lens of horizontal and vertical integration is necessary to evaluate how DRR can be framed.

Scholars Lafferty and Hovden[53] discuss the idea of horizontal integration. They argue that horizontal integration occurs when a central authority has put in place a comprehensive cross-sectoral strategy for integration. The authors suggest the central authority could be a government, a government ministry, or an agency or commission that has been entrusted with an overarching responsibility[54] — for instance, on climate change disaster risk reduction or sustainable development. Often, national laws embrace horizontal integration through institutional arrangements and allocation of mandates. In such a case, horizontal integration tends to establish "authorities", "agencies", or "commissions" with powers to supervise compliance. As a case in point, the Kenyan EMCA set up a National Environmental Management Authority (NEMA) as the principal public agency "to exercise general supervision and coordination over all matters relating to the environment and to be the principal instrument of Government in the implementation of all policies relating to the environment".[55] In further evidence of horizontal integration, NEMA has a statutory mandate[56] to evaluate environmental impact assessment study reports and issue licences for development projects that require prior impact assessment.[57] This function manifests the horizontal integration role because EMCA sets the requirement for impact assessment, and NEMA (as a central authority) executes the environmental impact assessment requirement for other public and private economic activities.

Vertical integration of climate change and DRR, on the other hand, indicates the extent to which a particular governmental sector has adopted and sought to implement climate resilience, and reduction of vulnerability as part and parcel of the core objectives assigned to that sector. This approach is key to completing the sustainability continuum, such that

---

[53]William Lafferty and Eivind Hovden, "Environmental Policy Integration: Towards and Analytical Framework," *Environmental Politics*, 12(3) (2003), 1–22.

[54]Lafferty and Hovden (2003), *op. cit.*, supra note 52 at 14.

[55]Section 9(1).

[56]Section 58–60.

[57]These projects are listed in the Second Schedule to EMCA.

adaptation and risk reduction measures are prescribed through horizontal integration and mainstreamed into the responsibilities of various sectors and county governments. Legal scholar Hans Christian Bugge concurs, noting that institutional integration of sustainability objectives requires facilitating and convincing sectoral authorities to work with objectives that they do not consider or identify as primary tasks.[58]

Since the likelihood of disaster risk tends to encompass most areas of society, economy, and environment, it is important to practice mainstreaming disaster risk reduction. This would include an overall (horizontal integration) system exists, providing a comprehensive national law and institutional arrangements that:

- Identifies the linkages between disaster risk, climate vulnerability, and sustainable development;
- Creates a mandatory obligation on various social, economic, and environmental sectors to integrate DRR into their functions and mandates; and
- Requires incorporation of climate change-induced DRR into sectoral planning, budgeting, and implementation of functions.

Elsewhere in this book, fellow scholars have chronicled how countries in Asia, with experience in facing disasters, have put in place national and regional legal mechanisms for DRR but face difficulties at the point of implementation. This is a key point, and therefore, such a Kenyan regulatory framework must have budgeting provisions such that the Medium Term Expenditure Framework (MTEF), which is a three-year budgeting cycle in the public section, includes disaster risk reduction expenditure associated with each sector, or level of government. This may require amendments to the 2012 Public Finance Management Act. In addition, it is necessary to have compliance mechanisms in place. Penal rules are helpful in this regard, in order to ensure public officers face strict liability for failure to take action as required by law, during a disaster management, or in implementing preventive measures.

---

[58] See, Hans Christian Bugge, "1987–2007: 'Our Common Future' Revisited." In Hans Christian Bugge and Christina Voigt (eds.), *Sustainable Development in International and National Law* (Oslo: Europa Law Publishing, 2007), p. 9.

16.5.2. *Specific areas of governance where law could implement mainstreaming of disaster risk reduction*

As noted above, a mainstreaming approach provides the various coordinating and sectoral agencies of the Kenyan national and county governments with the tools to effectively respond to the complex challenges of climate change risks in order to eliminate or minimize the vulnerability and hazard that escalates adverse impacts to disaster levels. This is mainly because mainstreaming includes the integration of climate change disaster risk reduction policy responses into national, county, and sectoral planning, and budgeting and management processes, providing a full continuum for preventive and disaster management options by each sector, with national oversight and budgeted funding. Horizontal integration, on the one hand, involves providing overarching national guidance, either through a national law, policy, or institutional framework for management of climate change-induced disaster risk. Vertical integration, on the other hand, requires clarity on how the various stakeholders will incorporate and manage climate change-induced disaster risk in their regular functions. Out of necessity, vertical integration of climate change functions and disaster risk management is best undertaken by specialized sector agencies of government, and in the case of Kenya, by local levels of government:

(1) The role of local (county) governments

Kenya has a long history of local government. Until 2013, the country was governed through a unified central government and local authorities that were also controlled by the national executive.[59] As a result of Constitutional change in 2010, a devolved system of government was introduced, taking effect after the first general elections held in March 2013. This system of devolved government means that Kenya has two levels of government: a national government and 47 semi-autonomous county governments. In terms of demarcation of functions, the Constitution is clear that the national government has the responsibility for making policy on disaster management, while county governments have the task of implementing disaster management. This approach endorses the mainstreaming approach. It is an important

[59] See Local Government Act, Cap 265, Laws of Kenya (now repealed).

approach because through the Constitution, the county governments posses the broad range of functions that are necessary for design and implementation of climate change adaptation, and disaster risk reduction interventions necessary to secure sustainable development. These include agriculture, planning, and development control approvals, soil and water conservation, and implementation of public participation in governance. The latter is key because it is the central concept around the development of a culture of public safety for disaster risk management. Therefore, in seeking vertical integration of disaster risk reduction and climate change adaptation, county governments provide a key legal option through which mainstreaming can be implemented through various functions, including the budgeting process. This is supportive of the principle of subsidiarity in governance.

(2) The instrumental function of environmental impact assessment
Environmental impact assessment (EIA), monitoring, and audit are identified by the Constitution as critical for prevention of environmental harmful practices.[60] In practice, the EIA continuum (from EIA to audit), is a crucial legal tool for ensuring and maintaining sustainability, by identifying various risks to environment and to society, and requiring mitigation measures. When implemented properly, environmental audits should ensure complete compliance with the conditions of an EIA licence. The range of projects and activities that must undergo an EIA is quite broad. This means that if climate change vulnerability was made one of the factors that an EIA study considered when examining each project, it would be possible to stop economic activities that pose a risk of disasters, or impose mitigation measures that can be monitored through environmental audits. At a more strategic and planning level, climate change vulnerability risk assessment should be undertaken when strategic environmental assessments (SEA) are undertaken on proposed policies, programs, or legislation. The strategic nature of SEAs is important because it can inform the economic planning and budgeting processes of governments, and put in place DRR contingencies ahead of time.

[60] Article 69(1).

16.6. Conclusion

There is a need to ensure a linkage between climate change vulnerability, disaster risk, and sustainable development is explicitly brought out as a risk to the national development goals of Kenya. This is because the country is prone to the negative impacts of climate change, including disasters that undermine efforts at sustainable development. The conceptual approach of a low-carbon climate resilient development has been endorsed by the national policy in Kenya and by the IPCC. This approach represents the frameworks for sustainable development and adaptation as being at the core of climate resilient development. Adaptation to climate change is very important to Kenya, as it is a path to reduction of vulnerability to adverse impacts of climate change (which if unchecked might escalate to disasters) and an avenue for framing equitable ways of overcoming threats to livelihoods — for sustainable development of people. The pursuit of the mainstreaming approach in governance of disaster risk reduction is therefore important, but the link between climate change, disaster risk, and sustainable development must be appreciated. The mainstreaming approach proposed here comprises vertical and horizontal integration. It provides a route through which sustainable development and climate change can be linked, and the entry of related disasters mitigated or reduced altogether.

Chapter 17

Land Tenure Systems as a Challenge for Disaster Recovery: Adapting to Extreme Weather Events after Typhoon Haiyan

Daniel Fitzpatrick and Caroline Compton†*

With anthropogenic climate change, there is a high probability that intensification of extreme weather events will combine with disaster vulnerability to increase the impacts of natural disasters in the Asia-Pacific region. As a result, adaptation to climate change includes proposals to "build back better" after natural disasters. Yet building back better is not a one-off or *sui generis* technical response to natural hazards as it involves adjustments to human systems that have evolved over long periods of political conflict and environmental stress. This paper argues that there is a link between human systems for managing land and adaptation to extreme weather events in the Philippines. In the Philippines, the government plans to relocate as many as 200,000 people away from the foreshore in the aftermath of Typhoon Haiyan. However, the Philippines government has provided little guidance

*Professor, ANU College of Law, Australian National University; Australian Research Council Future Fellow. Website: https://law.anu.edu.au/staff/daniel-fitzpatrick. Professor Fitzpatrick is grateful for Future Fellowship funding assistance from the Australian Research Council (FT110101065).

†PhD Student, ANU College of Law, Australian National University. Ms. Compton is grateful for funding from Oxfam International to undertake field research in Typhoon Haiyan-affected areas. Oxfam International published a policy brief in August 2014.

on security of land tenure as an element of relocation planning, or as part of recovery efforts for people wishing to remain in their pre-disaster locations. While the current relocation program offers the Philippines the opportunity to provide secure land tenure rights to large numbers of vulnerable coastal dwellers, a failure to provide secure rights will almost inevitably result in households returning to hazardous areas, or continuing to reside in unsafe zones because they have been excluded from the relocation program. Moreover, outside relocation areas, there are risks that farmers and farm workers will miss out on a historic opportunity to obtain secure land tenure through agrarian reform due to the loss of land records as a result of Typhoon Haiyan.

17.1. Introduction

Land issues provide a powerful example of the link between human activity and natural disasters. Population growth has forced increasing numbers of people to settle in areas at risk from cyclones, floods, eruptions, or earthquakes. Global climate change further heightens the vulnerability of many settlements to natural disaster risks. In 2012, approximately 32% of city-dwellers in developing regions lived in slums characterized by informal or illegal land tenures, and poor land use planning. A further 862,000,569 of the world's population lived in areas classified as urban slums.[1] There is little doubt that the vulnerability of a settlement to disaster, and its capacity to recover from a disaster, is closely connected to the quality of systems for land use, governance, and tenure.

Natural disasters are not a necessary result of natural hazards. A natural hazard has disastrous effects when human systems fail to cope with its social, economic, and physical impacts.[2] A key cause of human vulnerability to natural hazards is a poor quality system for governing land.

[1]UN-HABITAT, "Global Urban Indicators Database 2012." Retrieved 30 May 2014, from http://ww2.unhabitat.org/programmes/guo/guo_indicators.asp, accessed 30 May 2014.

[2]UNDP, "Reducing Disaster Risk: A Challenge for Development." Retrieved 10 April 2015, from http://www.undp.org/bcpr. (A natural disaster is "a serious disruption triggered by a natural hazard causing human, material, economic or environmental losses, which exceed the ability of those affected to cope".)

Writing for a UN-HABITAT publication, Fitzpatrick identifies a number of indicators of poor quality land governance that are relevant to the Philippines. These include:

- *Institutional exclusion.* Large numbers of landholders are denied access to formal systems of land tenure, administration, and planning.
- *Informal land rights.* Large numbers of landholders are deemed to hold informal or extralegal rights to land. There are large numbers of informal settlements in urban areas.
- *Unplanned development.* Land use planning is out of touch with actual settlement patterns and inconsistent across different levels of government. There is little community consultation in the process of land use planning.
- *Institutional incapacity.* Government land institutions do not serve the needs of all members of the population. Land officials have a low level of technical skill, and inadequate incentives for efficient, transparent, and accountable land management.
- *Inequality of land ownership.* There is a high degree of inequality in the distribution of land ownership or a history of social conflict over land ownership patterns. Large numbers of people either do not own land or own land that is insufficient for agricultural livelihoods.
- *Inadequate quality and coverage of land data.* Significant amounts of land are not covered by government land information systems. Data in the land information systems are poorly recorded and/or out-of-date. The boundaries between different types of land, including land claimed by the state, are not surveyed or defined with sufficient precision.
- *Inadequate laws and legal institutions relating to land.* Laws relating to the definition, acquisition, and transfer of rights to land bear little relation to the social practices of poor landholders or the landless. The rules for determining rights to land are not clear and subject to excessive forms of discretionary decision-making by government officials.

Around the world, governments are proposing coastal buffer zones and other land use planning measures to respond to increased risks of storm surges and sea level rise. In the Philippines, the government plans to relocate as many as 200,000 people away from the foreshore in the aftermath of Typhoon Haiyan.[3] On average, the Philippines experiences 20 typhoons a year,[4] and discussion of post-Haiyan rehabilitation has included adaptive measures to the likelihood of future Typhoon disasters.[5] Yet, the Philippines government has provided little guidance on security of land tenure as an element of relocation planning, or as part of recovery efforts for people wishing to remain in their pre-disaster locations. While the current relocation program offers the Philippines the opportunity to provide secure land tenure rights to large numbers of vulnerable coastal dwellers, a failure to provide secure rights will almost inevitably result in households returning to hazardous areas, or continuing to reside in unsafe zones because they have been excluded from the relocation program. Moreover, outside relocation areas, there are risks that farmers and farm workers will miss out on a historic opportunity to obtain land tenure security through agrarian reform due to the loss of land records as a result of Typhoon Haiyan.

This paper identifies land tenure as a key challenge to "building back better" after natural disasters. Its core argument is that building back better is not a one-off or *sui generis* technical response to natural hazards. It involves adjustments to human systems that have evolved over long periods of political conflict and environmental stress. Section 17.2 discusses historical sources of land tenure insecurity in the Philippines, including the colonial categorization of large areas of land as state-owned land, the emergence of inequality of rural land ownership as a result of land grants to *hacienda* estates, and the dramatic growth of informal settlements in urban areas. Section 17.3 sets out field data on the impact of Typhoon Haiyan on groups without security of land tenure. These groups include informal urban settlers, coconut and rice farmers, and fisherfolk. Section 17.4 highlights correlations between poverty, land tenure insecurity, and disaster

[3]Aya Lowe, "Typhoon Haiyan Survivors in Tacloban Face Upheaval as City Tries to Rebuild," *The Guardian*, 8 May 2014. Retrieved 8 November 2014, from http://www.theguardian.com/world/2014/may/08/typhoon-haiyan-survivors-tacloban-philippines.

[4]G.P. Yumul *et al.*, "Extreme Weather Events and Related Disasters in the Philippines, 2004–08: A Sign of What Climate Change Will Mean?," *Disasters*, 35(2) (2011), 362–82.

[5]Climate Change Commission of the Philippines, "Typhoon Haiyan: Philippines Prepares Climate Change Plans for Worse to Come." Retrieved 10 April 2015, from http://climate.gov.ph/newsandupdates/typhoon-haiyan-philippines-prepares-climate-change-plans-worse-come.

vulnerability in areas affected by Typhoon Haiyan. It also identifies the ways in which long-standing problems of land governance constrain efforts to build back better after the disaster.

17.2. Historical Sources of Land Tenure Insecurity in the Philippines

The causes of land tenure insecurity in the Philippines date back to processes of colonization under Spain and the United States of America. The Spanish declared sovereign authority of the Philippines as a result of acts of conquest, treaty, and alleged discovery. Law 1, title 12, book 4 of the *Recopilacion de Leyes de las Indias* vested all land "held without proper and true deed of grant" in the Spanish Crown. Except for land reserved for public and common use, and after "distributing to the natives what may be necessary for tillage and pasturage", all land was "free and unencumbered" for disposal by the Crown. Under this "Regalian doctrine", the state became the sole source of proprietary rights to land. Claims to rights based on occupation or use of land could not have a proprietary character without approval and grant by the state. The Supreme Court confirmed the Regalian doctrine in Valenton *v.* Murciano, 3 Phil. 537, 543 [1904]:

> While the State has always recognized the right of the occupant to deed if he proves a possession for a sufficient length of time, yet it has always insisted that he must make that proof before the proper administrative officers, and obtain from them his deed, and until he did that the State remained the absolute owner.

The legal denial of rights based on occupation and use, and the vesting of title in the Spanish Crown, continued by virtue of the "Maura Law" of 1893,[6] which mandated the registration of all agricultural lands. Without registration, agricultural lands reverted to the property of the state.[7]

---

[6] *Maura Law* (1893). The law was named after its author, Antonio Maura y Montañer, the then-Minister of Colonies of Spain.

[7] "... title to all agricultural lands which were capable of adjustment under the Royal decree of ... 1880, but the adjustment of which has not been sought at the time of promulgation of this decree ... will revert to the State. Any claim to such lands by those who might have applied for adjustment of the same but have not done so at the above-mentioned date, will not avail themselves in any way nor at any time." — *Maura Law* (1893), Art 4.

The Spanish administration utilized its claims of sovereign title over land to grant large areas of land to the Catholic Church and to hacienda estates engaged in the plantation sector. Agricultural tenancies became the predominant form of landholding in rural areas as part of a sharecropping system. By the time of transfer of sovereignty to the United States of America in 1898, inequality of landownership had emerged as a major source of social unrest. Yet, the American administration also retained the legal principle that the state was the sole source of proprietary interests in land.[8] The various Constitutions of the Philippines have confirmed the so-called Regalian doctrine of state title to land.[9] The introduction of the Torrens system of land administration in 1903 further diminished the legal status of rights based on occupation and use through its basic requirement that registration of title, rather than contracts with a legal owner, are the sole basis for the acquisition of legal rights to land.

17.2.1. *Land inequality and agrarian land reform*

The 1935 Constitution included a provision reserving to the state the right to regulate the relationship between landlord and tenant,[10] which was a response to continuing tensions over inequality of land ownership, including an uprising in Central Luzon in 1935. Other legal changes during this period also focused on the landlord–tenant relationship. These included the introduction of the Rice Tenancy Law,[11] which mandated for the first time a 50/50 share of crops between landlord and tenant. By 1955, land

---

Those with pending application were given one year to secure title. See P. J. Brosius *et al.*, (eds.), *Communities and Conservation: Histories and Politics of Community-Based Natural Resource Management* (US: Rowman Altamira, 2005).

[8]"[T]he Public Land Act operated on the assumption that title to public lands in the Philippine Islands remained in the government; and that the government's title to public land sprung from the Treaty of Paris and other subsequent treaties between Spain and the United States": see separate opinion of Puno J; *Cruz v. DENR* 347 *SCRA* 128, Sec. II (2000).

[9]See the discussion in separate opinion of Puno J; *Cruz v. DENR* 347 *SCRA* 128, Sec. II (2000).

[10]Constitution (1935), Art. XIV, Sec. 6: "The State shall afford protection to labor, especially to working women, and minors, and shall regulate the relations between the landowner and tenant, and between labor and capital in industry and in agriculture. The State may provide for compulsory arbitration."

[11]Rep. Act no. 4054 (1933).

reform projects were beginning their current fócus on redistribution,[12] which has continued in various guises today. The 1987 Constitution includes a statement of the social function of land,[13] and requires the state to implement a land redistribution program "founded on the right of farmers and regular farmworkers who are landless, to own directly or collectively the lands they till".[14]

The progress of land reform has been slow. The basic causes of delay include shortfalls of political will and technical expertise, landlord resistance to redistribution of both land and crop income, and a corresponding lack of resources on the part of farmer beneficiaries.[15] There are also structural issues such as the disconnection between legal concepts of tenancy, and the nature of social relations between farmworkers and land owners. For example, it takes time and expertise for Department of Agrarian Reform staff to translate, both culturally and linguistically, legal concepts that may bear little relation to the lived experience of tenant farmers.[16]

17.2.2. *The growth of informal urban settlements*

Since independence, the Philippines has experienced rapid rates of population growth and rural–urban migration. Between 1948 and 2010, the population increased from 19.2 to 92.34 million people.[17] In the mid-1970s, some 36% of the population lived in urban areas.[18] By 2010, 66.4% of the

[12]Rep. Act no. 1400 (1955).

[13]"The use of property bears a social function, and all economic agents shall contribute to the common good" (Constitution, Art. xii. Sec. 6).

[14]"The State shall, by law, undertake an agrarian reform program founded on the right of farmers and regular farmworkers who are landless, to own directly or collectively the lands they till or, in the case of other farmworkers, to receive a just share of the fruits thereof" (Constitution, Art. viii, Sec. 4).

[15]F. Hirtz, "The Discourse that Silences: Beneficiaries' Ambivalence towards Redistributive Land Reform in the Philippines," *Development and Change*, 29 (1988), 247.

[16]*Ibid.*

[17]Philippines Statistics Authority, "Population." Retrieved 22 May 2014, from http://www.nscb.gov.ph/secstat/d_popn.asp.

[18]Agnes R. Quisumbing and Scott McNiven, "Migration and the Rural-Urban Continuum: Evidence from the Rural Philippines," FCND Discussion Paper 197, 2005.

population resided in urban areas,[19] with Manila and the National Capital Region growing from 5.9 million residents in 1980 to 11.9 million residents by 2010.[20] This rapid process of urbanization has been accompanied by the growth of informal settlements. By 2007, it was estimated that over 550,000 households had informal tenure arrangements.[21] The inhabitants of these settlements lack security of land tenure primarily because their land is classified as either owned by private individuals or the state. The land was vacant at the time of informal settlement either because it was designated as unsafe for residential development (state land) or has been set aside by land speculators for future investment (private land). There are also substantial costs associated with obtaining title to land, paying land tax as an owner, and recording the purchaser or lease of land in land title records.

17.2.3. *Laws on security of land tenure*

The political pressures of land inequality and informal urban settlements have led to a number of legislative measures and programs to promote security of land tenure. These include the Urban Development Housing Act (UDHA), the Comprehensive Agrarian Reform Law, and the Fisheries Code.

The Urban Development and Housing Act of 1992 (UDHA) provides options for "underprivileged and homeless" citizens in urban areas to improve their security of land tenure through subsidised housing programs. The options include freehold ownership, usufruct, community ownership, leases, and options to purchase. UDHA does not apply to lands covered by

---

[19] Philippines Statistics Authority, "Urban Population of the Philippines." Retrieved 28 May 2014, from http://www.tradingeconomics.com/philippines/urban-population-percent-of-total-wb-data.html. Note that current World Bank and Philippines National Statistics Authority currently record an urban population of 49.12% for 2012, including a drop in the proportion of the population that is urban of nearly 20% between January 2010 and January 2011.

[20] Philippines Statistics Authority, "Population." Retrieved 22 May 2014, from http://www.nscb.gov.ph/secstat/d_popn.asp.

[21] J. Cruz, "Estimating Informal Settlers in the Philippines," 11th National Convention on Statistics, 2010.

the Comprehensive Agrarian Reform Law.[22] The Comprehensive Agrarian Reform Law of 1988 established the Comprehensive Agrarian Reform Program (CARP) to redistribute agricultural land from large landowners to landless farmers and farm workers.[23] The process involves a Notice of Coverage procedure that initiates compulsory acquisition of land for redistribution. Agrarian reform beneficiaries receive ownership of the land subject to a lien to secure payment for the land, and restrictions on sale or transfer of the land for a period of 10 years.[24] The Fisheries Code requires the Department of Agriculture to create settlement areas on public land near to fishing grounds for fisherfolk. The Department of Agriculture must coordinate with Local Government Units to identify areas of public domain for the settlements. Ownership of the land in the settlement area is to be vested in the fisherfolk.[25]

These laws establish mechanisms to improve security of tenure for the poor in the Philippines. However, their implementation has been challenged by a high degree of fragmentation of land administration and planning functions across multiple government agencies. A key agency is the Lands Management Bureau, situated within the Department of Environment and Natural Resources, which is responsible for the survey, management, and disposal of alienable and disposable lands.[26] Once forest land is delineated, the remaining alienable and disposable land is then classified according to its intended use. Use types include four categories of land:

(1) Agricultural
(2) Commercial, industrial, or for similar productive purposes.
(3) Educational, charitable, and other similar purposes.
(4) Reservations for town sites, and for public and quasi-public uses.[27]

---

[22]Rep. Act no. 7279 (1992), Sec. 5(a).

[23]Rep. Act no. 9700 (2009), Secs. 3(c) and 4.

[24]CARP applies to land devoted to agricultural activity, including public domain land suitable for agriculture, but not to land classified as mineral, forest, residential, commercial, or industrial: Rep. Act no. 9700 (2009), Secs. 24, 26, 27.

[25]Rep. Act no. 8550 (1998), Sec. 108.

[26]Exec. Order no. 192 (1987) 83 O.G. 3020, Sec. 20.

[27]Com. Act no. 2874 of 1936, Sec. 9.

The classification of land is significant as it determines the type of tenure that may apply to the land. Table 1 illustrates the problem of fragmentation of governance and tenure arrangements in the Philippines land sector.

Table 1: Duplication and Overlaps in the Philippines Land Sector

Duplication/Overlap	Agencies Involved	Source of Duplication/Overlap
Primary classification of public land as A&D	DENR/ NAMRIANCIP	Conflict between responsibilities as determined in EO 292 of July 1987 and the Indigenous Peoples Rights Act.
Undertaking land surveys for titling purposes	DENT (LMB); DAR Potentially NCIP	Administrative code provides for both DENR and DAR to undertake surveys. IPRA 1987 gives NCIP responsibility for the identification, *delimitation*, and recognition of ancestral domains.
Approval of subdivision survey for titling purposes	LMB; LRA	Property registration degree (PD 1539 of 1978 as amended) permits either LMB or LRA to approve such plans.
Award of original private rights on A&D land	DENR (patents); DAR (CLOAs); Courts (decrees); NCIP (CADTs)	Two titling processes (administrative and judicial) both mandated by law. Legislation authorising different forms of ownership rights in land by administrative processes.
Maintenance of independent, uncorrelated versions of cadastral maps/records	DENL (LMB); LRA	A consequence to the two agencies involved in the titling process. The practice is neither mandated or required by law.
Compilation of land maps and information	Multiple agencies	A reflection of differing agency needs for land information. Some necessary overlap occurs.
Land valuation and related mapping for tax purposes	BIR; Local government units	Different valuation methods mandated by different property taxation laws.

Source: N. Bangsal, "Land Administration System: Functional and Efficiency Implications," Philippines Congress Policy Advisory No. 2008–05, 2008.

A major governance challenge concerns the central role of Local Government Units (LGUs) in land administration and land use planning. LGUs have statutory responsibility for:

- Compulsory land acquisition for public purposes.[28]
- Land use planning, including the zoning of land.[29]
- Resettlement of persons living in unsafe areas.[30]
- Provision of land for "underprivileged and homeless citizens" subject to court orders for eviction.[31]

A number of studies conclude that LGUs do not have sufficient resources or expertise to undertake these tasks, particularly in the aftermath of Typhoon Haiyan.

17.3. The Impacts of Typhoon Haiyan

On 8 November 2013, Typhoon Haiyan (locally known as Yolanda) hit the Central Philippines.[32] In March 2014, Philippine National Disaster Risk Reduction and Management Council (NDRRMC) reported the following estimates of the effects of the disaster.

- Total families affected: 3,424,593 (16.8 million persons)
- Total deaths reported: 6,201
- Persons reported injured: 28,626
- People unaccounted for: 1,785
- Total families displaced: 890,895 (4.1 million persons)
- Families still housed in evacuation centers: 20,924 (101, 527 persons)
- Displaced families with host families/other housing alternatives: 869,971 (4 million persons)
- Total houses damaged: 1,140,332

---

[28] Rep. Act No. 7160 (1991), Sec. 19.
[29] *Ibid.*, Sec. 20(c).
[30] Rep. Act No. 7279 (1992), Sec. 28(8).
[31] Rep. Act No. 7279 (1992), Sec. 29.
[32] M. Weaver and Fishwick C, "Typhoon Haiyain Hits Philippines — As It Happened," *The Guardian*, 8 November 2013. Retreived 8 November 2014, from http://www.theguardian.com/world/2013/nov/08/typhoon-haiyan-hits-philippines-live-updates.

- Houses completely damaged: 550,928
- Houses partially damaged: 589,404.[33]

In December 2013, the United Nations undertook a multicluster needs assessment (Phase II). The needs assessment provided data on community services, markets, livelihoods, food security, health, water and sanitation, nutrition, and protection. It noted that 49% of people living in evacuation centers were in the poorest quartile of the population.[34]

17.3.1. *Land tenure in Haiyan-affected areas*

The regions affected by Typhoon Haiyan are among the poorest areas of the Philippines. Region VIII (covering Leyte and Eastern Samar) experienced the largest rise in poverty rates among farmers, migrants, the unemployed, and individuals residing in urban areas between 2006 and 2009.[35] The overall poverty rate in the region increased from 41.5% in 2006 to 45.2% to 2012 (compared to declining rates of 26.6% in 2006 and 25.2% in 2012 for the Philippines generally).[36] Region VIII also experienced a rise in the poverty rates of women, from 33.7% in 2006 to 37.4% in 2012 (compared to the decline from 21% in 2006 to 19.7% in 2012 for the Philippines generally).[37]

There are large numbers of renters and informal settlers in the Philippines. According to the 2007 Census, there were 62,187 renter households, and 11,462 informal settler households out of a total of 804,991 households in Region VIII alone. A further 251,480 households lived

---

[33]Government of the Philippines, NDRRMC Update Sitrep No. 106 Effects of Typhoon "Yolanda" (Haiyan), 6 March 2014.

[34]See "Multi-Cluster Needs Assessment (Phase 2)." Retrieved 10 April 2015, from https://philippines.humanitarianresponse.info/document/multi-cluster-needs-assessment-phase-2.

[35]Philippines Statistics Authority, "Fishermen Still Poorest Sector in 2009." Retrieved 14 April 2014, from http://www.nscb.gov.ph/pressreleases/2012/PR-201206-SS2-01_pov2009.asp.

[36]Philippines Statistics Authority, "Poverty Series and Charts." Retrieved 12 April 2014, from http://www.nscb.gov.ph/poverty/dataCharts.asp. Table 2: Annual Per Capita Poverty Threshold, Poverty Incidence and Magnitude of Poor Population, by Region and Province: 1991, 2006, 2009, and 2012.

[37]Philippines Statistics Authority, "Poverty Series and Charts," *op. cit.*

rent-free on land with the consent of the landowner.[38] In total, therefore, 32% of the population of Region VIII fell into the category of "landless". While official numbers are not known, many of these informal settlers live in urban areas such as Tacloban city.

There are correlations between rights to land and rural poverty in the Philippines. For example in Eastern Samar, where there are approximately 6,300 coconut farmers, it is estimated that 59.4% of the population live in poverty, whereas over 50% of coconut farmers lack secure rights to land.[39] Only 2% of land in Eastern Samar covered by the Comprehensive Agrarian Reform Program Extension with Reforms [CARPER] has been distributed to agrarian reform beneficiaries.[40] Across Region VIII generally, the proportion of farmers classified as poor increased from 37.3% in 2003 to 46.7% in 2009 (compared to declining from 37.2% in 2006 to 36.7% in 2009 for the Philippines generally).[41]

The relationship between poverty and tenure status applies to fisherfolk as well as farmers. Most fishing settlements are situated on land classified as public foreshore land, and are subject to the threat of eviction. According to a survey prepared by the Philippines Partnership for the Development of Human Resources in Rural Areas, 87% of coastal respondents identified fisherfolk in their communities who were informal settlers, and 69.6% identified fisherfolk who were at risk of eviction.[42] In 2009, the poverty rates for fisherfolk were estimated at 41.4% compared to a national poverty

[38]Table 7: Households by Tenure Status of the Lot and Region: 2007. Retrieved 14 April 2014, from http://www.census.gov.ph/sites/default/files/attachments/Philippines_Table%25207.pdf.

[39]Food Security and Agriculture Cluster Philippines, "Rebuilding Coconut Production after Typhoon Yolanda," 2013 quoting Department of Agrarian Reform 2013 figures on Published, Served and Registered Notices of Coverage.

[40]The largest portion of undistributed land is used for coconut farming (30% or 262,524 ha). There are similar figures for rural areas of Leyte.

[41]Philippines Statistics Authority, "Poverty Series and Charts." Retrieved 14 April 2014, from http://www.nscb.gov.ph/poverty/dataCharts.asp. Table 7a: Poverty Incidence for Farmers, by Region: 2003, 2006, and 2009.

[42]NGOs for Fisheries Reform, Letter to Hon. Jeffrey Ferrer, Committee Chairperson, Special Committee on Land Use, House of Representatives, 9 February 2011.

rate of 26.5%.[43] In Region VIII, 29% of fisherfolk were classified as poor in 2006, rising rapidly to 45.7% in 2009 (compared to 35% in 2006 and 41.4% in 2009 for fisherfolk in the Philippines generally).[44]

Even for those that may claim ownership of land, there were shortfalls in accurate documentation before Typhoon Haiyan. Large numbers of landowners do not have documents of title to land in the Philippines.[45] Our field consultation suggests that the numbers of Haiyan-affected landowners with land title documents were relatively few in comparison with groups without land title documents. For example, in Barangay 7, Quinapundon, Eastern Samar, informants described a range of land rights, including purchase, inheritance, and processes of borrowing land. While some residents had land tax documents, not one had formal certificates of ownership. The Mayor of Quinapundon described a common situation where land documentation such as tax receipts remained in the names of parents of current occupiers. The Mayor indicated that the lack of formal land title documents was a result of errors in the cadastral maps of the local government area. The Local Government Unit did not wish to distribute formal title documents for fear of community and social unrest, and because the costs of having land remapped were prohibitive.

17.4. The Effects of the Disaster on Land Tenure Security

The following section sets out fieldwork findings on the impacts of Typhoon Haiyan on groups with insecure forms of land tenure. The topics covered are (1) increased risks of evictions, (2) access to shelter assistance for informal settlers, (3) delays to agrarian reform, and (4) inappropriate site selection and planning for relocation. The data illustrate the relationship

[43]Philippines Statistics Authority, "Fishermen Still Poorest Sector in 2009." Retrieved 14 April 2014, from http://www.nscb.gov.ph/pressreleases/2012/PR-201206-SS2-01_pov2009.asp.

[44]Philippines Statistics Authority, "Poverty Series and Charts," *op. cit.*

[45]In 2008, the Department of Environment and Natural Resources (DENR) estimated that 69% of all land available for titling in the Philippines had been titled (47% through administrative titling and 22% through judicial titling). Department of Natural Resources and the Environment, "Compendium of Basic Environment and Natural Resources Statistics for Operations and Management (Second Edition) (2002–2008)." Retrieved 10 April 2015, from http://www.denr.gov.ph/images/stories/combined_compendium_enr_stat_ 2000_2008.pdf.

between land tenure insecurity and vulnerability to future disasters. The findings are based on Oxfam-funded fieldwork undertaken in February 2014, involving interviews in Cebu — Bantayan Island (Santa Fe, Bantayan and Madridejos), Leyte — Tacloban City, Tanauan and Palo, and Eastern Samar — Guiuan and Quinapondan.[46]

17.4.1. *Evictions of the poor and vulnerable*

Our interviews provided evidence that people who had insecure forms of land tenure prior to the disaster are most likely to face the threat of forced eviction after the disaster. For example, in Guiuan, Eastern Samar, officials from the Department of Agrarian Reform stated that the typhoon has provided further opportunities to evict farmers from land so it can be redeveloped for tourism. There is a powerful economic incentive for landowners to evict coconut farmers as it takes 10 years for a coconut tree to become productive. It is estimated that up to 90% of coconut trees were lost in typhoon affected areas. One of the basis for evictions from agricultural tenancies is an argument that there are legal obligations on tenants to cultivate the land.

Cases of evictions are connected to prohibitions on return to the land by displaced persons. In Tacloban City and Tanuan, Leyte, Eastern Samar, and Bantayan there are cases where informal settlers were prohibited from returning to their former homes. In Tanuan, Leyte, the prevention of return by informal settlers is connected to uncertainty as to the status of unsafe zones. In Bantayan, Cebu, there are people who have lived on the land for 25 years that are now unable to return because the Local Government Unit recognises another person as the "owner" due to their payment of real estate tax, even though the land is classified as a public foreshore land. In all areas there are examples of landowners preventing informal settlers from reconstruction of permanent structures, including essential toilet facilities.

The Philippines Constitution requires that evictions take place "in accordance with law and in a just and humane manner".[47] The Urban Housing and Development Act (UDHA) provides further safeguards against evictions. It requires that evictions be limited to circumstances where

[46]Leyte and Eastern Samar are within Region VIII, Bantayan Island is in Region VII.
[47]Const., Art. VIII, Sec. 10.

informal dwellings are sited on hazardous or dangerous land, and there is a court order for the demolition of dwellings. The UDHA also requires a 30-day notice of eviction, a period of consultation, and provision of an alternative site for residence. The Commission on Human Rights of the Philippines states that the mandatory anti-eviction safeguards of the UDHA apply to evictions arising from proposals for relocation away from unsafe zones after Typhoon Haiyan.[48] However, the UDHA only applies to protect "underprivileged and homeless citizens" that are eligible for socialized housing programs in urban or urbanizable areas. The UDHA does not apply to agricultural areas subject to the Comprehensive Agrarian Reform Program. While the UDHA provides a mechanism to secure land tenure for those at risk of post-disaster eviction, there is no current evidence of its application in Haiyan-affected areas. In order for the UDHA to apply, there is a need to link post-disaster housing programs managed by the Department of Social Welfare and Development (DSWD) with the mechanisms for socialised housing programs under the National Housing Authority.

17.4.2. *Access to shelter for informal settlers*

As yet there are no specific shelter guidelines for post-Haiyan recovery and rehabilitation. The DSWD Omnibus Guidelines on Shelter Assistance after natural disasters provide the basic framework for shelter assistance in Haiyan-affected areas. The guidelines establish eligibility criteria based on poverty in relation to DSWD permanent housing programs (Core Shelter Assistance [CSA] and Modified Shelter Assistance [MSA]). However, the guidelines have the potential to exclude renters and informal settlers from CSA and MSA programs because they require that beneficiaries have a guarantee of ownership or permanent or long-term occupation of at least 10 years of the land to be used for shelter. This requires Local Government Units to provide guarantees of tenure security for renters and informal settlers in order to ensure their eligibility for shelter assistance. In other words, there is a policy sequencing challenge involving multiple government agencies as, without priority measures to guarantee ownership

---

[48]Commission on Human Rights of the Philippines, "Human Rights Standards on Housing, Land and Property Rights of Populations Affected by Typhoon Yolanda," 40 (CHR-A2014-001, 27 February 2014).

or long-term occupation of land, renters, and squatters will not be eligible for permanent shelter assistance from DSWD.

Without permanent shelter assistance, it is likely that renters and informal settlers will either form residual caseloads of the homeless, or return to build ad hoc shelters on unsafe land. In the Philippines, a return to unsafe lands occurred after Typhoon Ketsana in September 2009[49] and Typhoon Bopha in December 2012.[50] In Indonesia after the 2004 Indian Ocean tsunami, temporary shelter arrangements could not be decommissioned as there was an initial failure to provide access to housing programs for renters and informal settlers. Without housing programs targeted at renters and informal settlers, these groups formed residual caseloads that prevented transitions from humanitarian response to early recovery.

The debate over relocation after Typhoon Haiyan has contributed to uncertainty and delay in the development of shelter policy. A major source of uncertainty is eligibility for permanent shelter in a context of relocation. In interviews, LGU officials suggested a number of possible options for determining eligibility for permanent shelter in sites for relocation, including:

- That renters would be ineligible as "only owners have lost something, renters can just rent somewhere else";
- That informal settlers who were been living outside the danger zone, whose houses had been destroyed and who were being refused permission to rebuild, would be ineligible; and
- That anyone eligible for emergency accommodation (in bunkhouses) that refused to take up said offer would be required to waive their right to all future shelter assistance, including permanent shelter.

---

[49]Purple Romero, "Between Typhoons and Relocations, A Life Defined by outside Forces," *Informal City Dialogues*. Retrieved 10 April 2014, from http://nextcity.org/informalcity/entry/between-typhoons-and-relocations-a-life-defined-by-outside-forces; "Philippines: One Year after Ketsana, Challenges Remain," *IRIN News*. Retrieved 10 April 2014, from http://www.irinnews.org/report/90573/philippines-one-year-after-ketsana-challenges-remain.

[50]Tessa Kelly, "Legal and Regulatory Issues: Typhoon Bopha," March 2013. Retrieved 10 April 2015, from http://www.shetercentre.org/sites/default/files/regulatory_issues_in_60pha_response.pdf.

In Tacloban City, the eligibility criteria for relocation will likely include informal settlers along the foreshore. However, there are reports from municipal staff that renters from unsafe zones will potentially not be eligible for relocation. In Tanauan, Leyte there are reports that informal settlers will not be eligible for relocation, even though they have not been allowed to return to their land. In Bantayan, Cebu it seems that informal settlers will be included in relocation measures.

In Tacloban City, the urban context means there may not be enough land available to relocate as many as 30,000 inhabitants of foreshore settlements. At the time the research was conducted, only enough land to resettle 10,000 people had been identified. In Eastern Samar, government officials indicated that there was no public land available for relocation, and that funding was required to acquire private land. At the same time, there are reports of substantial increases in the price of land in Guiuan. All these cases indicate widespread uncertainty as to where permanent shelters will be located.

17.4.3. *Delays in agrarian reform*

The physical effect of Typhoon Haiyan on the administration of the Philippines Agrarian Reform Program is significant. The loss of documents in the Leyte Provincial Agrarian Reform and Department of Environment and Natural Resources offices included (1) three boxes of "Notice of Coverage" documents that were ready to be redistributed to landowners, and (2) documentation supporting the certificates of coverage that had been issued prior to the typhoon. The Leyte Provincial Agrarian Reform Programme Officer indicated that the loss of documentation meant that it would be impossible for the office to meet the 30 June 2014 deadline established under the extended Comprehensive Agrarian Reform Program. The loss of documentation also had the potential to leave already distributed certificates vulnerable to legal challenge. Restoration of the land reform process will require close coordination with the Department of Environment and Natural Resources (for the resumption of cadastral mapping activities) and the Register of Deeds (for the reconstitution of lost land titles). This will be a challenge for Leyte as the provincial Registry of Deeds has been destroyed and there has been substantial damage to the cadastral maps and other records held by the Department of Environment and Natural Resources.

In addition to Notices of Coverage, the Department of Agrarian Reform is mandated to improve the terms and conditions of leases for agricultural land that is under the minimum size limit for compulsory acquisition, or that has not yet been acquired through Notice of Coverage procedures. In 2014, the Department of Agrarian Reform issued Administrative Order no. 2 on tenanted coconut lands affected by natural disasters or calamities. The Order provides that no tenant shall be ejected or dispossessed as a result of non-payment of rent due to a natural disaster or calamity, except where the dispossession is authorised by an appropriate body or court. The challenge is to ensure that coconut farmers are now aware of Administrative Order no. 2, and that its safeguards against forced eviction are enforced against landowners.

Typhoon Haiyan also caused extensive damage to irrigation systems in a number of rice-producing areas. In one set of interviews, tenant farmers and a departmental staff member in Quinapundon indicated that some rice farmers were required to make a payment of a designated amount of rice irrespective of the success of the harvest, rather than a percentage of the crop as mandated by law. These leases had an annual term, and allowed for the eviction of tenant farmers following one late rent payment. The Agrarian Reform staff member said that it was not the department's role to renegotiate leases, only to adjudicate those leases that were already in place and any role in enforcing the legal standards for agricultural leases would "create . . . trouble" within the community. The farming community in question is located within an area that floods regularly. The loss of livelihoods, combined with evictions from land, provided examples where land tenure insecurity before the disaster contributes to vulnerability after the disaster.

17.4.4. *Rights to suitable land for fishing communities*

Many subsistence fishing communities in the Philippines are located on foreshore areas, leaving them particularly vulnerable to storm surges and sea level rise. Many such communities in Haiyan affected areas were devastated by the typhoon and following storm surge, as they lack a buffer between their dwellings and the sea. Any relocation of subsistence fishing communities must provide for safe land in an area that continues to facilitate their access to the sea. As noted, the Fisheries Code is the appropriate mechanism to ensure fishing communities access to livelihoods and tenure security in relocation sites. The Code requires the Department of Agriculture to create settlement areas on public land, near to fishing grounds, for fisherfolk.[51] Yet, historically there has been little or no implementation of this relocation provision. The Philippines House of Representatives Committee on Aquaculture and Fisheries has conducted an investigation into the failure to implement this aspect of the Fisheries Code. The Committee concluded that the failure to establish fisherfolk settlement areas was due to lack of (1) capacity in the Department of Agriculture, (2) coordination among government agencies, (3) legal authority in the Department of Agriculture with respect to public domain land, and (4) rules and regulations to implement Section 108. A further obstacle was the authority of Local Government Units over land zoning.[52] The House of Representatives has issued a Resolution (no. 2480) requesting that the President establish a Task Force on Fisherfolk Settlement.[53]

17.4.5. *Land use planning and the security of sites for relocation*

The debate over relocation after Typhoon Haiyan has placed a spotlight on the national issue of land use planning. Land use planning is not only a question of geo-hazard mapping of safe and unsafe land. Effective land use planning provides a mechanism to protect against forced evictions and to improve tenure security for the poor and vulnerable. For example, the zoning of land for agricultural purposes provides protection against evictions for commercial developments. Effective land use planning also informs affected persons where safe settlements should be placed, and allows community participation in site selection and planning. Community engagement in site selection increases investment in the relocation process, and decreases the likelihood that people will move back to sites designated

---

[51] Rep. Act 8550 (1998), Sec. 108.

[52] AAMBIS-Owa Party List, "Solons Urge Creation of Settlement Areas for Displaced Fisherfolk Families." Retrieved 10 April 2014, from https://www.facebook.com/notes/aambis-owa-party-list/solons-urge-creation-of-settlement-areas-for-displaced-fisherfolk-families/10151057169966980.

[53] Gary B. Lao, "Benhur Bats for Fisher Folk Settlement," *The Freeman*. Retrieved 15 April 2014, from http://www.philstar.com/cebu-news/770167/benhur-bats-fisher-folk-settlement.

"high risk". Land use plans that ensure both tenure security and livelihoods create an economic incentive for relocated individuals to remain in their relocated settlements.

Site selection for relocation provides a Haiyan-related example where land use planning would reduce the risk of future disaster. In Tacloban City, one identified site contains a large garbage dump, which cannot currently be in-filled for logistical reasons. Preparation of the site is going on around the dump. The Director of Agrarian Reform for Region VIII indicated that the department had been approached to provide sites for use as relocation sites, and that one site being considered for relocating fishing communities in Palo was 8–10 km from the coast, with hills between the land and coastline. The site also held the area jail. Previous disaster experience in the Philippines shows that people are more likely to return to unsafe lands where suitable sites for relocation are selected (see the examples of Typhoons Ketsana and Bopha above).

Effective land use planning provides a long-term structural response to the recurring challenges of land tenure security after Typhoon disasters in the Philippines. As large scale relocations are likely to be an ongoing feature of post-disaster rehabilitation programs in the Philippines, there is a need to ensure the necessary legislative framework is in place to ensure such relocations consistently to safe, and appropriate sites. House Bill (HB) no. 478 or the Zoning and Land Use Policy Act (ZLUPA) before the 16th Congress of Parliament provides a means to bring greater coherence and community participation to disaster risk reduction in the Philippines.

17.5. Conclusion

The problem of land tenure insecurity after Typhoon Haiyan illustrates the challenges of adaptation to extreme weather events. Insecure land tenure is a primary cause of poverty and disaster vulnerability in the Philippines. Yet there are no current mechanisms to integrate tenure security into recovery planning after Typhoon Haiyan. Without a focus on tenure security there are risks that post-Haiyan recovery will not "build back better", and thereby enhance adaptation to extreme weather events, for poor and vulnerable groups that were most affected by the disaster. The Philippines legal framework includes mechanisms to provide tenure security for the poor and vulnerable, including renters, fisherfolk, and informal settlers. However, the law has not been effective due to the embedded socio-historical causes of land tenure insecurity in the Philippines, and the excessive fragmentation of land administration functions across agencies and levels of government. All these circumstances highlight the challenging relationship between climate change adaptation and adjustments to human systems such as the governance of land.

Chapter 18

The Role of ASEAN in Disaster Management: Legal Frameworks and Case Study of Typhoon Haiyan/Yolanda

Koh Kheng-Lian and Rose-Liza Eisma Osorio

18.1. Introduction

Typhoon Haiyan, also known as Typhoon Yolanda, devastated portions of central Philippines on 8 November 2013. It was the strongest typhoon ever recorded to make landfall (hereinafter referred to as 'Haiyan').[1] Of the seven worst hit provinces in the Philippines,[2] the most destruction was recorded in Leyte and Eastern Samar which have a combined population of 2.3 million.[3] The deadly storm, with sustained winds of 315 kilometers (195 miles) per hour and gusts of up to 378 kilometers (235 miles) per hour and storm surge of 7.5 meters (24.6 feet) high, killed more than 6,000 people.[4] According to the figures provided by the Philippine government, the United Nations Office for the Coordination of Humanitarian Affairs (OCHA) reported that 14.1 million people were affected, with more than 4.1 million displaced.[5]

---

[1] Oxfam Report: "A month after Yolanda, What Challenges Remain, What Needs to Be Done," 7 December 2013. Retrieved 17 July 2014, from http://interaksyon.com.

[2] Presidential Proclamation no. 682, dated 11 November 2013, declared a State of National Calamity, affecting seven major provinces in Central Philippines.

[3] Philippines: National Statistics Office (2010).

[4] OCHA & UNEP, "Typhoon Haiyan (Yolanda) Philippines Environmental Situational Overview," 14 January 2014.

[5] *Ibid.*

The Philippines is a member of ASEAN (Association of South East Asian Nations). This chapter considers the interactive role of ASEAN in disaster management and that of the Philippines.

To understand ASEAN's role in the Haiyan context, it is necessary to consider ASEAN at two levels: (1) ASEAN at the sub-regional level; (2) ASEAN at the national level — with its 10 members countries (AMCs), namely, Brunei Darussalam, Cambodia, Indonesia, Lao PDR, Malaysia, Myanmar, Philippines, Singapore, Thailand, and Vietnam.

It is also necessary to survey the ASEAN legal frameworks on disaster management and emergency response, its cooperation with the United Nations (UN), the AMCs, and other organizations (Section 18.2 and Appendix I), *infra*.

ASEAN (and the world) *had* awakened to the fact that many disasters in Asia (e.g., the Indian Ocean Tsunami 2004), need a 'whole-of-society' if not 'whole-of-the-world' approach. Indeed, the ASEAN Agreement on Disaster Management and Emergency Response, 2005 (AADMER)[6] was formulated after the 2004 tsunami hit a number of AMCs causing untold damage, e.g., Aceh (Indonesia) and Thailand. ASEAN was then 'non-existent' at the scene as it had only a very rudimentary soft law disaster framework, e.g., the ASEAN Declaration on Mutual Assistance on Natural Disasters 1976,[7] which only dealt with disaster warnings, exchange of information, and dissemination of medical supplies. The Indian Ocean Tsunami in 2004 was a clarion call at the highest level of ASEAN to draft AADMER. The tsunami also inspired the World Conference on Disaster Reduction in 2005[8] in Hyogo, Japan, which adopted the globally applicable Hyogo Declaration and the Hyogo Framework for Action, 2005–2015.[9] The Hyogo Framework stressed the need to build resilience, and strengthen and develop coordinated regional approaches for effective disaster response.

[6] Reproduced in Koh Kheng-Lian (compiler and ed.), *ASEAN Environmental Law, Policy and Governance, Selected Documents*, volume II (Singapore: World Scientific, 2013), p. 901.

[7] Reproduced in Koh (2013), *op. cit.*, p. 895.

[8] "Brief History of the WCDR Process," World Conference on Disaster Reduction, Hyogo, Japan, 18–22 January 2005. Retrieved 14 April 2015, from http://www.unisdr.org/2005/wcdr/wcdr-index.htm.

[9] "Hyogo Framework for Action (HFA)," The United Nations Office for Disaster Risk Reduction. Retrieved 14 April 2015, from http://www.unisdr.org/we/coordinate/hfa.

The kinds of disaster risks are numerous. Just to mention some direct damage and impacts: loss of lives, property, and livelihoods, and infrastructure destruction. As to 'indirect' effects, these include pollution, diseases, food and water shortages, looting, raping, etc. What kinds of emergency response are needed? What governance system should be put in place, what kinds of responsibilities arise from donors and parties in terms of, for example, relief supplies where there are delays due to customs clearance or simply an overwhelming amount of goods that choke up procedures causing delays — who pays for demurrage — the donor or the disaster party? What are the issues in the aftermath of the disaster, i.e., rehabilitation, recovery, and reconstruction can be long-drawn and bring about a different set of problems, not to mention funding. These are some of a litany of woes of disaster management that can confront ASEAN and/or donors.

Haiyan provides an opportunity to (1) examine the relevant ASEAN and other relevant frameworks on disaster management; (2) analyze the efficacy of these instruments and their implementation at the national and regional levels; (3) examine the comments made on ASEAN's role in the Haiyan context; (4) identify the adequacy of the laws, both at the regional and national levels; and (5) consider the lessons learned.

We will give a brief survey of the relevant instruments in order to better understand ASEAN's role. One of the joint writers, Rose-Liza Eisma Osorio, brings with her on-the-ground experience, as she was and still is involved in the recovery process of Haiyan. Her case study is reflected in Section 18.5.

18.2. ASEAN Legal Frameworks on ASEAN Disaster Management

ASEAN's role must be viewed holistically at two levels: (1) its regional level as a sub-regional organization. This includes its coordination with international and other organizations such as the United Nations, APEC (Asia-Pacific Economic Cooperation), ARF (ASEAN Regional Forum) (see Appendix I); (2) the national level, where ASEAN Member Countries (AMCs) are charged with the responsibility for the implementation of ASEAN instruments, in this case study, the Philippines.

The core mechanism of ASEAN's legal frameworks on disaster management is AADMER together with other legal frameworks set forth below. AADMER came into force on 24 December 2009. Hence, ASEAN's experience in disaster management is relatively recent.

The scope of "disaster management in AADMER covers a broad range of activities which include preventive actions such as building resilient societies and risk reduction prior to, during, and after the disasters. Also rehabilitation, recovery and reconstruction from the impact of the disasters (*infra*).

Haiyan is clearly a "disaster" under AADMER which is defined as "a serious disruption of the functioning of a community or a society causing widespread human, material, economic or environmental losses".

The following are the main ASEAN instruments relating to disaster management which are relevant in considering ASEAN's role in Haiyan:

- The ASEAN Regional Programme on Disaster Management, 2004–2010.[10]
- Declaration on Action to Strengthen Emergency, Relief, Rehabilitation, Reconstruction and Prevention on the Aftermath of Earthquake and Tsunami Disaster, 26 December 2004.[11]
- ASEAN Agreement on Disaster Management & Emergency Response, 2005 (AADMER).[12]

---

[10]Reproduced in Koh (2013), *op. cit.*, p. 899.

[11]"Declaration on Action to Strengthen Emergency, Relief, Rehabilitation, Reconstruction and Prevention on the Aftermath of Earthquake and Tsunami Disaster, 26 December 2004," Association of Southeast Asian Nations. Retrieved 14 April 2015, from http://www.asean.org/communities/asean-socio-cultural-community/item/declaration-on-action-to-strengthen-emergency-relief-rehabilitation-reconstruction-and-prevention-on-the-after math-of-earthquake-and-tsunami-disaster-of- 26-december-2004.

[12]Reproduced in Koh (2013), *op. cit.*, p. 901. Chairman's Statement of the First Meeting of the Conference of the Parties to the [AADMER], Jakarta, 16 March 2012. Retrieved 14 April 2015, from http://www.asean.org/news/asean-statement-communiques/item/chairman-s-statement-of-the-first-meeting-of-the-conference-of-the-parties-to-the-asean-agreement-on-disaster-management-and-emergency-response-jakarta-16-march-2012; Chairman's Statement of the Second Meeting of the Conference of the Parties to the [AADMER], Hanoi, 30 May 2013. Retrieved 14 April 2015, from http://ptv.ph/13-asean-corner/389-chairman-s-statement-of-the-second-meeting-of-the-conference-of-the-parties-to-the-asean-agreement-on-disaster-management-and-emergency-response.

- ➢ ASEAN Agreement on Disaster Management and Emergency Response: Work Programme — 2010–2015 (building disaster-resilient nations and safer communities).[13]
- ➢ SASOP Standard Operating Procedure for Regional Standby Arrangements and Coordination of Joint Disaster Relief and Emergency Response Operations, 2008.[14]
- ➢ Disaster Relief Fund (AADMER).

- ASEAN Socio-Cultural Community Blueprint, 2009–2015 (ASCC).[15]
- ASEAN Political–Security Community Blueprint, 2009–2015 (APSC).[16]
- Initiative for ASEAN Integration (IAI) Strategic Framework and IAI Work Plan 2 (2009–2015).[17]
- Cha-am Hua Hin Statement on East Asia Summit (EAS) Disaster Management.[18]
- Tripartite Memorandum of Cooperation (MoC) between ASEAN, the United Nations International Strategy for Disaster Risk Reduction (UNISDR) and the World Bank on Disaster Risk Reduction in the Region, 2009.[19]
- Joint Declaration on ASEAN-UN Collaboration on Disaster Management, 30 October 2010.[20]
- Agreement on the Establishment of the ASEAN Coordinating Centre for Humanitarian Assistance on Disaster Management 2011.[21]

---

[13]Reproduced in Koh (2013), *op. cit.*, p. 901.

[14]"Standard Operating Procedure for Regional Standby Arrangements and Coordination of Joint Disaster Relief and Emergency Response Operations (SASOP)," Association of Southeast Asian Nations. Retrieved 14 April 2015, from http://www.scribd.com/doc/112149284/Standard-Operating-Procedure-for-Regional-Standby-Arrangements-and-Coordination-of-Joint-Disaster-Relief-and-Emergency-Response-Operations-SASOP.

[15]Reproduced in Koh Kheng-Lian (compiler and ed.), *ASEAN Environmental Law, Policy and Governance, Selected Documents*, volume I (Singapore: World Scientific, 2009), p. 615.

[16]Reproduced in *ibid.*, p. 593.

[17]Reproduced in *ibid.*, p. 571.

[18]Reproduced in Koh (2013), *op. cit.*, p. 927.

[19]Reproduced in *ibid.*, p. 923.

[20]Reproduced in *ibid.*, p. 933.

[21]Reproduced in *ibid.*, p. 1325. The UN Office for Disaster Risk Reduction (UNISDR) was established in 1999 to facilitate the implementation of the international strategy for disaster reduction.

- ASEAN Coordinating Centre for Humanitarian Assistance on Disaster Management, 2011.[22]
    - ➢ ASEAN Coordinating Centre for Humanitarian Assistance (AHA Centre).
    - ➢ ASEAN Disaster Emergency Logistic System.
    - ➢ ASEAN Emergency Response Assessment Team (ERAT).
    - ➢ ASEAN Early Rapid Assessment Team.
    - ➢ Bali Declaration on ASEAN Community in a Global Community of Nations "ASEAN Concord III", 2011.[23]

Some highlights of the main provisions in the above instruments as applicable to Haiyan are given below.

(1) ASEAN Agreement on Disaster Management and Emergency Response, 2005 (AADMER)

Before AADMER, there was the Declaration on Mutual Assistance on Natural Disasters 1976. It was a very brief document to improve communication channels on disaster warnings, exchange of information, and dissemination of medical supplies. There was also the ASEAN Regional Programme on Disaster Management 2004–2010 which provides the framework for promoting regional cooperation on disaster management. However, nothing much was done to operationalize it.

After the tsunami struck on 26 December 2004 — killing over 230,000 people in 14 countries in the Indian Ocean, and inundating coastal communities with waves up to 30 meters (100 ft) high, the ASEAN Member Countries (AMCs) resolved to enhance cooperation, and crafted AADMER.

AADMER forms the core mechanism which has spawned other frameworks and organizational bodies (see *infra*). The scope is comprehensive, covering the period prior to, during, and after disasters. AADMER was later operationalized by further instruments, namely, the AADMER Work Programme, 2010–2015[24] and was incorporated into the Socio-Cultural

[22]Reproduced in Koh (2013), *op. cit.*, p. 1325.

[23]"Bali Declaration on ASEAN Community in a Global Community of Nations 'Bali Concord III'." Retrieved 14 April 2015, from http://www.asean.org/archive/documents/19th%20summit/Bali%20Concord%20III.pdf.

[24]"ASEAN Agreement on Disaster Management and Emergency Response. Work Programme 2010–2015." Retrieved 14 April 2015, from http://www.asean.org/images/

Community Blueprint (ASCC), the Political–Security Blueprint (APSC), the Initiative for ASEAN Integration (IAI) Strategic Framework and the IAI Work Plan 2 (2009–2015), all part of the Roadmap for an ASEAN Community 2009–2015.[25] The ASEAN Humanitarian Centre (AHA Centre) was established by the Agreement on the Establishment of the ASEAN Coordinating Centre for Humanitarian Assistance on Disaster Management, 2011 (AHA Agreement). AADMER also made provisions for the establishment of the AHA Centre to facilitate cooperation among the Parties. Article 20 of AADMER provides for ASEAN to enter into MoC with its Dialogue Partners, the United Nations and World Bank to collaborate efforts in tackling disasters.

Article 2: Objective

The objective of AADMER is to provide effective mechanisms to achieve substantial reduction of disaster losses in lives and in the social, economic, and environmental assets of the Parties, and to jointly respond to disaster emergencies through concerted national efforts and intensified regional and international cooperation.

AADMER: Article I, key definitions

The following are the main key terms, and they reflect the wide scope of AADMER, encompassing a range of activities.

> *'Disaster risk' means the probability of harmful consequences, or expected losses in terms of deaths, injuries, property, livelihoods, economic activity or damage to the environment resulting from interactions between natural or human-induced hazards and vulnerable conditions.*
>
> *'Disaster risk reduction' means a conceptual framework of elements considered with the possibilities to minimise vulnerabilities and disaster risks throughout a society, to avoid through prevention or to limit through mitigation and preparedness the adverse impacts of hazards, within the broad context of sustainable development.*
>
> *'Disaster emergency' means a situation where a Party declares that it is unable to cope with a disaster.*
>
> *'Party' means a Member State that has consented to be bound by this Agreement and for which the Agreement is in force.*

resources/ASEAN%20Publication/2013%20(12.%20Dec)%20-%20AADMER%20Work%20Programme%20(4th%20Reprint).pdf.

[25]Reproduced in Koh (2009), *op. cit.*, pp. 593 and 615.

'Receiving Party' means a Party that accepts assistance offered by an Assisting Entity or Entities in the event of a disaster emergency.

'Requesting Party' means a Party that requests from another Party or Parties assistance in the event of a disaster emergency.

(2) AADMER and ASEAN's Role in Haiyan

AADMER is a proactive regional framework for cooperation, coordination, technical assistance, and resource mobilization in all aspects of disaster management. It also affirms ASEAN's commitment to the Hyogo Framework of Action (HFA)[26] and is the first legally-binding HFA-related instrument in the world. It forms the backbone of disaster management, giving priority to disaster risk reduction, and views disasters as a development concern from a humanitarian response. In the context of the ASEAN Charter, it brings about a people-centered focus with expected participation from such various stakeholders, even though the primary responsibility lies with individual AMCs, and it also promotes the development of the ASEAN Community by 2015.

To understand ASEAN's precise role in Haiyan, it is important to note that AADMER applies the *principle of sovereignty* in Article 3 and also sets out the limitations and roles of parties to AADMER:

1. *The sovereignty, territorial integrity and national unity of the Parties shall be respected, in accordance with the Charter of the United Nations and the Treaty of Amity and Cooperation in Southeast Asia, in the implementation of this Agreement. In this context, each affected Party shall have the primary responsibility to respond to disasters occurring within its territory and external assistance or offers of assistance shall only be provided upon the request or with the consent of the affected Party.*
2. *The Requesting or Receiving Party shall exercise the overall direction, control, co-ordination and supervision of the assistance within its territory.*
3. *The Parties shall, in the spirit of solidarity and partnership and in accordance with their respective needs, capabilities and situations,*

[26] "Hyogo Framework for Action 2005–2015: Building the Resilience of Nations and Communities to Disaster," World Conference on Disaster Reduction, UN Doc. A/CONF.206/6, 18–22 January 2005.

> *strengthen co-operation and co-ordination to achieve the objectives of this Agreement.*
>
> ...
>
> *6. The Parties, in addressing disaster risks, shall involve, as appropriate, all stakeholders including local communities, non-governmental organisations and private enterprises, utilising, among others, community-based disaster preparedness and early response approaches.*

The principle of sovereignty holds sway. This principle is not unique to ASEAN but is also contained in the UN Charter. Unlike the situation in Cyclone Nargis in 2008 (note that AADMER had not entered into force), where ASEAN was at first confronted by Myanmar's military ruler's refusal to grant access on grounds of its sovereignty (though subsequently, it permitted access to international organizations through ASEAN's coordination), there was no such problem of access in Haiyan, as the Philippines was the Requesting Party. Another important point under AADMER is the *primary responsibility* of the Requesting Party to exercise overall direction, control (including border controls and customs), coordination, and supervision of the assistance within its territory. The principle of sovereignty also carries with it the responsibility of the Requesting Party to ensure the safety of those rendering assistance when they are in its territory. ASEAN has therefore to work within the confines of AADMER. It does not have a 'free hand', as it were — this is not always understood (see *infra*).

It should also be noted that all 10 AMCs have ratified AADMER. However, if each AMC were to render assistance to the Philippines in Haiyan, it does not have to go through ADDMER or the AHA Centre, see *infra*. It can deal directly with the country requiring assistance.

To operationalize AADMER, the actions to be taken are also incorporated in the ASCC and APSC. A distinction should be appreciated between the two.

(1) *Distinction between ASCC and APSC approach*

ASEAN Socio-Cultural Community (ASCC), Section II, B 7

The objective of the ASCC approach is to strengthen effective mechanisms and capabilities to prevent and reduce losses in lives, and in social, economic, and environmental assets of AMCs, and to jointly respond to disaster emergencies through concerted national efforts and intensified regional and international cooperation.

ASCC focuses on pre-disaster preventive measures to build a disaster-resilient nation, and also assists during disasters to facilitate operations with national agencies and international organizations. When a disaster *morphs* into a serious one, APSC comes into operation, but it is mutually reinforcing with ASCC.

Finally, ASCC is aligned with the ASEAN Charter's people-centered approach in integration and community building, reflected in "forging a common identity and building a caring and sharing society . . . where the well-being, livelihood, and welfare of the peoples are enhanced".

ASEAN Political-Security Community Blueprint, 2009–2015 (APSC): Section II, B 5: Strengthen ASEAN Cooperation on Disaster Management and Emergency Response

Haiyan attracted the APSC because of the magnitude of its destruction. APSC focuses on ASEAN's political and security cooperation at a higher plane. It proceeds, inter alia, on a shared responsibility for comprehensive security. This involves formal communication at diplomatic levels such as access, security, and sovereignty issues, as in AADMER.

Some of the 'Actions' envisaged in this section are:

- Enhance joint effective and early response at the political and operational levels in activating the disaster management arrangements.
- Enhance civilian-military coordination.
- Work towards effective interface on disaster management between ASEAN-related bodies such as the ASEAN Regional Forum (ARF), ASEAN Plus Three, and East Asia Summit (EAS).

The 'security' aspects of disaster management operationalized under APSC bring a political and security cooperation to a higher plane. Some of the platforms to enhance ASEAN's disaster management capacities under APSC are the ASEAN Regional Forum (ARF), ASEAN Plus Three and East Asia Summit (EAS).[27]

---

[27]Reproduced in Koh (2009), *op. cit.*, p. 593, Section B5(v).

The relevant 'Actions' to be undertaken in Haiyan include:

- Joint effective and early response.
- Enhance civilian-military coordination in providing effective and timely response.
- Develop ASEAN Regional Forum (ARF) strategic guidelines for humanitarian assistance and disaster relief operations.

One of the most crucial issues of disasters is to gain access to victims of disasters who are cut off from relief operations. Access has to be quick as the lack of the basic necessities of life — food, water, and medicine — would increase the death toll. The military is more well-equipped to gain access than a civilian mechanism. However, military intervention is not without its sensitivities, given questions of military secrets of the Requesting Party. Be that as it may, the APSC has provision for 'civil-military' aid for timely response.

As noted, military intervention is a very sensitive matter and ASEAN and its member countries (at the time when Haiyan occurred) preferred to focus on humanitarian assistance together with the ASEAN Regional Forum (ARF) and the ASEAN Defence Ministers Meeting Plus process (ADMM-Plus).[28] However, it was up to each individual AMC to send military aid, and the Singapore Armed Forces from the Singapore Defence Attaché did send a team to the Philippines.

In the case of humanitarian assistance under APSC Section B.3., it calls for *strengthening* of such assistance. The actions taken under this section include intensifying cooperation with the United Nations and the role and contributions of relevant international organizations on humanitarian assistance, civil military dialogue and coordination in humanitarian assistance, and expanding the role and contribution of women in field-based humanitarian operations.

(2) *AADMER Work Programme, 2010–2015*

Soon after the entry into force of AADMER on 24 December 2009, its Work Programme was adopted at the 15th Meeting of the ASEAN

[28]Rory Medcalf, "Typhoon Haiyan and the Geopolitics of Disaster Relief," *The Interpreter*, 11 November 2013.

Committee on Disaster Management (ACDM). It is a rolling plan, with many strategic priorities in Phase 1 (2010–2012) which includes ongoing activities initiated in the previous cycle (2004–2010) as well as those that are already in the pipeline. Phase 2 (2013–2015) activities are designed to sustain the momentum and further strengthen disaster management capacities, mechanisms, and systems that have been developed and/or established.[29]

The objectives are:

- *Develop and implement a Regional Risk Assessment Programme that establishes protocols and best practices to complement already existing Member State initiatives;*
- *Develop and implement the necessary policy and procedural mechanisms to ensure that AHA Centre is able to receive and consolidate data from National Focal Points (NFPs) and disseminate to each Party the analysed data and risk level associated with identified hazards;*
- *Implement the necessary mechanisms and systems to ensure AHA Centre is able to conduct analysis on possible regional-level implications; and*
- *Develop and implement the necessary protocols to allow risk assessment outputs relating to regional hazards be made available to all ASEAN working groups.*

The Work Programme consists of four strategic components: (1) disaster preparedness and response; (2) risk assessment, early warning, and monitoring; (3) prevention and mitigation; and (4) recovery. It also outlines several building blocks in order for these four components to be sustainably implemented. These building blocks include training, knowledge management, and strengthening and forming new partnerships.

The Work Programme seeks to translate AADMER's spirit and intent into concrete actions and initiatives to be implemented in order to attain

[29]This first reprint of the Work Programme includes the List of Flagship Projects for AADMER Work Programme Phase 1 (2010–2012) as Annex 7, which was adopted by the 16th ACDM Meeting.

the ASEAN vision of disaster-resilient nations and safe communities by 2015.

It is designed to support the national agenda and complement AMCs' efforts to attain the vision of disaster-resilient nations. Its guiding principles are consistent with the principles of the ASEAN Charter and the ASEAN Community Blueprints towards an ASEAN Community by 2015. The Work Programme primarily adheres to a strategic regional approach to all aspects of disaster management and outlines activities that will add more value to and leverage the expected benefits of ongoing and planned national initiatives of AMCs. It operationalizes the strategic objectives and actions outlined in the ASEAN Socio-Cultural Community Blueprint (in particular Section B.7) and the ASEAN Political-Security Community Blueprint (in particular Sections B.5 and B.6) as part of the Roadmap for an ASEAN Community for the period 2009–2015.

This strategic component of the Work Programme aims to reduce loss of life and damage to property from natural and human-induced disasters through the identification of hazards and risks prior to impacts and by increasing warning time. The activities include improved regional risk assessment and early warning activities with a focus on cross-boundary issues that require intercountry collaboration, thereby providing regional benefits, more inclusive disaster planning and mitigation efforts, as well as targeted response and recovery activities.

With constant development and environmental change, risk is evolving and manifesting in new areas and among new populations. Identifying high risk populations and geographic areas and understanding underlying hazards, vulnerabilities, and disaster management capacities are critical to effectively plan for, respond to, and recover from disasters. Risk assessment activities are being undertaken in the region.

(3) *Agreement on the Establishment of the ASEAN Coordinating Centre for Humanitarian Assistance on Disaster Management, 2011.*[30]

Pursuant to Article 20.1 of AADMER, the ASEAN Coordinating Centre for Humanitarian Assistance on Disaster Management (AHA Centre) was

[30]Reproduced in Koh (2013), *op. cit.*, p. 1325.

launched at the 19th ASEAN Summit on 18 November 2011, in Bali, Indonesia, following the signing of the Agreement on the Establishment of the ASEAN Coordinating Centre for Humanitarian Assistance on Disaster Management, 17 November 2011 (AHA Agreement). It aims to be the regional hub for information and knowledge for disaster management. It serves as the center point for mobilization of resources to disaster-affected areas and acts as the coordination engine to ensure ASEAN's fast and collective response to disasters within the ASEAN region. The AHA Centre works on the basis that the Requesting Party will act first to manage and respond to disasters within its territory. Its functions include receiving and analyzing data on risk assessment and making recommendations; facilitating the establishment and maintenance of arrangements for disaster relief and emergency response; and facilitating joint emergency response. A number of ASEAN Dialogue Partners, in particular Australia, Canada, Japan, New Zealand, the United States, and recently the European Union, have also been assisting the establishment of the AHA Centre. AMCs when providing assistance can do so directly and need not work through the AHA Centre, as in AADMER, *supra*.

To reduce loss of life and assets resulting from disasters, AADMER COP (consisting of ministers or secretaries of the government bodies in charge of disaster management and risk reduction) and the ASEAN Community on Disaster Management (ACDM) operationalize AADMER and collaborate with relevant regional organizations such as the ASEAN Partnership Group (AGP)[31] (see Appendix I, and Section 18.3, below).

[31] Consisting of international NGOs and many other regional and international organizations, and supported by the ASEAN Secretariat and its Disaster Management and Humanitarian Assistance Division, see ASEAN Organizational Chart on Disaster; see Potential Policies and Areas of Cooperation on Disaster Management and Emergency Response between ASEAN and Canada. Retrieved 14 April 2015, from http://www.mfa.go.th/asean/contents/files/asean-media-center-20130715-161750-794058.pdf.

18.3. ASEAN Disaster Management Cooperation: International, Regional, and National Mechanisms

The Chart (see Appendix I) shows a congestion of relevant regional and international organizations, agencies, committees, National Focal Points, etc., on disaster management. There are some overlapping jurisdictions. Governance is both horizontal and vertical, and extends beyond ASEAN members and the region to cover the United Nations, World Bank, ASEAN Dialogue Partners, and other international organizations.

The AHA Centre facilitates cooperation and coordination among the parties with relevant United Nations and international organizations in promoting regional collaboration in disaster management and emergency response.[32] The AADMER Conference of Parties (COP)[33] together with its ASEAN Committee on Disaster Management (ACDM) coordinates with relevant international bodies. The ACDM comprises the National Focal Points.

The AHA Agreement launched a Disaster Emergency Logistic System on 7 December 2012 at the Royal Malaysian Air Force Base in Subang, Malaysia. The establishment of the Disaster Emergency Logistic System is supported by the Government of Japan through the Japan–ASEAN Integration Fund. A significant new body established by Singapore after the Haiyan disaster is the Changi Regional Humanitarian Assistance and Disaster Relief Coordination Centre, in Singapore (*infra*).

In Haiyan, supplies were sent to the devastated areas from Subang. ASEAN ADMM-Plus, ASEAN's Dialogue Partners, ASEAN+3, East Asia Summit (EAS), ASEAN Regional Forum (ARF), and Asia–Europe Meeting (ASEM) are platforms for political-security strategic planning, including disasters.

More recently, new entities have developed to augment the existing array to tackle disasters. Among them are the ASEAN Defence Ministers'

---

[32]Agreement on the Establishment of the ASEAN Coordinating Centre for Humanitarian Assistance on Disaster Management. Reproduced in Koh (2013), *op. cit.*, p. 1325, Art. 3 (1).

[33]ASEAN Agreement on Disaster Management and Emergency Response, Art. 21, reproduced in Koh (2013), *op. cit.*, at p. 914.

Meeting (ADMM) (on non-traditional security, humanitarian assistance, and disaster relief) and Civil Society Organisations (CSOs).

The ASEAN Secretariat has a Disaster Management and Humanitarian Assistance Division that supports COP, ACDM, and the ASEAN Disaster Management and Emergency Relief Fund. The Secretary-General of ASEAN is the ASEAN Humanitarian Assistance Coordinator.

18.4. Typhoon Haiyan: Differing Accounts on ASEAN's Contributions

There are contrasting accounts of ASEAN's role in Haiyan. We will examine Jeffrey Wright in "Wherefore Art Thou ASEAN? Typhoon Haiyan's Teachable Moment",[34] and Steven Keithley in "ASEAN Slowly Gets up to Speed on Haiyan".[35] They gave rather scathing accounts of ASEAN's response in Haiyan. In contrast, an on-the-ground account is given by Lee Khiam Jin, in "ASEAN's Response to Typhoon Haiyan".[36] It is useful to quote them *in extenso* for analysis.

First, let us consider Wright's account:

(1) "*Wherefore Art Thou ASEAN? Typhoon Haiyan's Teachable Moment*"

What is the nuance of this caption, "Wherefore Art Thou, ASEAN?".

Is the caption inspired by *Romeo and Juliet*: "O Romeo, Romeo! *wherefore art thou* Romeo?" (Shakespeare: *Romeo and Juliet*). The "wherefore" here means *why* rather than *where*. What Juliet is asking, in allusion to the feud between her Capulet family and Romeo's Montague clan, is "Romeo,

[34]Jeffrey Wright, "Wherefore Art Thou ASEAN? Typhoon Haiyan's Teachable Moment," *The Internationalist*, 18 November 2013. Retrieved 14 April 2015, from http://blogs.cfr.org/patrick/2013/11/18/wherefore-art-thou-asean-typhoon-haiyans-teachable-moment/. Jeffrey Wright is Research Associate, International Institutions and Global Governance.

[35]Steven Keithley, "ASEAN Slowly Gets up to Speed on Haiyan," *The Diplomat*, 23 November 2013. Retrieved 14 April 2015, from http://thediplomat.com/2013/11/asean-slowly-gets-up-to-speed-on-haiyan/. Steven Keithley studies international history at Georgetown University.

[36]Lee Khiam Jin, "ASEAN's Response to Typhoon Haiyan," *Penang Monthly*, 4 March 2014. Lee Khiam Jin is head of the Corporate Affairs and Programme Division, ASEAN Coordinating Centre for Humanitarian Assistance on Disaster Management (AHA Centre), and formerly the Economic Affairs officer of the Information and Communication Technology and Disaster Risk Reduction Division, UNESCAP.

why are you a Montague?". Their love is impossible because of their family names and she asks him to change his allegiance, or else she will change hers.

> Juliet:
> *O Romeo, Romeo! wherefore art thou Romeo?*
> *Deny thy father and refuse thy name;*
> *Or, if thou wilt not, be but sworn my love,*
> *And I'll no longer be a Capulet.*

If taken in the Romeo context, "why" or literally "where", it does mean, "ASEAN why are you not being effective?". Or, "where are you?". Whichever nuance, it does express frustration that ASEAN is not effective!

Haiyan generated worldwide interest with many comments, not the least from the media. The writers note that much of what has been written lacks examination or knowledge of the relevant instruments and ASEAN's exact role (see *supra* and *infra*) in the Haiyan context. Or, if there were references, analysis was lacking, including the operationalization of the various frameworks, before Haiyan (hence, this chapter gives a comprehensive account of the legal frameworks, *supra*).

Wright stated:

> *Disasters also reveal the capacities of actors to respond effectively. In the case of Typhoon Haiyan, it is clear that the region's most prominent organization possesses no such resilience. The Association of Southeast Asian Nations (ASEAN) carries a mandate to respond to all forms of [security] threats. But it lacks the political will and resources to fulfil its Charter obligations . . .*
>
> *ASEAN's role in the Philippines has been limited to basic information-sharing functions. Prior to Haiyan making landfall, the Coordinating Centre for Humanitarian Assistance — ASEAN's disaster management unit — monitored the storm's movements and deployed logistics personnel to Manila and Tacloban to help coordinate relief efforts with national authorities.*
>
> *Now in the typhoon's aftermath, ASEAN lies in the shadows altogether, wielding neither the power nor funds to play a substantial part in the humanitarian response.*
>
> *In the absence of a strong ASEAN, the Philippine recovery is simply more fragile and harder to execute, relying on a range of actors to fill the void.*

Keithley also decried ASEAN's inefficacy.

(2) "*ASEAN Slowly Gets up to Speed on Haiyan*"

> *About two weeks into Haiyan ASEAN's initial response was said to have been sluggish. 'Will it now step up?'. Yet the Association of Southeast Asian Nations has been conspicuously absent. Despite the immense destruction inflicted upon one of its one own, ASEAN has not stepped up and led. Instead, that mantle has been seized by the United States and the U.N. Office for the Coordination of Humanitarian Affairs. Worse, beyond providing reporting and shuffling staffers through the ASEAN Coordinating Centre for Humanitarian Assistance (AHA), ASEAN has done comparatively little in any capacity.*
>
> *It is not insignificant that Philippine President Benigno Aquino III failed to mention ASEAN during his first address about the relief. Although the director of the AHA announced an aid package amounting to [US]$500,000 on the same day, November 12, Washington notified Manila that it would send [US]$20 million together with an aircraft carrier, the USS George Washington, to the region. Add in multimillion dollar pledges from Australia, Japan, the United Kingdom and others, and it is easy to see the relative meagreness of ASEAN's contribution, disturbing given Filipino membership in ASEAN. Some observers have pointed to structural weaknesses within the organization to try to explain this perceived failure, but the organization demonstrated during Nargis that it does possess the ability to institutionalize effective mechanisms for emergency management.*
>
> *However, it is still early. The devastation in Tacloban, Daanbantayn, Guiuan and the numerous other cities impacted by Haiyan will take months, if not years, to repair. So ASEAN could still show the international community that its response to Cyclone Nargis was not an exception to the organization's longer history of weak and ineffective disaster management. There are even signs that this might be happening. In an interview with BBC World News, Secretary-General Le Luong Minh stated that 'What we provided is only initial, there is more support to be given'. In the days since its initial [US]$500,000 contribution, the AHA has started to supervise the arrival of more personnel and supplies. Under the ASEAN banner, Brunei and Indonesia have sent C-130 and CN-235 transport planes full of food, medicine and other important items. The AHA is even in the process of constructing a permanent office in Tacloban. The prospects, therefore, are not entirely negative. Nonetheless, there is*

> *as yet no evidence of more powerful mechanisms being utilized, namely, the ASEAN Disaster Management and Emergency Relief Fund formalized by the AADMER. If that fund can follow the same trajectory as the Fund for ASEAN after its creation in 1969, the benefits could be immense, not only for the Philippines, but for the credibility of ASEAN.*
>
> *Johann Wolfgang von Goethe wrote in Faust, 'The time has come to prove by deeds/that man will not quake before the pit'. ASEAN stands before its own pit right now. Only time will tell if Nargis was just a bright spot in an otherwise depressing history, or the beginning of an significant improvement in the capabilities of the Association of Southeast Asian Nations.*

(3) *Comments and analysis of Jeffrey Wright and Steven Keithley*

Apart from making general statements regarding ASEAN's inefficacy, Wright raises a number of questions which touch on the legal frameworks which he does not address. These frameworks have been set out in Sections 18.2 and 18.3, *supra.*

Wright also points out ASEAN's role and what he perceived was ineffective both from the evidence he gathered and from the ASEAN legal frameworks on disaster — such as he mentioned.

We would like to comment on and analyze the following points he raised: ASEAN's response to 'security' threats. He said:

> *In the case of Typhoon Haiyan, it is clear that the region's most prominent organization possesses no such resilience. The Association of Southeast Asian Nations (ASEAN) carries a mandate to respond 'to all forms of [security] threats.' But it lacks the political will and resources to fulfil its Charter obligations . . .*
>
> *What is the role of ASEAN in providing regional security, and what is ASEAN's place in the broader architecture of Asian security? How does ASEAN best leverage the competencies of its partners and allies? Can the organization improve its capacity to respond to emerging threats, including national disasters such as Typhoon Haiyan and the risk of infectious disease in its aftermath? How can states close the development gap in the region?*

It is noted that from 21–24 October 2013, a few weeks before Haiyan, ASEAN Member States together with the AHA Centre and the

Pacific Disaster Center (PDC) conducted the annual ASEAN Regional Disaster Emergency Response Simulation Exercise (ARDEX-13) in Hanoi, Vietnam. This exercise is one of a regular tests of ASEAN's preparedness mechanisms under the ASEAN Standby Arrangements. During the exercise, the ASEAN Disaster Monitoring and Response System[37] at the AHA Centre was used for multi-hazard monitoring, early warning, decision, and information sharing. ASEAN's contributions to Haiyan include sending relief supplies, being at the scene to give humanitarian assistance — it was very much guided by the AADMER and APSC/ASCC spirit. Was ASEAN really as Keitheley said: "conspicuously absent". The fact that the Secretary-General of ASEAN was not there is not to be taken that ASEAN was ineffective (see *infra*).

Keithley also quoted from Faust:

> *Johann Wolfgang von Goethe wrote in Faust, 'The time has come to prove by deeds/that man will not quake before the pit'. ASEAN stands before its own pit right now. Only time will tell if Nargis was just a bright spot in an otherwise depressing history, or the beginning of a significant improvement in the capabilities of the Association of Southeast Asian Nations.*

[37]Officially launched on 10 January 2013 — "U.S. Supported System Brings a New Era in ASEAN Disaster Management and Response." Retrieved 14 April 2015, from http://www.usaid.gov/asia-regional/press-releases/us-supported-system-brings-new-era-asean-disaster-management-and; "Chairman's Statement of the 22nd ASEAN Summit, 'Our People, Our Future Together'," para. 12, 24 and 25 April 2013. Retrieved 14 April 2015, from http://www.asean.org/news/asean-statement-communiques/item/chairmans-statement-of-the-22nd-asean-summit-our-people-our-future-together; "ASEAN Defence Ministers Meeting (ADMM)." Retrieved 14 April 2015, from http://www.asean.org/communities/asean-political-security-community/category/asean-defence-ministers-meeting-admm; "Brunei Ready for Hosting ASEAN's Humanitarian Assistance, Disaster Relief Exercise Meetings." Retrieved 14 April 2015, from http://news.xinhuanet.com/english/world/2013-05/28/c_132415044.htm; "Joint Communique, 46th ASEAN Foreign Ministers' Meeting," Bandar Seri Begawan, Brunei Darussalam, 29–30 June 2013, paras. 30–31. Retrieved 14 April 2015, from http://www.asean.org/images/2013/news/joint%20communique%20of%20the%2046th%20asean%20foreign%20ministers%20meeting%2046th%20amm%20-%20final%20-%2030%20june%202013.pdf; "ASEAN Tests Regional Disaster Response Readiness Prior to Typhoon Haiyan," Pacific Disaster Centre's News and Media update, 27 November 2013. Retrieved 14 April 2015, from http://www.pdc.org/news-n-media/pdc-updates/ASEAN-Tests-Regional-Disaster-Response-Readiness-Prior-to-Typhoon-Haiyan/. See also an account of ASEAN's role by Lee Khiam Jin, *infra*.

ASEAN is essentially a coordinating agency to facilitate the work of the United Nations for the Coordination of Humanitarian Affairs (OCHA),[38] and not a regional organization for humanitarian affairs, replacing OCHA or the United Nations World Food Programme (WFP).[39]

While ASEAN has, at the regional level, carried out emergency preparedness exercise and, indeed, had one in Hanoi, the responsibility for implementation of AADMER and with it ASCC "early response" preparedness is left to the national disaster management offices of each AMC, i.e., the Philippines. This is not to say that ASEAN could not do more. Indeed, after Haiyan, ASEAN did speed up efforts to facilitate joint responses together with regional and international cooperation.[40]

The ASEAN Regional Programme on Disaster Management, 2004–2010; the Joint Declaration on ASEAN UN Collaboration on Disaster Management, 31 October 2010; the AADMER COP and the ASEAN Committee on Disaster Management (ACDM); and the strategies under the AADMER Work Programme and the AHA Centre will facilitate the national disaster agency in the Philippines. Assistance was in fact given, as reported by Lee Khiam Jin, *infra*. Given the magnitude of Haiyan, the contributions by ASEAN (with its limitations on funding and capacity) are not expected to overcome all the myriad of problems. Besides, there were other donors. There was confusion regarding governance resulting in delays (e.g., who is to do what) etc., see, *infra*, Sections 18.5 and 18.8.

In the case of military aid, ASEAN focused more on humanitarian aid as the US had sent 66 military aircrafts and 12 marine vessels to the disaster areas to clear roads and transport aid workers. The US subsequently reduced its presence making way for civilian and commercial assistance. Jeremy

---

[38]United Nations Office for the Coordination of Humanitarian Affairs (OCHA) website, http://www.unocha.org/.

[39]World Food Programme (WFP) website, http://www.wfp.org/.

[40]Yan, "Chinese Troop Joins ASEAN Humanitarian Assistance, Disaster-Relief Exercise," *Xinhuanet*, 29 April 2014. Retrieved 14 April 2015, from http://news.xinhuanet.com/english/china/2014-04/29/c_133296424.htm. The Chinese military has sent a 25-member troop to join an ASEAN humanitarian assistance and disaster-relief exercise, which kicked off Monday in eastern Thailand's Chachoengsao province. The exercise, jointly hosted by Thai and Malaysian militaries, is designed to enhance the capacity for joint humanitarian assistance in the ASEAN region. A total of 13 countries including China, the United States, Japan, France, Canada, and some ASEAN members will participate in the exercise.

Konyndyk, Director for Foreign Disaster Assistance at the US Agency for International Development (USAID), said the US military had started to reduce its presence to allow civilian aid agencies to step up efforts. "What we have seen, particularly over the past week, is now civilian and private-sector commercial capacity has started coming back up again and that is taking the burden off of the military actors," Konyndyk told *Reuters* in an interview, "You don't want the military playing that role in the long run, they are an interim bridging capacity there, but in the long run, that really needs to be civilian role."[41]

Several ASEAN members have also deployed defence assets, bilaterally, to support the relief operation. Considering the high-profile attention given to HADR (humanitarian assistance and disaster relief) cooperation within ASEAN, the ARF, and the ADMM-Plus, there was expectation that ASEAN could have responded with more alacrity. But the unexpected magnitude of Haiyan was overwhelming.

In regard to funding, although there is the ASEAN Disaster Risk Financing (DRF), there is limited funding from member states. Also, funding has been eroded by other disasters that have been occurring in the region. In any case, ASEAN has no intention to set up a fund to finance disasters as such — *the fund is merely for operations of AHA training and coordination*, etc. The private sector and other organizations should be involved in financing disasters. ASEAN organized a Forum on ASEAN Disaster Risk Financing and Insurance under the AADMER Work Programme on 8–10 November 2011. It was an initiative of ASEAN, World Bank (WB), and United Nations International Strategy for Disaster Reduction (UNISDR). ASEAN has entered into MoC with WB and UNISDR. Among the participants were the finance and insurance sectors which assisted in developing a disaster risk financing roadmap for ASEAN. During discussions, some options for consideration for disaster risk financing and

---

[41] "Typhoon Haiyan: US Military Scales down Aid Efforts in Philippines," *The Straits Times*, 23 November 2013. Retrieved 14 April 2015, from http://www.straitstimes.com/the-big-story/typhoon-haiyan/story/typhoon-haiyan-us-military-scales-down-aid-efforts-philippines-20#sthash.ftwFJdZY.dpuf. See also Euan Graham, "Super-Typhoon Haiyan: ASEAN's Katrina Moment?," Center for Strategic and International Studies, 20 November 2013. Retrieved 14 April 2015, from http://csis.org/publication/pacnet-82-super-typhoon-haiyan-aseans-katrina-moment.

insurance were considered. These include key challenges such as limited data and tools to assess the economic and fiscal impact; disaster and risk financing and insurance strategies are yet incomplete at both national and regional levels; lack of mechanisms to disburse funds post-disaster; private catastrophe risk insurance markets remain underdeveloped; and there is limited regional cooperation among AMCs on disaster risk financing and insurance although it was noted that political support for cooperation is beginning to show up.

Disaster risk financing (DRF) schemes are a means of transferring risk and must be integrated with disaster risk management while encouraging the private sector to be involved. It was pointed out at the forum that some countries are reluctant to pursue DRF as this might minimize their chances of getting aid from donors. To successfully implement a DRF into disaster reduction risk framework, there must be political and technical understanding of risk identification and analysis, capacity building of disaster risk management, and collaboration with the private sector. Having regard to the complexities of disaster risk funding and the recent entry point of ASEAN into disaster management and humanitarian response (about four years), ASEAN was not fully capable of handling Haiyan. Indeed, other countries have also experienced extreme difficulties, such as Katrina (US). While Haiyan and other serious disasters will galvanize a 'whole-of-society' or 'whole-of-the-world' approach, much still needs to be done.[42]

Given the magnitude of the disaster in Haiyan and the many donors, it is not easy to assess the actual impact of ASEAN's contributions. But there is evidence that more than just a monitoring role, ASEAN together with some of its member countries did provide some basic supplies and services, including humanitarian relief and also in some instances, military aid.

Wright notes:

> *The fact remains that there are larger questions about the state of ASEAN cooperation in disasters. What is the role of ASEAN in providing regional security, and what is ASEAN's place in the broader architecture of Asian security? How does ASEAN best leverage the competencies of its partners*

[42]"Towards an ASEAN Roadmap for Disaster Risk Financing and Insurance," ASEAN Disaster Risk Financing and Insurance Forum Proceedings (8–10 November 2011), ASEAN Secretariat, Jakarta.

such as the UN, World Bank and Dialogue partners? Can the organization improve its capacity to respond to emerging threats, including national disasters such as Typhoon Haiyan and the risk of infectious disease in its aftermath? How can states close the development gap in the region?

Wright's queries about the role of ASEAN have been discussed earlier in both its legal frameworks, as well as in cooperation with the UN and other organizations. As for the other queries, we would like to bring to attention the on-the-ground account by Lee Khiam Jin, in "ASEAN's Response to Typhoon Haiyan" (4 March 2014):

Typhoon Haiyan was a natural disaster on a catastrophic scale. How did Asean react to this, and how good were its humanitarian assistance and disaster relief efforts?

UN Under-Secretary-General for Humanitarian Affairs and Emergency Relief coordinator Baroness Valerie Amos launched a US$301mil flash appeal from Manila to cover relief efforts for the first six months, and urged the international humanitarian community to scale up its operations to provide lifesaving aid.

As a result of wide coverage of the disaster, generous amounts of relief items such as personal hygiene kits, food, clean water, shelter, family tents and basic medicines were mobilised from international societies, including Asean countries, to the most seriously affected provinces such as Leyte and Easter Samar...Meanwhile, the World Bank has offered the Philippines a US$500mil emergency loan to help it construct buildings able to withstand high winds and severe flooding.

Asean's emergency response mechanism tested.

But how did the Philippines' own neighbours respond? Guided by the AADMER spirit "to provide effective mechanisms to achieve substantial reduction of disaster losses in lives and in the social, economic and environmental assets of the Parties, and to jointly respond to disaster emergencies through concerted national efforts and intensified regional and international cooperation", the AHA Centre had actually deployed one staff member before the landfall of the typhoon to establish an emergency satellite-based communication system to support the Office of Civil Defence at Tacloban City, Leyte Province. Two members of the ASEAN Emergency Response Assessment Team (Erat) from Brunei and one personnel from the AADMER Partnership Group were also among the earliest Asean emergency responses to Typhoon Haiyan. In the weeks following the disaster, there was not a single day that went by without the presence of AHA Centre personnel supporting the Philippine government in facilitating and delivering relief assistance from Asean member states.

Emergency response to such a large-scale natural disaster is a test of the efficacy of ASEAN's mechanism. All the ASEAN member states contributed to the disaster-affected people in the Philippines with relief supplies, cash and medical personnel either through the AHA Centre or a bilateral initiative. The AHA Centre also mobilised relief items stockpiled at the Disaster Emergency Logistic System for Asean based at the Subang Royal Air Base, namely office prefabs and supplies kits, generators, forklifts and mobile storage units, transporting them with a C-130 Hercules military transport aircraft chartered from the government. Items such as rice, water and tarpaulins, on the other hand, were procured in Manila.

The Malaysian government, through Sasop, mobilised tons of food supplies, blankets, tarpaulins, medicine and medical personnel, equipment to build field hospitals, and a search and rescue team in addition to a donation of US$1 million to the Philippine government. All these assistances were deployed via the C-130 military aircraft, which made three round trips at the expense of the Malaysian government.

Meanwhile, Brunei mobilised tons of rice, instant noodles, biscuits and drinking water in five batches of shipment transported by CN235 planes and navy cargo ships. AHA Centre personnel were tasked to facilitate the receiving and handover of the relief items to the Philippine government according to Form 4: Offer of Assistance, as stipulated in SASOP. The Indonesian government contributed US$1 million and an additional US$1 million worth of relief items. Singapore mobilised two personnel from the Singapore Civil Defence Force to support the ICT and provided relief supplies via Singapore's Red Cross society. The Thai government, on the other hand, provided substantial relief supplies, donated US$2.7 million and collected an additional US$1.7 million by raising funds. Vietnamese, Cambodian and Myanmar governments each provided US$100,000, while Laos contributed US$50,000 to support the emergency response.

Challenges ahead

Although ASEAN's combined relief assistance was less than the international society's pledge, Asean member states have demonstrated solidarity when it comes to disaster emergency response. There were comments that ASEAN was invisible during Haiyan relief operations, but The Economist [see Penang Monthly], in its reporting, had this fair remark: "Disaster relief is not a talent show or beauty contest. In fact, AHA Centre's efforts, like those of the Philippine government and the world as a whole, have been swamped by a typhoon more destructive than anyone had thought possible".

The utilisation of military assets in humanitarian assistance and disaster relief (HADR) is a sensitive issue, particularly in the West. Nonetheless, the heavy dependency on C-130 and CN235 aircrafts as well as navy ships in the relief operations demonstrated that Asean will continue to count on its military for HADR operations.

In 2013 alone, there were four major regional exercises conducted in ASEAN, such as ARDEX-13 and the ASEAN Defence Minister Meeting (ADMMPlus), which previously in 2010 had already agreed on five areas of practical cooperation to pursue (maritime security, counter-terrorism, disaster management, peacekeeping operations and military medicine). These exercises and formation of themed task forces demonstrated ASEAN's commitment to develop standby mechanisms and assess its efficiency in utilising ASEAN military assets and capacities in HADR.

The UN, with its specialised agencies, is able to address post-disaster needs especially during the reconstruction and rehabilitation phase. For instance, the World Food Programme (WFP) has provided food supplies to over 2.6 million people in the Philippines, the United Nations Children's Fund (UNICEF) restored clean water and sanitation supplies, the World Health Organization (WHO) supported the local government to open health clinics and hospitals, and the UN Refugee Agency (UNHCR) provided shelter materials including tents and plastic sheeting to aid reconstruction work. This is an area where ASEAN sectoral bodies and committees can be organised to deliver coherent services and assistance amidst major disasters.

To advance the AADMER and ASEAN mechanism to meet the challenges of future large-scale HADR operations, ASEAN is to develop proactive media and public relations strategies, not only to make its emergency relief actions visible to the public, but also to attract interested parties, including the private sector, to contribute to its ASEAN Disaster Management and Emergency Relief (ADMER) fund.

Relief assistance provided by respective ASEAN member states should be strongly encouraged to mobilise based on the existing SASOP mechanism, widely used web-based crisis information systems such as the recently-launched ASEAN WebEOC system and the geospatial application for risk assessment, monitoring and early warning, as well as strengthening cooperation with the UN Office for the Coordination of Humanitarian Affairs (UNOCHA) and other regional humanitarian actors. The ASEAN mechanism in HADR will continue to be assessed for the better of this vulnerable community that 600 million people call home.

Lee's account and analysis of ASEAN's contributions puts ASEAN in a much more positive role, even if there are gaps. There are lessons to be learned.

The crucial role that ASEAN has played in Haiyan is that which is provided by its relatively new AHA Centre in its coordinating role under Article 20 of AADMER:[43]

The AHA Centre works on the basis that the Party will act first to manage and respond to disasters. In the event that the Party requires assistance to cope with such a situation, in addition to direct request to any Assisting Entity, it may seek assistance from the AHA Centre to facilitate such request.

To this end, it performs the following important functions:

- Receive and consolidate data as analyzed by and recommendations on risk level from the National Focal Points (Article 5.4);
- On the basis of such information, disseminate to each Party, through its National Focal Point, the analyzed data and risk level arising from the identified hazards (Article 5.4);
- Where appropriate, conduct analysis on possible regional-level implications.

18.5. The Haiyan Aftermath

(1) *An analysis of policies, institutions, and governance in the Philippines' and ASEAN's response to climate change and natural disasters*

The Haiyan disaster has been mentioned in the Introduction. As one of the most vulnerable countries to climate change, the Philippines had, before

[43]The ASEAN Coordinating Centre for Humanitarian Assistance on disaster management (AHA Centre) aims to facilitate cooperation and coordination among the Parties, and with relevant United Nations and international organizations, in promoting regional collaboration.

Haiyan, adopted major laws directly dealing with natural disasters and climate change. These are Republic Act no. 9729, otherwise known as the Climate Change Act of 2009, and Republic Act 10121, also known as the Philippine Disaster Risk Reduction and Management Act of 2010. Both laws have defined adaptation as "the adjustment in natural or human systems in response to actual or expected climatic stimuli or their effects, which moderates harm or exploits beneficial opportunities".[44]

The Philippine Climate Change Act is a landmark legislation which adopts strategic goals in order to build national and local resilience to climate change-related disasters. Basically, it advances the policies to strengthen, integrate, consolidate, and institutionalize government initiatives to achieve coordination in the implementation of plans and programs to address climate change in the context of sustainable development. This was later on amended in 2012 by Republic Act 10174 that, among others, established the People's Survival Fund to provide long-term finance streams to enable the government to effectively address the problem of climate change.

The Philippine Climate Change Commission (CCC), an independent and autonomous agency, serves as the sole policy-making body of the government tasked to coordinate, monitor, and evaluate the programs and action plans of the government relating to climate change. The CCC is mandated to create the official National Strategic Framework and Program on Climate Change as the basis for climate change planning, research and development, extension, monitoring of activities, and climate financing, to protect vulnerable and marginalized communities from the adverse effects of climate change.[45]

In accordance with its mandate, the CCC formulated the National Framework Strategy on Climate Change: 2010–2022 in order to build a roadmap that will serve as the basis for a national program on climate change. Its primary goal is to build the adaptive capacity of communities and increase the resilience of natural ecosystems to climate change, and optimize mitigation opportunities towards sustainable

[44]See Rep. Act 9729 and Rep. Act 10121.

[45]Rep. Act 9729, Sec. 8 (2009).

development. Although it identifies specific actions to mitigate effects of climate change, it falls short of identifying adaptation strategies by merely focusing on enhancing the availability and quality of vulnerability and adaptation assessments to serve as the country's scientific basis for formulating appropriate climate change adaptation strategies. At the local level, climate-smart local planning and the implementation of pilot measures to enhance adaptation to climate change have been initiated in a few areas while strengthening capacity of the CCC in its task of policy development, coordination, and monitoring, among others, is still ongoing.

Pursuant to the Philippine Disaster Risk Reduction and Management Act of 2010, the National Disaster Risk Reduction and Management Council is the central authority for disaster management in the Philippines, running from national to municipal levels, with government ministries, the armed forces, emergency services, and civil society all represented. Its structure designates four cabinet secretaries as leads for different aspects of the disaster management cycle: Preparedness, Response, Prevention and Mitigation, and Rehabilitation and Recovery.

The said law has laid the basis for a paradigm shift from disaster preparedness and response to disaster risk reduction and management. One of its state policies provides for the development, promotion, and implementation of a comprehensive National Disaster Risk Reduction and Management Plan that aims to strengthen the capacity of the national government and the local government units (LGUs), together with partner stakeholders, to build the disaster resilience of communities, and to institutionalize arrangements and measures for reducing disaster risks, including projected climate risks, and enhancing disaster preparedness and response capabilities at all levels. It also seeks to ensure that disaster risk reduction and climate change measures are gender responsive, sensitive to indigenous knowledge systems and cultures, and respectful of human rights.

Pursuant to these basic laws on disaster risk reduction management and climate change adaptation, all local chief executives as chairpersons of the Local Disaster Coordinating Councils are urged to mainstream climate change adaptation and disaster risk reduction measures into local policies, plans, budgets, and investment programs as a priority concern.[46] These legal measures are aligned with Sections 4a and 4d of the ASEAN Agreement on Disaster Management and Emergency Response (AADMER), which task parties thereto to cooperate in developing and implementing measures to reduce disaster losses including, but not limited to, identification of disaster risk, development of monitoring, assessment, and early warning systems. To bolster this cooperative undertaking, ASEAN member countries have pledged to take necessary legislative, administrative, and other measures.

Since 1991, the Philippine Inter-Agency Committee on Climate Change (IACCC) was established by virtue of Presidential Administrative Order no. 220, making the Philippines among the first countries to set up such a national administrative committee. This initiative was implemented right after the 2007 Bali United Nations Framework Convention on Climate Change (UNFCCC) meeting (COP13) where an agreement was reached on "the urgent need to take further meaningful action to reduce emissions from deforestation and forest degradation".[47] The Climate Change Commission was given the tasked to coordinate existing climate change initiatives such as Reducing Emissions from Deforestation and Forest Degradation (REDD+) — plus other similar mechanisms.[48]

In recent years, Executive Order no. 43, Series of 2011, has set in place a Cabinet Cluster in the Executive Department focusing on Climate Change Adaptation and Mitigation (CCAM). The CCAM Cluster shall take the lead in pursuing measures to adapt to and mitigate the effects of climate change on the Philippine archipelago; and undertake all the necessary preparation for both natural and man-made disasters, such as, but not limited to, typhoons, earthquakes, tsunami, floods, landslides, civil disturbance, and terrorism. The Cluster is chaired by the Secretary, Department of Environment and Natural Resources, with the Climate Change Commission performing secretariat functions.

---

[46]Memorandum Circular no. 2008–69.

[47]"Bali Delegates Agree to Support Forests-For-Climate (REDD) Plan," 16 December 2007. Retrieved 28 October 2014, from http://news.mongabay.com/2007/1215-redd.html.

[48]Executive Order no. 881, s. 2010.

(2) *Philippine government's response in the aftermath of Haiyan devastation*

At the aftermath of Typhoon Haiyan, most of the basic services like power, telecommunications, and water supplies could not be delivered as local authorities were themselves visibly shaken by the gravity of the emergency situation. The local populace was reporting about the scarcity of food and access to the area was not possible with transport facilities heavily damaged or destroyed. Emergency responders were also crippled and unable to provide assistance due to the logistical difficulties. The reported delays of relief efforts were generally attributed to these factors, including the decentralized authority, a shortage of available government workers, looting, and heavy rains.[49]

In the Philippines, local governments, pursuant to the Local Government Code of 1991, are tasked to provide basic services and safeguard the general welfare of the inhabitants within their territorial jurisdiction, including harmful effects of natural disasters and acts which endanger the environment. Local governments are likewise the primary authorities for disaster risk reduction and management in particular. Each local government is mandated to implement the plans crafted by the local DRRM office with guidance from national agencies. However, full implementation of Climate Change and DRRM laws is clearly lacking in most LGUs in the Philippines. No local CCAM/DRRM plans are in place, which visibly impairs the ability to provide effective emergency humanitarian services in times of climate-induced disasters.

When Haiyan struck, humanitarian response to a disaster of such scale required active government leadership and coordination at the local level and supplemental assistance from national government agencies. While local authorities were initially coping with the natural disaster, several cabinet secretaries from the national government were dispatched to oversee relief operations in different places. A host of national and international relief efforts also became involved to support such immediate actions. But there was initially no clear coordination for the whole response. The lack of coordination between national government and local officials was evident

[49]Max Fisher, "47 Statistics That Explain Typhoon Haiyan," *Washington Post*, 12 November 2013.

when the local mayor of the most badly-hit-city, Tacloban in Leyte, accused the Department of Interior and Local Government (DILG) Secretary of "politicking" over Yolanda.[50]

On-the-ground coordinated response could not be led by local governments because they were almost literally overwhelmed by the disaster. National government action tried to fill in the gaps but came almost a month after. The Office of the Presidential Assistant for Rehabilitation and Recovery (OPARR) was created to develop post-Yolanda recovery activities and a funding tracking system on 2 December 2013. The OPARR, as the lead agency to head rehabilitation and reconstruction efforts, was viewed as the government's strategy to ensure fast track transition from relief efforts to rehabilitation and rebuilding of affected areas.

However, conflicting policy measures from the OPARR were observed. One example is the observance of "no build zone" on legally mandated salvage zones (three meters in urban areas, 20 meters in agricultural areas, and 40 meters in forest areas).[51] Despite the existence of this law, the OPARR nevertheless sought for exemption from such rule showing the lack of foresight and disregard to adaptation requirements for climate change thus exposing the hazard-prone communities to more climate-related disasters.[52]

A "no build zone" policy in the Yolanda-affected areas would have prohibited building any structure within 40 meters from the coastline. But Panfilo Lacson of OPARR says that a blanket application of the "no build zone" will not address exceptional circumstances and may be impractical for certain areas.[53] For instance, Lacson points out that areas with fishing industries or tourism-oriented businesses still need to build structures within 40 meters from the coastline, but other areas are located in high elevations which are not susceptible to storm surge.[54] In the meantime, with no definite

[50]Rigoberto Tiglao, "You Are a Romualdez, and the President Is an Aquino," *Manila Times*, 10 December, 2013.

[51]Presidential Decree 1067, otherwise known as the Water Code of the Philippines.

[52]Briefing Room: Presidential Assistant for Rehabilitation and Recovery, "PARR: 'No Build Zone' Policy Not Recommended in Yolanda-Affected Areas," Official Gazette, 14 March 2014.

[53–54]*Ibid.*

resolve to enforce the "no build zone" policy in sight and the slow pace of rehabilitation work especially in Tacloban City, homeless survivors have started to rebuild their homes in areas formerly declared as "no build zones".

A year later, OPARR came up with its Comprehensive Rehabilitation and Recovery Plan (CRRP), an 8,000 page document outlining the government's commitment to implement over 25,000 rehabilitation and recovery projects worth PHP167.9 billion.[55] With the CRRP, the Philippine government has crafted two plans so far. The Philippine government's planning agency, National Economic Development Authority (NEDA), has earlier crafted the Reconstruction Assistance for Yolanda (RAY): 'Build Back Better' Plan in December 2014. Under the RAY-BBB Plan, the "government seeks to enable new modalities to encourage and facilitate the active involvement of the private sector with options including expansion of public-private partnership arrangements for major investment programs".[56]

Despite its promising features, a lawmaker criticized that the BBB Plan does not say how and when this will be done; hence, it may even fall short of achieving pre-Yolanda economic levels.[57] In addition, rebuilding the agricultural sector seems to be a low priority for the national government despite the National Economic Development Authority's (NEDA) admission that "majority of those who lost their incomes and sources of livelihood are the marginalized rural workers — farmers, fisherfolk, and coconut farmers".[58]

A recent report — Implementation for Results — under NEDA's RAY program was released by the same agency "to provide a framework to ensure close alignment between the objectives of the recovery and rehabilitation program and the Philippine Development Plan".[59] This was essentially

[55]"Island Left out," *Cebu Daily News*, 17(272), 10 November 2014.

[56]Dennis Carcamo, "Bayan Questions Gov't Rehab Plan for Yolanda-Hit Areas," *Philippine Star*, 8 January 2014.

[57–58]*Ibid.*

[59]NEDA, Document: Reconstruction Assistance on Yolanda, "Implementation for Results," Official Gazette, 29 September 2014. Retrieved 14 April 2015, from http://www.gov.ph/2014/09/29/document-reconstruction-assistance-on-yolanda-implementation-for-results/.

aimed at "accelerating and intensifying the recovery and rehabilitation process" which seeks to respond to continuing questions regarding the government's plans for the Yolanda-hit areas.

Other agencies and organizations also started to implement projects without necessarily aligning its objectives with the aforementioned plans. For instance, the Department of Environment and Natural Resources (DENR) started its own P1-B mangrove rehabilitation project.[60] The Department of Agriculture (DA) — Bureau of Fisheries and Aquatic Resources (BFAR) awarded fishing boats to fisherfolk beneficiaries.[61] Whilst the Department of Social Work and Development (DSWD) provided cash-for-work to survivors of the typhoon,[62] emergency cash assistance were given out by Tzu Chi and Oxfam Foundation. The Tzu Chi Foundation, Philippines has spent over PHP1.2 billion in cash assistance to survivors of Typhoon Yolanda so far.[63] According to Oxfam, where markets have been revived, particularly in urban centers, cash assistance helped families to buy food, and local traders to recover.[64]

At the request of the Philippine government and in a supporting role, the United Nations established the Disaster Assessment and Coordination (UNDAC) and the UN Office for the Coordination of Humanitarian Affairs (OCHA) teams.[65] With the UN Humanitarian Country Team already in

---

[60]Manila Bulleting, "DENR-8 Starts P1-B Mangrove Rehab Project," 27 April 2014. Retrieved 25 November 2015, from https://ph.news.yahoo.com/denr-8-starts-p1-b-mangrove-rehab-project-151013364.html.

[61]Official Gazette, "DA-BFAR Awards Fishing Boats to Fisherfolk-Beneficiaries," 28 February 2014. Retrieved 25 November 2014, from http://www.gov.ph/2014/02/28/da-bfar-awards-fishing-boats-to-fisherfolk-beneficiaries/.

[62]Official Gazette, "DSWD Implements Cash-for-Work Program for Typhoon Victims," 22 January 2014. Retrieved 25 November 2014, from http://www.gov.ph/2011/10/10/dswd-implements-cash-for-work-program-for-typhoon-victims/.

[63]Tara Quismundo. "Group Aims to Spread Love amid Devastation," *Philippine Daily Inquirer*, 10 February 2014.

[64]OXFAM, "Typhoon Haiyan: The Response so Far and Vital Lessons for the Philippines recover," 7 December 2013.

[65]Lum and Margesson, "Typhoon Haiyan (Yolanda): U.S. and International Response to Philippines Disaster," Congressional Research Service, 25 November 2013. Retrieved 17 June 2015, from http://www.crs.gov.

place, OCHA has assisted in coordinating international actors on the ground and to enlist donor support.[66] Humanitarian relief sectors are typically established during humanitarian crises to enable the United Nations to coordinate partners, prioritize resources, and facilitate planning.[67] Relief sectors have been organized into 13 relief sectors or "clusters" led by various agencies which include: Camp Coordination and Camp Management; Early Recovery; Education; Emergency Shelter; Emergency Telecommunications; Food Security and Agriculture; Health; Livelihood; Logistics; Nutrition; Protection; Water, Sanitation and Hygiene; and Communications and Communities.[68]

One important issue for the Philippine economic recovery in the months and years ahead will be whether decentralized governance and corruption will limit the efficiency of their stimulus efforts. In response to widespread fears about government corruption, the Aquino administration launched a web portal that allows the public to track international disaster funds, the Foreign Aid Transparency Hub (FAITH).[69]

While local and international NGOs and civil society hold relief operations in typhoon Yolanda-affected areas, the NGOs likewise called for government leadership in reconstruction of disaster areas.[70] According to NGOs led by Oxfam, strong government leadership is needed for the recovery of coconut farmers and fisher folk struggling to rebuild communities as "many of the lower-income municipalities are struggling to recover after Yolanda and are heavily dependent on private sector companies, the United Nations, or NGOs for rehabilitation support".[71] The groups further noted that "government attention and investment in these areas is required to ensure people do not fall further into poverty".[72]

[66–68] *Ibid.*

[69] Zinia de la Pena, "'Faith' for Foreign Aid," *Philippine Star*, 18 November 2013. Retrieved 25 November 2014, from http://www.philstar.com/headlines/2013/11/18/1258028/faith-foreign-aid.

[70] D.J. Yap, "NGOs Call for Government Leadership in Reconstruction of Disaster Area," *Philippine Daily Inquirer*, 15 March, 2014. Retrieved 25 November 2014, from http://newsinfo.inquirer.net/585870/ngos-call-for-government-leadership-in-reconstruction-of-disaster-area.

[71–72] *Ibid.*

How would you link the laws set out in Section 18.5(1), to the state of response by the government as discussed in Section 18.5(2)? It will help set a meaningful foundation for your conclusions later.

Also, below in Section 18.6(3), are you able to succinctly create a link on how the government related with the ASEAN mechanism at that moment of crisis?

(3) *Emergent opportunities for an ASEAN regional response to humanitarian crises*

At about the time when international support was extended to the country, ASEAN moved to help relief operations in areas hit by Haiyan.[73] Within the framework of the AADMER, the ASEAN Secretary-General Le Luong Minh stated that they are working closely with the AHA Centre to ensure ASEAN's effective response to the disaster by facilitating ASEAN cooperation and coordination through the ASEAN Coordinating Centre for Humanitarian Assistance on disaster management (AHA Centre).[74]

Thereafter, the ASEAN Secretary called on the international community to provide fresh support for the post-disaster recovery and rehabilitation needs of communities affected by Haiyan in the Philippines. The call was made at a high level conference on Assistance for the Recovery of Yolanda-Affected Areas (ARYA).[75] The ASEAN Secretariat indicated that the conference was able to generate fresh commitments for ARYA through the sharing of expertise and experience

[73] Amy R. Remo, "ASEAN Moves to Help Relief Operations in Areas Hit by Yolanda," *Philippine Daily Inquirer*, 11 November 2013. Retrieved 25 November 2014, from http://globalnation.inquirer.net/90283/asean-moves-to-help-relief-operations-in-areas-hit-by-yolanda/. See, "President Aquino's Address to the Nation," 6 February 2015. Retrieved 25 November 2014, from http://www.gov.ph/2015/02/06/english-speech-of-president-aquino-addressing-the-nation-on-february-6-2015/.

[74] *Ibid.*

[75] "ASEAN Gathers Fresh Support from International Community for Haiyan/Yolanda Recovery in the Philippines." Retrieved 14 April 2015, from http://www.dfa.gov.ph/index.php/2013-06-27-21-50-36/dfa-releases/3833-asean-gathers-fresh-support-from-international-community-for-haiyan-yolanda-recovery-in-the-philippines.

on building disaster-resilient communities, technical, technological, and financial contributions from various sources.[76] The conference was part of ASEAN's strategic plan to support recovery efforts in the Philippines. A capacity-building workshop for municipalities affected by the typhoon was conducted after the conference to raise their capacity for disaster recovery. ASEAN also called on the international community to adopt municipalities affected by Haiyan.[77] ARYA was spearheaded by ASEAN in collaboration with the OPARR and the Department of Foreign Affairs of the Philippines.[78]

There are emerging opportunities to enable a strategic response throughout the ASEAN region. One of the fundamental issues after the typhoon was the breakdown of local authority resulting in the inability to deliver basic emergency services. In the light of absence of local governance, the magnitude of the calamity heightened the need for the national government to step in. However, national government authorities also failed to immediately respond to assist in response in areas severely affected by the natural disaster.[79]

In just over six months following the landfall of Haiyan in the Philippines, rebuilding efforts in regions impacted by the storm continued to be hampered by a lack of government coordination and supplies. The lack of coordination is evident in the distribution of relief goods and cash for work beneficiaries. For many local inhabitants that remained dependent on overseas aid, distribution of relief goods were hampered by bureaucracy.[80]

Government officials in the Philippines admitted they lacked coordination and effective responses to Haiyan, which devastated wide swathes of the country.[81] The admission comes after locals, still struggling to rebuild six months later, charged the government with failing to provide

[76–78] *Ibid.*

[79] "In Face of Lackluster Government Response, Typhoon Haiyan Survivors Ask the World 'Please Send Food!'," *Catholic Online* (News Consortium), 19 May 2014. Retrieved 25 November 2014, from http://www.catholic.org/news/international/asia/story.php?id=55461.

[80] Liv G. Campo, "Donation for Yolanda Survivors Stuck at CIP," *The Freeman*, 3 April 2014; "Relief Goods Stuck," *Sunstar Cebu*, 18 May 2014.

[81] Supra note 41.

even basic, and coordinated relief to victims.[82] As the UN assessment report of nine municipalities in Leyte showed, "at the end of November, food appears to be effectively distributed in some [areas], but not effectively or evenly distributed in others . . . the more remote communities are not notified adequately or are required to walk in for what remains available".[83]

The OPARR which was expected to take on the challenges of consolidating efforts of the local, national, and international in the recovery efforts has been hampered by inadequate resources. Thus, absence of an integrated response by various agencies is quite evident.

Overall, the lack of general capacity to manage a disaster of this proportion provides regional bodies, such as ASEAN, with the opportune time to extend humanitarian assistance pursuant to current regional agreements. The Philippines has called for "focused international effort" to meet climate threats.[84] The critical role of ASEAN to extend support to address the threats posed by climate change, and to build communities that are resilient in the face of disaster cannot be overemphasized.

Haiyan left millions of people in need of urgent assistance. But it is also pushing millions of poor people into deepening debt and destitution — more exposed to the next disaster.[85] With as many as 50 storms hitting the country each year, steady population increases in disaster-prone areas, combined with a geography consisting of islands and poor infrastructure, the Philippines is vulnerable to humanitarian crises.[86] With this reality, the Philippines authorities and other AMCs together with the ASEAN must increase efforts to work together to address inadequate capacities and resources throughout the region, and also, to reduce the growing risk of disasters driven by climate change that the Philippines and other countries in the region will face.

---

[82] *Ibid.*

[83] Supra note 26.

[84] International Federation of Red Cross and Red Crescent Societies (IFRC), 10 June 2014.

[85] Oxfam Report: "A Month after Yolanda, What Challenges Remain, What Needs to Be Done," 7 December 2013. Retrieved 14 April 2015, from http://interaksyon.com.

[86] Supra note 27.

18.6. AADMER Article 17 — Rehabilitation; AADMER Work Programme

(1) *Rehabilitation and reconstruction*

One of the provisions of AADMER Article 17 is:

> *2. Rehabilitation and Reconstruction*
>
> *a. Support and emphasize the importance of national rehabilitation and reconstruction programs given the devastation of the basic infrastructures and services in the affected countries. The reconstruction and rehabilitation phase should link seamlessly with the humanitarian relief efforts.*
> *b. Call on the international community, in particular the donor countries, the World Bank, the Asian Development Bank, the Islamic Development Bank, European Investment Bank, and related international financial institutions to provide the necessary fund for the viability and sustainability of those programs.*
> *c. Establish a partnership, upon the request and the leadership of the country concerned, involving donor countries and regional as well as international financial institutions, to support the respective national programs of the affected countries.*
> *d. Welcome the initiative of several countries on the moratorium of payments of the external debt of the affected countries to augment their national capacity to carry out the rehabilitation and reconstruction efforts.*
> *e. Promote and encourage private sector participation in and contribution to the rehabilitation and reconstruction efforts.*

Although the above provisions apply to earthquakes and tsunamis, nonetheless, it could equally apply to other types of disaster like Haiyan. However, research must be conducted to determine the extent, if any, the above provisions have been put into operation in the Haiyan situation.

(2) *AADMER Work Programme, 4 — Recovery*

The recovery component aims to strengthen the capacity of member states to:

- Conduct damage and loss assessment within a month after the disaster occurs;

- Develop an effective recovery action plan for rehabilitation and reconstruction within three months after a disaster occurs;
- Mobilize resources from the local, regional, and international community to support implementation and the recovery process;
- Develop a transition plan and link post-disaster recovery process into sustainable development one year before the end of the recovery period.

The recovery principle is to promote a proactive process for early to long-term recovery even before a disaster occurs. This is to be done at all levels of government. Pre-disaster recovery planning helps recovery faster. The recovery component contributes to the achievement of the overall goals of disaster resilience and sustainable development.

Under recovery, ASEAN should assist in planning the reconstruction process. It can facilitate mobilizing from all over the world in rebuilding in the aftermath of Haiyan. Indeed, the role of humanitarian disaster relief has changed and UNISDR is organizing a conference.

For reconstruction is a mega-disaster and lessons can be learned from Aceh. It had put in place a unique governance structure which was non-cumbersome, non-bureaucratic, and used principles of transparency that incorporated an anti-corruption entity as part of its structure. A single agency was used to coordinate reconstruction. The 2004 Indian Ocean tsunami recovery success is attributed to this post-disaster institutional design. Lessons can also be learned from Cyclone Nargis in terms of mobilizing all stakeholders from top to bottom.

Other lessons include the newly expected role of the military which may cause misapprehension among states and hinder humanitarian, civil, and military coordination. There are also questions of human rights as where assistance is favored for the rich at the expense of the poor. Questions of governance and corruption may disrupt recovery, as also will recurrence of disasters even before full recovery.

As the Philippines was recovering from Haiyan, a massive cyclone, Hagupit (locally known as Typhoon Ruby) struck in early December 2014. It came from the same part of the western Pacific Ocean as Haiyan did 13 months before, and made landfall in Eastern Samar. As it neared the Eastern Seaboard of the Philippines, thousands of people living in islands on the path of the super typhoon were evacuated, with many seeking shelter in churches, schools, and other makeshift evacuation centers. Preparations were underway for potentially life-threatening winds, storm surge, and flash floods as Typhoon Ruby gained strength with maximum sustained winds at 180 mph.

Unlike Typhoon Haiyan, the country marshalled all appropriate government agencies to prepare for the potentially life-threatening disaster. With hourly weather updates from the weather agency, PAGASA, the military placed on high alert, and thousands of families evacuated by local governments, it was evident that the government was well-prepared this time. Adaptation measures had been put in place. Although the tasks of rebuilding after Typhoon Haiyan still continue, the increasing frequency of abnormal weather events now call for reinforcing community resilience. Basic supplies and services — such as food, water, and communication lines have to be in place and ready to be deployed at a moment's notice. Early warning systems together with necessary infrastructure support are necessary aspects of disaster preparedness and climate adaptation. Such are the realities now in countries often visited by strong typhoons like the Philippines and in countries around the ASEAN region.

18.7. Scaling up Disaster Management after Haiyan

The magnitude of Haiyan was such that ASEAN could not meet all its potential obligations. ASEAN has since Haiyan held a number of meetings to scale up its disaster management strategies and has moved to a higher level of response for the future.[87]

The Haiyan experience has raised ASEAN's urgency to strengthen its disaster mechanism including the East Asia Summit Rapid Disaster

[87]"ASEAN to Lead Conference on Renewed Yolanda Assistance." Retrieved 14 April 2015, from http://www.dfa.gov.ph/index.php/2013-06-27-21-50-36/phl-embassies-and-consulates/3579-asean-to-lead-conference-on-renewed-yolanda-assistance; "ASEAN Gathers Fresh Support from International Community for Haiyan Recovery in the Philippines," 18 August 2014. Retrieved 14 April 2015, from http://www.asean.org/news/asean-secretariat-news/item/asean-gathers-fresh-support-from-international-community-for-haiyan-recovery-in-the-philippines; "ASEAN Sec-Gen Pushes Members to Continue Giving for Rehab of Yolanda-Hit Areas," *Kicker Daily News*, 16 August 2014. Retrieved 14 April 2015, from http://m.kickerdaily.com/tag/yolanda-affected/.

Response.[88] The East Asia Summit (EAS) is a regional leaders' forum for strategic dialogue and cooperation for closer regional cooperation. Under the EAS, India is developing a virtual knowledge center for EAS countries to share knowledge and best practices in disaster management. Australia has also developed an emergency management. Australia serves as a temporary repository information and EAS countries can contribute to the information and documentation on the site. A High Level Conference for the Recovery of Yolanda Affected AREAS (ARYA) was held in Manila on 14 August 2014[89] to further consolidate regional recovery initiatives. It also welcomed the roundtable of 14 × 7 point of contacts of the National Response Agencies of EAS participating countries and the inauguration of the Virtual Knowledge Portal (VKP) which was hosted by India from 4–5 December 2014.[90]

Another recent initiative is the establishment of the Changi Regional HADR (Humanitarian Assistance and Disaster Relief) Coordination Centre (RHCC) housed at the Changi Command and Control Centre (CC2C) in Singapore. Singapore has responded rapidly to disasters, following the ASEAN Defence Ministers Meeting in Singapore in September 2014.[91] RHCC leverages on ready facilities available at the CC2C for large-scale multinational operations and exercises. It will:

a. Focus on supporting a disaster affected state's military in coordinating assistance provided by foreign militaries.
b. Support and complement other existing mechanisms such as the ASEAN Coordinating Centre for Humanitarian Assistance on

[88]"East Asia Summit Rapid Disaster Response," *Disaster Resilient Australia*. Retrieved 14 April 2015, from https://www.emknowledge.gov.au/connect/east-asia-summit-rapid-disaster-response/. See also 2014 EAS Indonesia-Australia Rapid Disaster Response Workshop, 11–12 June 2014. There are 18 EAS member countries including India and Australia.

[89]ASEAN, "ASEAN, High-Level Conference on ARYA," (n.d.). Retrieved 14 April 2015, from http://www.asean.org/communities/asean-socio-cultural-community/category/asean-high-level-conference-on-arya.

[90]India is a dialogue partner of ASEAN. See Chairman's Statement of the Fourth East Asia Summit (EAS)'s Foreign Ministers Meeting on 10 August 2014, Nay Pyi Taw. Retrieved 14 April 2015, from http://www.mofa.gov.mm/wp-content/uploads/2014/08/Chair-Statement-on_4th_East_Asia_Summit_FMM_Final1.pdf, at para. 13.

[91]Singapore MINDEF, "Fact Sheet: Changi Regional HADR Coordination Centre (RHCC)," 12 September 2014. Retrieved 14 April 2015, from http://www.mindef.gov.sg/imindef/press_room/official_releases/nr/2014/sep/12sep14_nr2/12sep14_fs.print.img. html.

Disaster Management (AHA Centre) and UN Office for the Coordination of Humanitarian Affairs (OCHA), which focus on supporting a disaster affected state's national disaster management authorities in coordinating the overall relief effort.

Prior to predictable events such as a typhoon, the Changi RHCC will:

a. *Fuse information from regional disaster and early warning centres:*
b. *Distribute the situation picture to the partner militaries through its OPERA Command and Control Information System (CCIS) web-portal.*

This will allow partner militaries to prepare their response mechanisms.

a. *If the affected state agrees, the RHCC will concurrently also conduct initial planning and scale up for a possible multinational military HADR operation.*
b. *If disaster occurs:*

   - *The Changi RHCC will continue to broadcast the updated situation picture, including contributions of international aid, to partner militaries to facilitate decision-making, and minimise duplication and gaps in the provision of foreign military assistance.*
   - *This can be done via the OPERA CCIS web-portal as well as the RHCC's network of international liaison officers and linkages with operations centres of regional militaries.*
   - *The Changi RHCC will also coordinate and facilitate the deployment of foreign military assistance into theatre.*
   - *If required, the RHCC can forward deploy a mobile coordination unit to support the affected state's military in coordination efforts on the ground.*
   - *Once the disaster relief efforts reach a steady state and the affected state's military is able to coordinate the foreign military assistance provided without external support, the Changi RHCC will draw down its operations.*

Should a major disaster happen in the region, the affected country will be able to tap on RHCC.

It will complement the work of existing coordination centers working with civilian agencies, such as OCHA and AHA Centre. For example, instead of AHA coordinating with different nations' militaries, RHCC can be a 'one-stop' and integrate with them.

18.8. Conclusion and Way Forward

This chapter provides a timely examination of legal frameworks on disaster management which includes reconciliation, recovery, and reconstruction as applied in Haiyan. While it has demonstrated some of the strengths and inadequacies of ASEAN, its AMCs (including the Philippines), and other relevant organizations, the full extent of these have not been researched into, as yet. However, sufficient is known to point the way forward to be better equipped for future disasters not only in ASEAN but elsewhere.

Haiyan demonstrates that the Philippine government and other players need to work closely with local communities and involve them in the process so as to empower them understand their vulnerabilities and to conduct needs assesment. There should be a more focused international effort to cooperate and coordinate their activities.

There are lessons to be learned in Haiyan and ASEAN needs to do more to realize its vision of disaster reduction and developing resilient nations, as this is the first line of defence against disasters. It needs to improve the capacities for regional risk assessment, collect data, and provide effective and efficient early warning systems. Other areas include:

- Assisting AMCs to mainstream disaster risk reduction into national development policies and plans, and in formulating and implementing risk reduction measures that link climate change adaptation and key sectors to ensure sustainable development.
- Enhancing technical and organizational capacities — facilitating coordination with the UN, WB, its Dialogue Partners, and mobilizing their support.

There is a need for the world community to cooperate on risk reduction. This was discussed at the Third UN World Conference on Disaster Risk Reduction (World Conference), held from 14 to 18 March 2015 in Sendai City, Miyagi Prefecture, Japan.[92] Earlier, the United Nations

---

[92]"About the Third UN World Conference on Disaster Risk Reduction," UN World Conference on Disaster Risk Reduction, Japan, 14–18 March 2015. Retrieved 14 April 2015, from http://www.wcdrr.org/conference. See also, UN Resolution adopted by the General Assembly on 20 December 2013, Report of the Second Committee (A/68/438/Add.3), A/RES/68/211, 29 January 2014.

General Assembly Resolution for 2013[93] on International Strategy for Disaster Reduction stated that the World Conference would focus on a forward-looking and action-oriented outcome document and would have the following objectives which include:

- *To consider the experience gained through the regional and national strategies/institutions and plans for disaster risk reduction and their recommendations as well as relevant regional agreements within the implementation of the Hyogo Framework of Action.*
- *To adopt a post-2015 framework for disaster risk reduction.*
- *To identify modalities of cooperation based on commitments to implement a post-2015 framework for disaster risk reduction.*
- *To determine modalities to periodically review the implementation of a post-2015 framework for disaster risk reduction.*

The Conference and its preparatory process had participants from all relevant stakeholders, including parliaments, civil society, the International Red Cross and Red Crescent Movement, non-governmental organizations, national platforms for disaster risk reduction, focal points for the Hyogo Framework for Action, local government representatives, scientific institutions and the private sector, as well as organizations of the United Nations system and intergovernmental organizations.

A changing philosophy in disaster management and humanitarian assistance would involve the collaboration and complementarity of efforts rather than competition.[94] This is an important factor as it has often been perceived that rendering assistance in disaster situations is like a competition, with some countries trying to outdo the others. As a result, there is little or no coordination and this may lead to confusion and duplication of efforts.

In the case of Haiyan, for example, it would have been desirable if countries had worked through ASEAN rather than individually. There are

---

[93] "Resolution Adopted by the General Assembly on 20 December 2013." Retrieved 14 April 2015, from http://www.unisdr.org/files/resolutions/ARES68211E.pdf. See, Summary of the Third World Conference on Disaster Risk Reduction: 14–18 March 2015. Retrieved 14 April 2015, from http://www.iisd.ca/download/pdf/enb2615e.pdf; See, "Achieving Justice and Human Rights in an Era of Climate Disruption," International Bar Association, July 2014.

[94] See "Addressing the Changing Landscape of Humanitarian Action: Encouraging Complementarity Rather than Competition," 2013. Retrieved 22 June 2015, from https://www.wiltonpark.org.uk/addressing-the-changing-landscape-of-humanitarian-action/.

lessons in Haiyan for a 'whole-of-the world' approach, requiring closer cooperation and coordination among countries.

The military has mainly contributed to disaster relief operations by providing air transport for relief items and personnel; medical assistance; and expert personnel (in civil–military coordination and liaison, needs assessment, and logistics). The role of the military has become crucial in disaster situations. Also, governments are playing an increasing role in humanitarian assistance. This brings us to the question of how they are to coordinate their efforts with other organizations such as the Overseas Development Institute's Humanitarian Policy Group and the International Committee of the Red Cross (ICRC).

It has been widely acknowledged that embracing non-traditional humanitarian actors, including the private sector, will change the dynamics of humanitarian action (e.g., under APSC) in the future, and the question which remains is, how will a new humanitarian architecture make better use of this diversity? Humanitarian action needs a rethink in order to reflect the complexity of humanitarian engagement on the ground, and to expand the traditional mechanisms of coordination, cooperation, and funding. Transferring knowledge and experience, establishing equal partnerships, and sustaining dialogue are just some potential solutions to current and future challenges to humanitarian action. The landscape and operating environment are becoming increasingly complex and there is much greater demand for effective coordination between the growing number of actors, both recipient and donor. Focusing on similarities could be a potent beginning for long-term strategic partnerships, and especially in the case of new global actors, there have been calls for a change in mutual perceptions of 'traditional' and 'rising' global actors.

All said about ASEAN's limited response, it has made some contributions in terms of giving relief and financial assistance, including from its AMCs, which are all Parties to AADMER. Having regard to the enormity and the unexpected ravages of Haiyan, still more should have been done. But how much more in the circumstances? ASEAN is scaling up its role for future disasters.

The AHA Centre was established in November 2011 and there are many challenges ahead — it has the potential of playing a more robust coordinating role and facilitating risk reduction and recovery from not only its member states but also its Dialogue Partners, the UN, and the World Bank. The world should wake up to the fact that as more climate change disasters occur, big and small (the latter often not reported and neglected), a coordinated 'whole-of-the-world' approach should be adopted.

The Mid-Term Review of the ASEAN Socio-Cultural Community Blueprint (2009–2015) (Executive Summary)[95] emphasizes on:

- Communication Strategy: ASEAN should expedite the release of the ASEAN community-wide communication strategy/plan.
- Financial support for disaster management.

It also reiterates "Our People Our Future Together". This is the time for ASEAN to further demonstrate its solidarity as it did when it drafted AADMER.

On 10 November 2014, the ASEAN Secretariat launched the following: (1) *Typhoon Yolanda Book, 'Y' (It Happened),*[96] and (2) *The ASEAN Book, Weathering the Perfect Storm.*[97]

These publications are part of the Typhoon Yolanda Documentation Project spearheaded by the National Disaster Risk Reduction and Management Council and supported by ASEAN.

The first book is a narrative reflection of Typhoon Yolanda (Haiyan) and the Philippines' disaster risk reduction and management (DRRM) system. It documents the experiences of people and communities affected by the disaster, including good practices and case stories. Building from these lessons and experiences, it aims to inform a number of key stakeholders

---

[95]"Mid-Term Review of the ASEAN Socio-Cultural Community Blueprint (2009–2015)," 14 April 2014. Retrieved 14 April 2015, from http://www.asean.org/resources/item/mid-term-review-of-the-asean-socio-cultural-community-blueprint-2009-2015.

[96]"Typhoon Yolanda Book, 'Y It Happened', Launched in Commemoration of Typhoon Yolanda's One Year Anniversary," UNDP. Retrieved 14 April 2015, from http://www.ph.undp.org/content/philippines/en/home/presscenter/articles/2014/11/13/typhoon-yolanda-book-y-it-happened-launched-in-commemoration-of-typhoon-yolanda-s-one-year-anniversary/.

[97]"Statement of the Secretary-General of ASEAN for the Launching of Typhoon Yolanda Book 'Y (It Happened)' and ASEAN Book 'Weathering the Perfect Storm'," Association of Southeast Asian Nations. Retrieved 14 April 2015, from http://www.asean.org/news/asean-secretariat-news/item/statement-of-the-secretary-general-of-asean-for-the-launching-of-typhoon-yolanda-book-y-it-happened-and-asean-book-weathering-the-perfect-storm.

including the members of the national and international communities, national government agencies, local government units and civil society organizations — how they can collectively work together to reduce the risks of disasters, especially to the most vulnerable.

Le Luong Minh, Secretary-General of ASEAN, in his speech at the launching of the books on 10 November 2014, Pasig City, Philippines said:[98]

> *The two publications, the Typhoon Yolanda Book "Y (It Happened)" and the ASEAN Book "Weathering the Perfect Storm", could provide useful learning platforms for disaster management practitioners and the humanitarian community. These publications underscore the reality that in responding to a large-scale disaster, new modalities of work processes to cope with the situation are required. We would like to thank the Government of the Philippines as well as all stakeholders in the region for sharing their views and experiences and documenting them in these two publications to further improve ASEAN's efforts in responding to disasters and building a disaster-resilient and safer community.*

The first book is also an account and evaluation of the DRRM institutional set-up, mechanisms, and policies of the Philippines. It is anchored on the four thematic areas of the DRRM Law: Prevention and Mitigation, Preparedness, Response, and Rehabilitation and Recovery. It seeks to identify gaps under the existing DRRM structure vis-a-vis the Philippine Disaster Risk Reduction and Management Act (R.A. 10121) and the National DRRM Framework Plan, as well as to ascertain and share key lessons, learnings, and challenges that will serve as inputs to the TY Yolanda Comprehensive Rehabilitation and Recovery Plan. "*Y (It Happened)*" is a hopeful contribution to the continuing discourse on enhancing the disaster risk reduction and management practices, systems, and policies in the Philippines and in the wider international DRRM and humanitarian community, particularly in ASEAN.

We would like to end with a broader dimension to halt climate change to ensure that climate related disasters such as Haiyan will not occur, at least with such intensity and frequency. ASEAN should speak with a more concerted and forceful voice in climate change negotiations.

---

[98] *Ibid.*

As for the global community, a new architecture should be that of a 'whole-of-the-world' approach. All this requires cooperation, capacity building, and research.[99]

Appendix 1: ASEAN Disaster Management Cooperation: International, Regional, and National Mechanisms

AADMER (2005)	ASEAN Secretariat	ASEAN Insurance Council	ASEAN Dialogue Partners: Regional, International Cooperation Mechanisms, etc.
• AADMER Work Programme, 2010–2015 • AADMER COP • ASEAN-ERAT • ACDM • ARPDM • DELSA • Japan–ASEAN International Fund • APG • AAHA (2011) ➢ AHA Centre (Jakarta) ➢ United Nations Humanitarian Response Depot (UNHRD) Subang Base (Malaysia) • AMCs NDMOs • JICA	• ASEAN Socio-Cultural Community Department ➢ Disaster Management and Humanitarian Assistance Division ➢ ASEAN Disaster Management Relief and Emergency Administration ➢ DMRS	• ANDREWS • Natural Disaster Risk Standard Report	• ASEAN Dialogue Partners: Australia, Canada, EU, US, NZ, etc. • Regional Cooperation Mechanisms: ➢ ASEAN + 3 ➢ NEAT ➢ ADRC • APEC-TFEP ➢ EPWG • ARF ➢ DiREx • RHCC (Singapore) • OCHA • WB • UNIDSR

[99]The writers would like to thank Dr Raman Letchumanan for his comments but they remain responsible for any errors. Dr Raman was formerly Head of the Environment Division, Secretariat of the Association of Southeast Asian Nations; he is currently Senior Fellow, Centre for Non-Traditional Security Studies, S. Rajaratnam School of International Studies, Nanyang Technological University.

Acronyms:

AADMER:	ASEAN Agreement on Disaster Management and Emergency Response
AAHA:	ASEAN Agreement on Humanitarian Assistance
ACDM:	ASEAN Committee on Disaster Management
ADRC:	ASEAN Disaster Reduction Centre
AHA Centre:	ASEAN Coordinating Centre for Humanitarian Assistance on Disaster Management
AMCs:	ASEAN Member Countries
ANDREWS:	ASEAN Natural Disaster Research Work Sharing
APEC:	Asia-Pacific Economic Cooperation
APG:	AADMER Partnership Group (a group of international NGOs collaborating for an ASEAN people-centered implementation of AADMER)
ARF:	ASEAN Regional Forum
ARPDM:	ASEAN Regional Programme on Disaster Management
ASEAN + 3:	ASEAN, China, Japan, and South Korea
COP:	Conference of the Parties
CSOs:	Civil Society Organizations
DELSA:	Disaster Emergency Logistics System for ASEAN
DiREx:	Disaster Relief Exercise
DMRS:	ASEAN Disaster Monitoring and Response System
EPWG:	Emergency Preparedness Working Group
ERAT:	Emergency Rapid Assessment Team
JICA:	Japan International Cooperation Agency
NDMOs:	National Disaster Management Offices. These centers include Brunei Darussalam: National Disaster Management Centre, Ministry of Home Affairs; Cambodia: National Committee for Disaster Management (NCDM); Indonesia: National Disaster Management Agency (BNBP); Lao PDR: National Disaster Management Office, Ministry of Labour and Social Welfare; Malaysia: National Security Council, Prime Minister's Department; Myanmar: Relief and Resettlement Department, Ministry of Social Welfare,

	Relief, and Resettlement; Philippines: National Disaster Risk Reduction and Management Council and Administrator (NDRRMC); Singapore: Singapore Civil Defence Force (SCDF); Thailand: Department of Disaster Prevention and Mitigation (DDPM), Ministry of Interior Thailand; Vietnam: Department of Dyke Management, Flood, and Storm Control (DDMFSC) — Standing Office of CCFSC
NEAT:	Network of East Asian Think Tanks
NTS:	Non-Traditional Security
OCHA:	United Nations for the Coordination of Humanitarian Affairs
RHCC:	Changi Regional Humanitarian and Disaster Relief Coordination Centre, Singapore
TFEP:	(APEC) Task Force of Emergency Preparedness
UNISDR:	United Nations Office for Disaster Risk Reduction, to facilitate the implementation of the International Strategy for Disaster Reduction (ISDR)

CPSIA information can be obtained at www.ICGtesting.com
Printed in the USA
BVOW06*1608270815

413833BV00005B/13/P

9 789814 689731